Study Guide

for

Stewart, Redlin, and Watson's

Algebra and Trigonometry

Second Edition

D0074066

Douglas Shaw
University of Northern Iowa

BROOKS/COLE
CENGAGE Learning™

Australia • Brazil • Japan • Korea • Mexico • Singapore • Spain • United Kingdom • United States

Study Guide for Stewart, Redlin, and Watson's Algebra and Trigonometry, Second Edition
Douglas Shaw

Cover Image: Bill Ralph

For product information and technology assistance, contact us at
Cengage Learning Customer & Sales Support, 1-800-354-9706

For permission to use material from this text or product,
submit all requests online at **www.cengage.com/permissions**
Further permissions questions can be emailed to
permissionrequest@cengage.com

ISBN-13: 978-0-495-01358-7

ISBN-10: 0-495-01358-7

Brooks/Cole
20 Davis Drive
Belmont, CA 94002
USA

Cengage Learning is a leading provider of customized learning solutions with office locations around the globe, including Singapore, the United Kingdom, Australia, Mexico, Brazil, and Japan. Locate your local office at **www.cengage.com/global**

Cengage Learning products are represented in Canada by Nelson Education, Ltd.

To learn more about Brooks/Cole, visit **www.cengage.com/brookscole**

Purchase any of our products at your local college store or at our preferred online store **www.cengagebrain.com**

Printed in the United States of America
2 3 4 5 6   14 13 12 11 10

ED216

Preface

Do you want to get an A in this course? If so, you already know how, because it isn't that well kept of a secret: Read the textbook thoroughly, do the homework carefully, pay attention in class, and make sure to ask questions when you don't understand something. Oh — and practice, practice, practice. I have taught thousands of students at this point in my career. (Wait a minute, can that be right? Yes, it is. Thousands. Wow.) Every semester, some of them resolve to take this very ordinary advice, and they usually do extremely well in my class.

This book is meant to supplement and not to replace *Algebra and Trigonometry, Second Edition*. For each section, it provides a list of the main points, along with important definitions, and then gives you practice problems to solve. Lots of them. The exercises are in two-column format, so you can cover up the solution column, work out the problems on a separate sheet of paper, and then check your solutions. There is always a temptation to look at the solutions right away, before making a solid attempt to work out the problems yourself. But stay strong — you learn by doing. The solution column is there to allow you to check your work, and to explain things if you get stuck.

Hearty thanks go to John Banks, who wrote the previous edition; Andy Bulman-Fleming, who is the best typesetter in the business (he recently wrote me an email pointing out *fifteen things* that he did not like about a particular one of my sentences); Jordan Meyer, who is an ace proofreader and a good friend; and Stacy Green, whose editorial guidance is always tops.

During this project, both Andy and I wound up with new additions to our families. This book is dedicated to little Frances Lily Shaw and to Walter McKellar Bulman-Fleming; may you live lives full of adventure and joy.

Douglas Shaw

Contents

11 Analytic Geometry   277

12 Sequences and Series   301

13 Counting and Probability   317

P Prerequisites

P.1   Modeling the Real World

▼ Concepts

(A) **Using and finding models.**

▼ Hints and Tips

- The first step of solving a modeling problem is to understand the question. That seems like an obvious thing to say, but there is sometimes a temptation to start writing equations right away, without taking the time to read the question carefully and to think about it.

▼ Review

(A) **Modeling with Functions.**

In algebra we use letters to stand for numbers. This allows us to describe patterns, which we express in a formula.

Exercise: The amount of power which can be generated by wind is given by the formula $P = 0.65 \times s^3$, where P is the power in watts and s is the speed of the wind in m/s. How much power is generated by the following wind speeds?

Answer:

(a) 2 m/s

We substitute $s = 2$ into the model and solve:
$P = 0.65 \times s^3 = 0.65 \times (2)^3 = 5.2$ watts.

(b) 4 m/s

We substitute $s = 4$ into the model and solve:
$P = 0.65 \times s^3 = 0.65 \times (4)^3 = 41.6$ watts.

(c) 6 m/s

We substitute $s = 6$ into the model and solve:
$P = 0.65 \times s^3 = 0.65 \times (6)^3 = 140.4$ watts.

Exercise: A VCR plays 12.5 cm of tape per minute.

Answer:

(a) How much tape is played in 8 minutes?

   Hint: Use the distance formula $D = RT$.

We substitute the known values and solve: $D = RT \iff$
$$D = \left(\frac{12.5 \text{ cm}}{\text{minute}} \right) (8 \text{ minutes}) = 100 \text{ cm.}$$

(b) How much tape is played in T minutes?

$$D = RT \quad \Leftrightarrow \quad D = \left(\frac{12.5 \text{ cm}}{\text{minute}} \right)(T \text{ minutes}) = 12.5T \text{ cm}$$

(c) A movie is 2 hours 15 minutes long. How much videotape is used by the movie?

2 hours 15 minutes $= 135$ minutes and

$$D = RT = \left(\frac{12.5 \text{ cm}}{\text{minute}} \right)(135 \text{ minutes}) = 1687.5 \text{ cm}.$$

Finding patterns is an important skill that is developed by practice. This leads to writing formulas that can then be used to solve other problems.

Exercise: The water in a hot tub is drained and replaced with fresh water. When the hot tub is restarted the temperature of the water is $68°$. One hour later, the water temperature is $79°$.

Answer:

(a) Find a formula that models the temperature of the water h hours after the hot tub is restarted.

We use the model that the temperature of the water in the hot tub is equal to the initial temperature plus the increase in temperature each hour times the number of hours. Let T be the temperature of the water in the hot tub h hours after it is restarted. The initial temperature is $68°$ and during the first hour the temperature increases by $79° - 68° = 11°$. Thus we get the model $T = 68 + 11h$.

(b) Use the model developed in part (a) to determine when the temperature reaches $101°$.

We use the model $T = 68 + 11h$ with $T = 10$ and solve for h: $T = 68 + 11h \quad \Leftrightarrow \quad 101 = 68 + 11h \quad \Leftrightarrow \quad 33 = 11h \quad \Leftrightarrow \quad 3 = h$. So it will take 3 hours for the temperature to rise to $101°$.

◤ Core Exercises

5, 11, 25, 39

P.2   Real Numbers

Concepts

(A) **Subsets of the real numbers both named ("the natural numbers") and unnamed ("$\{x \mid x \geq 2\}$").**

- Converting decimals to fractions.
- Adding, subtracting, multiplying and dividing fractions.
- Expressing intervals in various notations.

(B) **Properties of real numbers.**

(C) **Properties of negative numbers.**

- Adding, subtracting, multiplying and dividing positive and negative numbers

(D) **The real number line.**

- Locating points and sets on the real number line.

(E) **Unions and intersections of sets.**

- Finding the union and intersection of two given sets.

(F) **Absolute value and distance.**

- Finding the absolute value of a given real number.
- Finding the distance between two given real numbers.

Definitions

- **Real numbers:** The set of all numbers that can be expressed in decimal notation.
- **Natural numbers:** The numbers with which you count: $1, 2, 3, 4, 5, \ldots$.
- **Integers:** The natural numbers, together with their negatives and zero: $\ldots, -3, -2, -1, 0, 1, 2, 3, \ldots$
- **Rational numbers:** All numbers that can be expressed as $r = \dfrac{m}{n}$, where m and n are integers and $n \neq 0$. Their decimal representation always terminates, or repeats in a cycle (1.125, $3.12121\overline{212}$).
- **Irrational numbers:** All real numbers that are not rational, that is, all numbers that cannot be expressed as $\dfrac{m}{n}$ where m and n are integers and $n \neq 0$. Their decimal representation is always nonterminating and nonrepeating.
- **Commutative property of real numbers:** $a + b = b + a$, $ab = ba$.
- **Associative property of real numbers:** $(a + b) + c = a + (b + c)$, $(ab)\, c = a\, (bc)$.
- **Distributive property of real numbers:** $a\, (b + c) = ab + ac$, $(b + c)\, c = ab + ac$.
- **Additive identity for real numbers:** The number 0: $a + 0 = a$.
- **Multiplicative identity for real numbers:** The number 1: $1a = a$.
- **Quotient of a over b:** $\dfrac{a}{b}$, $b \neq 0$.

- **Numerator of a fraction** $\dfrac{a}{b}$: The "top" part, a.

- **Denominator of a fraction** $\dfrac{a}{b}$: The "bottom" part, b.

- **Set:** A collection of objects.

- **Union of two sets:** $A \cup B$ is the set of all elements that are in A or B or both.

- **Intersection of two sets:** $A \cap B$ is the set of all elements that are in both A and B.

- **Empty set:** $\varnothing$ or $\{\}$, a set containing no element.

- **Absolute value of a real number:** $|a|$ is the distance between a and 0, and is always a positive number (or 0, if $a = 0$).

- **Distance between a and b:** $|a - b|$ or, alternatively, $|b - a|$.

- **a is less than b:** $a < b$, in which case $b - a$ is positive.

- **a is greater than b:** $a > b$, in which case $b - a$ is negative.

Questions to Ask Your Teacher

Which notation for intervals do you prefer, bracket notation or set-builder notation?

Review

(A) **Subsets of the real numbers both named ("the natural numbers") and unnamed ("$\{x \mid x \geq 2\}$").**

There are many different types of numbers that make up the **real number** system. Some of these special sets are shown below.

Symbol	Name	Set
$\mathbb{N}$	Natural (counting)	$\{1, 2, 3, 4, \ldots\}$
$\mathbb{Z}$	Integer	$\{\ldots, -2, -1, 0, 1, 2, \ldots\}$
$\mathbb{Q}$	Rational	$\{r = p/q \mid p, q \text{ are integers}, q \neq 0\}$
$\mathbb{R}$	Real	Numbers that can be represented by a point on a line

Every natural number is an integer, and every integer is a rational number. For example: $3 = \frac{3}{1} = \frac{6}{2} = \cdots$. But not every rational number is an integer and not every integer is a natural number. Real numbers that cannot be expressed as a ratio of integers are called **irrational**. π and $\sqrt{2}$ are examples of irrational numbers. Every real number has a decimal representation. If the decimal representation of a number has a sequence of digits that repeats forever, it is a rational number.

• **Converting decimals to fractions.**

We convert a rational number from its decimal representation to a fraction representation by following these steps:

1. Set $x =$ the repeating decimal.
2. Multiply x and the decimal representation by enough powers of 10 to bring one repeating sequence to the left of the decimal point.
3. Multiply x and the decimal representation by enough powers of 10 so that the first repeating sequence starts immediately after the decimal point.
4. Subtract the results of Step 2 from the results of Step 1. This creates an equation of the form *integer* $\times x =$ *another integer*.
5. Divide both sides by the coefficient and reduce.

Exercise: Convert each repeating decimal to its fractional representation.

(a) $0.\overline{4578}$

Answer:

Let $x = 0.\overline{4578}$. The repeating sequence has 4 digits in it and no digit before the repeating sequence. Start by multiplying both sides by 10^4:

$$10{,}000x = 4578.\overline{4578}$$
$$1x = 0.\overline{4578} \qquad \textit{Subtract}$$
$$9999x = 4578$$
$$x = \frac{4578}{9999} = \frac{1526}{3333} \qquad \textit{Divide}$$

So a fractional representation of $0.\overline{4578}$ is $\dfrac{1526}{3333}$.

(b) $3.12\overline{3}$

Let $x = 3.12\overline{3}$. The repeating sequence has 1 digit in it and there are 2 digits before the repeating sequence starts, so we multiply both sides by 10^3: $1000x = 3123.\overline{3}$. Next we need to multiply both sides by 10^2 to bring the 2 nonrepeating digits to the other side of the decimal point: $100x = 312.\overline{3}$. Thus we get:

$$1000x = 3123.\overline{3}$$
$$100x = 312.\overline{3} \qquad \textit{Subtract}$$
$$900x = 2811$$
$$x = \frac{2811}{900} = \frac{937}{300} \qquad \textit{Divide}$$

So a fractional representation of $3.12\overline{3}$ is $\dfrac{937}{300}$.

- **Adding, subtracting, multiplying and dividing fractions.**

We add and subtract fractions by first putting them over a common denominator.

Exercise: Calculate $\dfrac{1}{2} + \dfrac{2}{3}$.

Answer: We rewrite the fractions with the common denominator, 6: $\dfrac{1}{2} + \dfrac{2}{3} = \dfrac{3}{6} + \dfrac{4}{6}$. Then we add the fractions:
$$\dfrac{3}{6} + \dfrac{4}{6} = \dfrac{7}{6}.$$

Exercise: Calculate $4 - \dfrac{5}{8}$.

Answer: We rewrite the fractions with the common denominator, 8: $\dfrac{32}{8} - \dfrac{5}{8}$. Then we subtract the fractions:
$$\dfrac{32}{8} - \dfrac{5}{8} = \dfrac{27}{8}.$$

We multiply fractions by multiplying both numerators and denominators. We divide fractions by multiplying by the reciprocal.

Exercise: Calculate $\dfrac{5}{6} \times \dfrac{4}{15}$.

Answer: We multiply numerators and denominators:
$\dfrac{5}{6} \times \dfrac{4}{15} = \dfrac{5 \times 4}{6 \times 15}$. Then we express the result in lowest terms:
$$\dfrac{5 \times 4}{6 \times 15} = \dfrac{20}{90} = \dfrac{2}{9}.$$

Exercise: Calculate $\dfrac{5}{21} \div \dfrac{3}{14}$.

Answer: We multiply by the reciprocal:
$\dfrac{5}{21} \div \dfrac{3}{14} = \dfrac{5}{21} \times \dfrac{14}{3}$. Then we multiply numerators and denominators: $\dfrac{5}{21} \times \dfrac{14}{3} = \dfrac{5 \times 14}{21 \times 3}$. Finally, we express the result in lowest terms: $\dfrac{5 \times 14}{21 \times 3} = \dfrac{70}{63} = \dfrac{10}{9}$.

- **Expressing intervals in various notations.**

We can express intervals in three ways. For example, the set of all points strictly greater than 3 and less than or equal to 5 can be written in set-builder notation: $\{x \mid 3 < x \le 5\}$, interval notation: $(3, 5]$, or drawn on a number line (as described further below):

Exercise: Express $[-1, 5]$ in set-builder notation.

Answer: $[-1, 5]$ is the set of all real numbers between -1 and 5, including -1 and 5: $\{x \mid -1 \le x \le 5\}$.

(B) Properties of real numbers.

The basic properties used in combining real numbers are:

Commutative Laws	$a + b \;\;= b + a$	$ab \;\;\;\;\;= ba$
Associative Laws	$(a + b) + c \;\;= a + (b + c)$	$(ab)\,c \;\;= a\,(bc)$
Distributive Laws	$a\,(b + c) \;\;= ab + ac$	$(b + c)\,a \;\;= ab + ac$

The number 0, called the **additive identity**, is special for addition because $a + 0 = a$ for any real number a. Every real number a has a negative, $-a$, that satisfies $a + (-a) = 0$. **Subtraction** is the operation that undoes addition and we define $a - b = a + (-b)$. We use the following properties to handle negatives:

$(-1)\,a = -a$	$-(-a) = a$	$(-a)\,b = a\,(-b) = -\,(ab)$
$(-a)\,(-b) = ab$	$-(a + b) = -a - b$	$-(a - b) = b - a$

The number 1, called the **multiplicative identity**, is special for multiplication because $a \cdot 1 = a$ for any real number a. Every nonzero real number a has an inverse, $1/a$, that satisfies $a \cdot (1/a) = 1$.

Exercise: Use the properties of real numbers to write the given expression without parentheses.	**Answer:**
(a) $4\,(3 + m) - 2\,(4 + 3m)$	$4\,(3 + m) - 2\,(4 + 3m) = 12 + 4m - 8 - 6m$ $\qquad\qquad\qquad\qquad\quad = 4 - 2m$ Remember to properly distribute the -2 over $(4 + 3m)$.
(b) $(2a + 3)\,(5a - 2b + c)$	$(2a + 3)\,(5a - 2b + c)$ $\quad = (2a + 3)\,5a + (2a + 3)\,(-2b) + (2a + 3)\,c$ $\quad = 10a^2 + 15a - 4ab - 6b + 2ac + 3c$

(C) Properties of negative numbers.

To combine real numbers involving negative numbers we use the following properties:

Property	Example
1. $(-1)\,a = -a$	$(-1)\,5 = -5$
2. $-(-a) = a$	$-(-5) = 5$
3. $(-a)\,b = a\,(-b) = -\,(ab)$	$(-5)\,7 = 5\,(-7) = -\,(5 \cdot 7)$
4. $(-a)\,(-b) = ab$	$(-4)\,(-3) = 4 \cdot 3$
5. $-(a + b) = -a - b$	$-(3 + 5) = -3 - 5$
6. $-(a - b) = -b - a$	$-(5 - 8) = 8 - 5$

• **Adding, subtracting, multiplying and dividing positive and negative numbers**

Exercise: Compute the following: || **Answer:**

(a) $3 - 5$ || $3 - 5 = -2$

(b) $3 - (-5)$ || $3 - (-5) = 8$

(c) $(-3)(4)$ || $(-3)(4) = -12$ (Property 3)

(d) $(-3)(-5)$ || $(-3)(-5) = 15$ (Property 4)

(D) **The real number line.**

The real numbers can be represented by points on a line as follows:

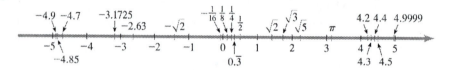

• **Locating points and sets on the real number line.**

Exercise: State whether the given inequality is true or false. || **Answer:**

(a) $-3.1 > -3$ || False; $-3.1 - (-3) = -0.1$.

(b) $2 \le 2$ || True; $2 - 2 = 0$

(c) $15.3 \ge -16.3$ || True; $15.3 - (-16.3) = 31.6$.

Exercise: Sketch the numbers $7, -2, 0$, and $\sqrt{2}$ on the real number line. || **Answer:**

Exercise: Sketch the following intervals on the real number line: || **Answer:**

(a) $[-2, 4]$ ||

(b) $(-3, -\sqrt{2}]$ ||

(c) $(-\pi, \pi)$ ||

(d) $\{x \mid x \ge 4\}$ ||

(E) Unions and intersections of sets.

The two key binary operations for sets are called **union** and **intersection**. The union of two sets is the set that consists of the elements that are in *either* set. The intersection of two sets is the set that consists of the elements in *both* sets. The **empty set**, $\varnothing$, is a set that contains no element.

- **Finding the union and intersection of two given sets.**

Exercise: Let $A = \{2, 4, 6, 8, 10, 12\}$, $B = \{3, 6, 9, 12\}$, and $C = \{1, 3, 5, 7, 9, 11\}$. Find $A \cup B$, $A \cup C$, and $B \cup C$.

Answer: $A \cup B = \{2, 3, 4, 6, 8, 9, 10, 12\}$, $A \cup C = \{1, 2, 3, 4, 5, 6, 7, 8, 9, 10, 11, 12\}$, and $B \cup C = \{1, 3, 5, 6, 7, 9, 11, 12\}$.

Exercise: Let $A = \{2, 4, 6, 8, 10, 12\}$, $B = \{3, 6, 9, 12\}$, and $C = \{1, 3, 5, 7, 9, 11\}$. Find $A \cap B$, $A \cap C$, and $B \cap C$.

Answer: $A \cap B = \{6, 12\}$, $A \cap C = \varnothing$, and $B \cap C = \{3, 9\}$.

Exercise: Let $A = \{x \mid x < 8\}$, $B = \{x \mid 3 \le x < 7\}$, and $C = \{x \mid 5 < x\}$. Find $A \cup B$, $A \cup C$, and $B \cup C$.

Answer: $A \cup B = \{x \mid x < 8\}$, $A \cup C = \{x \mid -\infty < x < \infty\} = \mathbb{R}$, and $B \cup C = \{x \mid 3 \le x\}$.

(F) Absolute value and distance.

The **absolute value** of a number a, denoted by $|a|$, is the distance from the number a to 0 on the real number line.

Remember that absolute value is always positive or zero. It is also defined as $|a| = \begin{cases} a & \text{if } a \ge 0 \\ -a & \text{if } a < 0 \end{cases}$ If a and b are real numbers, then the distance between the points a and b on the real line is $d(a, b) = |b - a|$. By Property 6 of negative numbers, $|b - a| = |a - b|$ (since the distance between a and b is the same as the distance between b and a).

- **Finding the absolute value of a given real number.**

Exercise: Find each absolute value.

Answer:

(a) $|-7|$

Since $-7 < 0$ we use the definition $|a| = -a$, so $|-7| = -(-7) = 7$.

(b) $|8|$

Since $8 \ge 0$ we use the definition $|a| = a$, so $|8| = 8$.

(c) $|\pi - 5|$

$|\pi - 5| = -(\pi - 5) = 5 - \pi$, since $\pi - 5 < 0$.

(d) $\left| \sqrt{10} - 3 \right|$

Since $\sqrt{10} > \sqrt{9} = 3$, we have $\sqrt{10} - 3 > 0$, so $\left| \sqrt{10} - 3 \right| = \sqrt{10} - 3$

• **Finding the distance between two given real numbers.**

Exercise: Find the distance between each pair of numbers.	**Answer:**
(a) 4 and 10	$|4 - 10| = |-6| = 6$. Remember to do the work on the inside of the absolute value bars first!
(b) 5 and -3	$|5 - (-3)| = |5 + 3| = |8| = 8$
(c) -2 and -9	$|-2 - (-9)| = |-2 + 9| = |7| = 7$

◤ Core Exercises

1, 5, 29, 33, 55, 57, 71, 75, 77, 79

◥ P.3   Integer Exponents

◤ Concepts

Ⓐ **Exponential notation.**

Ⓑ **Laws of Exponents.**

• Performing computations involving exponential notation.

Ⓒ **Scientific notation.**

• Performing computations involving scientific notation.

◤ Definitions

• $a^n = \underbrace{a \cdot a \cdot a \cdots a}_{n \text{ times}}$

• $a^0 = 1$

• $a^{-n} = \dfrac{1}{a^n}$

◤ Hints and Tips

• The next two sections may seem to you like a whole lot of things to memorize. And that is partially true. If you have ever played a board game like checkers, the first step was learning how the pieces move, how the checkers become kings when they reach the other end of the board, how jumping works, and so on. The rules of exponents and radicals form a large part of the basic rules of precalculus. If you take the time now to learn them solidly, you will find the rest of the course a lot easier, just as checkers is much easier to play if you know how the pieces move. The text proves a few of the rules; if you understand *why* the rules are true, that will make them easier to memorize and to retain.

▛ Review

Ⓐ Exponential notation.

We represent repeated multiplication ($6 \times 6 \times 6 \times 6 = 1296$) with exponential notation ($6^4 = 1296$).

The key exponent definitions are:

$$a^n = \underbrace{a \cdot a \cdot a \cdot \ldots \cdot a}_{n \text{ factors of } a} \qquad a^0 = 1, a \neq 0 \qquad a^{-n} = \frac{1}{a^n}, a \neq 0 \qquad a^{r/s} = \sqrt[s]{a^r} = \left(\sqrt[s]{a}\right)^r$$

Ⓑ Laws of exponents

The elegant thing about exponential notation is that it allows us to transform complicated expressions into simpler ones. The following exponent laws should be memorized:

$a^m a^n = a^{m+n}$	To multiply two powers of the same number, add the exponents.
$\dfrac{a^m}{a^n} = a^{m-n}, a \neq 0$	To divide two power of the same number, subtract the exponents.
$(a^m)^n = a^{mn}$	To raise a power to a new power, multiply the exponents.
$(ab)^n = a^n b^n$	To raise a product to a power, raise each factor to the power.
$\left(\dfrac{a}{b}\right)^n = \dfrac{a^n}{b^n}, b \neq 0$	To raise a quotient to a power, raise both numerator and denominator to the power.
$\left(\dfrac{a}{b}\right)^{-n} = \left(\dfrac{b}{a}\right)^n$	To raise a fraction to a negative power, invert the fraction and change the sign of the exponent.
$\dfrac{a^{-n}}{b^{-m}} = \dfrac{b^m}{a^n}$	To move a number raised to a power from numerator to denominator or from the denominator to numerator, change the sign of the exponent.

● **Performing computations involving exponential notation.**

Exercise: Simplify.

(a) $x^3 x^5$

(b) $w^{12} w^{-8}$

(c) $\dfrac{y^8}{y^{15}}$

(d) $(5x)^3$

(e) $\left(m^3\right)^7$

(f) $\dfrac{2^9}{2^7}$

Answer:

$x^3 x^5 = x^{3+5} = x^8$

$w^{12} w^{-8} = w^{12+(-8)} = w^4$

$\dfrac{y^8}{y^{15}} = y^{8-15} = y^{-7} = \dfrac{1}{y^7}$ or $\dfrac{y^8}{y^{15}} = \dfrac{1}{y^{15-8}} = \dfrac{1}{y^7}$

$(5x)^3 = 5^3 x^3 = 125x^3$

$\left(m^3\right)^7 = m^{3 \cdot 7} = m^{21}$

$\dfrac{2^9}{2^7} = 2^{9-7} = 2^2 = 4$ or $\dfrac{2^9}{2^7} = \dfrac{1}{2^{7-9}} = \dfrac{1}{2^{-2}} = 2^2 = 4$

Shortcut: In problems like 1(c) and 1(f) above, compare the exponents in the numerator and denominator. If the exponent in the numerator is larger, we simplify the expression by bringing the factor up into the numerator. If the exponent in the denominator is larger, we simplify the expression by bringing the factor down into the denominator.

Exercise: Simplify $\dfrac{27^3}{9^5}$.

Answer: Since both 27 and 9 are powers of 3, first express each number as a power of 3.

$$\frac{27^3}{9^5} = \frac{\left(3^3\right)^3}{\left(3^2\right)^5} = \frac{3^{3 \cdot 3}}{3^{2 \cdot 5}} = \frac{3^9}{3^{10}} = \frac{1}{3^{10-9}} = \frac{1}{3}.$$

Exercise: Simplify $\left(2x^7y^5\right)\left(3x^2y^3\right)^4$.

Answer:
$$\begin{aligned}
\left(2x^7y^5\right)\left(3x^2y^3\right)^4 &= \left(2x^7y^5\right)\left[(3)^4\left(x^2\right)^4\left(y^3\right)^4\right] \\
&= \left(2x^7y^5\right)\left(3^4 x^8 y^{12}\right) \\
&= (2)(81)\left(x^7 x^8\right)\left(y^5 y^{12}\right) \\
&= 162 x^{15} y^{17}
\end{aligned}$$

Exercise: Simplify $\left(\dfrac{x^4 y}{z^6}\right)^3 \left(\dfrac{xz^3}{y^4}\right)^5$.

Answer:
$$\begin{aligned}
\left(\frac{x^4 y}{z^6}\right)^3 \left(\frac{xz^3}{y^4}\right)^5 &= \frac{\left(x^4\right)^3 y^3}{\left(z^6\right)^3} \frac{x^5 \left(z^3\right)^5}{\left(y^4\right)^5} \\
&= \frac{x^{12} y^3}{z^{18}} \frac{x^5 z^{15}}{y^{20}} \\
&= \left(x^{12} x^5\right)\left(\frac{y^3}{y^{20}}\right)\left(\frac{z^{15}}{z^{18}}\right) \\
&= \frac{x^{17}}{y^{17} z^3}
\end{aligned}$$

Exercise: Eliminate negative exponents and simplify.

Answer:

(a) $\dfrac{3a^3 b^{-4}}{2a^{-2} b^{-1}}$

$$\frac{3a^3 b^{-4}}{2a^{-2} b^{-1}} = \frac{3}{2} a^{3-(-2)} b^{-4-(-1)} = \frac{3}{2} a^5 b^{-3} = \frac{3a^5}{2b^3}$$

(b) $\left(\dfrac{4w^3 v^{-4}}{2w^2 v^5}\right)^{-3}$

Method 1: First distribute the exponent -3, then simplify.
$$\begin{aligned}
\left(\frac{4w^3 v^{-4}}{2w^2 v^5}\right)^{-3} &= \frac{4^{-3} w^{3(-3)} v^{-4(-3)}}{2^{-3} w^{2(-3)} v^{5(-3)}} = \frac{2^{-6} w^{-9} v^{12}}{2^{-3} w^{-6} v^{-15}} \\
&= 2^{-6-(-3)} w^{-9-(-6)} v^{12-(-15)} \\
&= 2^{-3} w^{-3} v^{27} = \frac{v^{27}}{2^3 w^3} = \frac{v^{27}}{8w^3}
\end{aligned}$$

Method 2: Simplify inside the parentheses, then simplify.
$$\begin{aligned}
\left(\frac{4w^3 v^{-4}}{2w^2 v^5}\right)^{-3} &= \left(2w^{3-2} v^{-4-5}\right)^{-3} \\
&= \left(2wv^{-9}\right)^{-3} \\
&= 2^{-3} w^{-3} v^{-9(-3)} = 2^{-3} w^{-3} v^{27} \\
&= \frac{v^{27}}{2^3 w^3} = \frac{v^{27}}{8w^3}
\end{aligned}$$

There are additional strategies that could have been used to solve this exercise.

(c) $\dfrac{\left(5^3 m^7 n^2\right)^3}{\left(3^4 m^{-2} n^3\right)^4}$

$$\frac{\left(5^3 m^7 n^2\right)^3}{\left(3^4 m^{-2} n^3\right)^4} = \frac{5^{3(3)} m^{7(3)} n^{2(3)}}{3^{4(4)} m^{-2(4)} n^{3(4)}} = \frac{5^9 m^{21} n^6}{3^{16} m^{-8} n^{12}}$$

$$= \frac{5^9 m^{21-(-8)}}{3^{16} n^{12-6}} = \frac{5^9 m^{29}}{3^{16} n^6}$$

ⓒ Scientific Notation

Scientific notation is used to express very large numbers or very small numbers in a more compact form. The goal is to express the positive number x in the form $a \times 10^n$, where $1 \le a < 10$ and n is an integer. Calculators have a limited number of digits they can display. Any number greater than that must be displayed in scientific notation.

• **Performing computations involving scientific notation.**

Exercise: Write the following numbers in scientific notation.	**Answer:**

(a) 23,500,000,000

23,500,000,000
↑ ↑↑↑ ↑↑↑ ↑↑↑

We must move the decimal point ten places to the left. This gives us the representation 2.35×10^{10}.

(b) 0.000000000067

0.000000000067
↑↑↑↑↑↑↑↑↑↑↑

We must move the decimal point eleven places to the right. This gives the representation 6.7×10^{-11}.

Exercise: If $a \approx 4.1 \times 10^{-7}$, $b \approx 1.97 \times 10^9$, and $c \approx 3.24 \times 10^{-3}$, find the quotient $\dfrac{a}{bc}$.

Answer: $\dfrac{a}{bc} \approx \dfrac{4.1 \times 10^{-7}}{\left(1.97 \times 10^9\right)\left(3.24 \times 10^{-3}\right)}$

$\approx \dfrac{4.1}{(1.97)(3.24)} \times 10^{-7-9-(-3)}$

$\approx 0.64 \times 10^{-13} \approx 6.4 \times 10^{-14}$

Notice that the answer is stated to two significant digits because the least accurate of the numbers has two significant digits.

◤ Core Exercises

19, 47, 65, 75

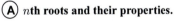

P.4   Rational Exponents and Radicals

◤ Concepts

Ⓐ **nth roots and their properties.**

• Simplifying and combining radicals.
• Rational exponents.
• Rationalizing denominators.

Definitions

- $a^{r/s} = \sqrt[s]{a^r} = \left(\sqrt[s]{a}\right)^r$

- $a^0 = 1$

- $a^{-r/s} = \dfrac{1}{a^{r/s}}$

- **Principal nth root:** If n is a positive integer, the principal nth root of a is the positive number b such that $b^n = a$.

- **Rationalizing the denominator:** If a fraction has a radical symbol in the denominator, we rationalize the denominator by rewriting the fraction in a form that does not have a radical.

Questions to Ask Your Teacher

Do you expect students to rationalize the denominators in all of their answers?

Review

(A) n**th roots and their properties.**

If n is a positive integer, the principal nth root of a, $\sqrt[n]{a}$, is the positive number b such that $b^n = a$. So we say that $\sqrt{16} = 4$, $\sqrt[3]{8} = 2$, $\sqrt[4]{81} = 3$, and so forth. Notice that we write $\sqrt{16}$ instead of $\sqrt[2]{16}$. We can take odd roots of both positive and negative numbers: $\sqrt[3]{-125} = -5$, because $(-5)^3 = -125$. We cannot take even roots of negative numbers, because any even power of any number is positive. Also notice that $\sqrt{16} = 4$ and $\sqrt{16} \neq -4$, even though both 4^2 and $(-4)^2$ equal 16.

The key rules of radicals are the following:

$$\sqrt{a^2} = |a|$$

If $a \geq 0$ and $b \geq 0$, then $\sqrt{a}\sqrt{b} = \sqrt{ab}$.   If $a \geq 0$ and $b > 0$, then $\sqrt{\dfrac{a}{b}} = \dfrac{\sqrt{a}}{\sqrt{b}}$.

$$\sqrt[n]{ab} = \sqrt[n]{a}\,\sqrt[n]{b} \qquad \sqrt[n]{\dfrac{a}{b}} = \dfrac{\sqrt[n]{a}}{\sqrt[n]{b}} \qquad \sqrt[m]{\sqrt[n]{a}} = \sqrt[mn]{a} \qquad \sqrt[n]{a^n} = \begin{cases} a & \text{if } n \text{ is odd} \\ |a| & \text{if } n \text{ is even} \end{cases}$$

- **Simplifying and combining radicals.**

Exercise: Simplify the following expressions.

Answer:

(a) $\sqrt{28}$

$\sqrt{28} = \sqrt{4}\sqrt{7} = 2\sqrt{7}$

(b) $\sqrt{50}\sqrt{8}$

$\sqrt{50}\sqrt{8} = \sqrt{50 \cdot 8} = \sqrt{400} = 20$ or
$\sqrt{50}\sqrt{8} = 5\sqrt{2} \cdot 2\sqrt{2} = 10\sqrt{4} = 20$

(c) $\dfrac{\sqrt{54}}{\sqrt{6}}$

$\dfrac{\sqrt{54}}{\sqrt{6}} = \sqrt{\dfrac{54}{6}} = \sqrt{9} = 3$

(d) $\sqrt[3]{16}$

$\sqrt[3]{16} = \sqrt[3]{8}\sqrt[3]{2} = 2\sqrt[3]{2}$

(e) $\sqrt[4]{81a^8b^{16}c^5}$

$\sqrt[4]{81a^8b^{16}c^5} = \sqrt[4]{81}\sqrt[4]{a^8}\sqrt[4]{b^{16}}\sqrt[4]{c^4}\sqrt[4]{c} = 3a^2c\sqrt[4]{c}$

Exercise: Simplify.	**Answer:**
(a) $\sqrt[4]{81}$	$\sqrt[4]{81} = \sqrt[4]{3^4} = 3$
(b) $\sqrt[5]{-32}$	$\sqrt[5]{-32} = \sqrt[5]{(-2)^5} = -2$. Odd roots of negative numbers are negative.
(c) $\sqrt[3]{32}$	$\sqrt[3]{32} = \sqrt[3]{2^5} = \sqrt[3]{2^3}\sqrt[3]{2^2} = 2\sqrt[3]{2^2} = 2\sqrt[3]{4}$

- **Rational exponents.**

The definitions of rational exponents are $a^{1/n} = \sqrt[n]{a}$ and $a^{m/n} = (\sqrt[n]{a})^m = \sqrt[n]{a^m}$.

Exercise: Simplify.	**Answer:**
(a) $8^{2/3}$	$8^{2/3} = (\sqrt[3]{8})^2 = 2^2 = 4$ or $8^{2/3} = \sqrt[3]{8^2} = \sqrt[3]{64} = 4$. Both give the same solution. However, most students find the first way easier.
(b) $9^{3/2}$	$9^{3/2} = (\sqrt{9})^3 = 3^3 = 27$.
(c) $(8x^3y^9)^{5/3}$	$(8x^3y^9)^{5/3} = 8^{5/3}x^{3(5/3)}y^{9(5/3)} = 2^{3(5/3)}x^{3(5/3)}y^{9(5/3)} = 2^5x^5y^{15} = 32x^5y^{15}$

- **Rationalizing denominators.**

Exercise: Rationalize the denominator in each of the following expressions.	**Answer:**
(a) $\frac{1}{\sqrt{2}}$	$\frac{1}{\sqrt{2}} = \frac{1}{\sqrt{2}} \cdot \frac{\sqrt{2}}{\sqrt{2}} = \frac{\sqrt{2}}{2}$
(b) $\sqrt{\frac{7}{3}}$	$\sqrt{\frac{7}{3}} = \frac{\sqrt{7}}{\sqrt{3}}$ Property 2 $= \frac{\sqrt{7}}{\sqrt{3}} \cdot \frac{\sqrt{3}}{\sqrt{3}}$ Rationalize by multiplying both numerator and denominator by $\sqrt{3}$ $= \frac{\sqrt{21}}{3}$ Simplify
(c) $\frac{1}{5\sqrt[3]{2}}$	$\frac{1}{5\sqrt[3]{2}}\left(\frac{\sqrt[3]{2^2}}{\sqrt[3]{2^2}}\right) = \frac{\sqrt[3]{4}}{10}$
(d) $\frac{3}{\sqrt{x}}$	$\frac{3}{\sqrt{x}}\left(\frac{\sqrt{x}}{\sqrt{x}}\right) = \frac{3\sqrt{x}}{x}$

Core Exercises

5, 55, 63, 69, 71

P.5   Algebraic Expressions

Concepts

(A) **Combining algebraic expressions.**

- Adding and subtracting polynomials.
- Multiplying algebraic expressions using the distributive property.
- Multiplying algebraic expressions using special product formulas.

Definitions

- **Variable**: A letter that can represent any number from a given set of numbers.
- **Domain of a variable**: The set of numbers that the variable is permitted to have.
- **Algebraic expression**: A set of variables and numbers combined using addition, subtraction, roots, and any other operations.
- **Polynomial**: An algebraic expression of the form $a_n x^n + a_{n-1} x^{n-1} + \cdots + a_1 x + a_0$, where x is a variable and the as are constants, with $a_n \neq 0$. (Example: $5x^4 + 3x^3 + x^2 - \frac{4}{27}x + 2$.)
- **Degree of a polynomial**: The highest power of the variable in the polynomial. (Example: The polynomial $5x^4 + 3x^3 + x^2 - \frac{4}{27}x + 2$ has degree 4.)
- **Monomial**: An algebraic expression of the form ax^k where a is a real number and k is a nonnegative integer. You may think of this as a polynomial with exactly one term. (For example: $3x^5$, $-5x^{11}$, or $\sqrt{2}x$.)
- **Binomial**: The sum of two monomials, or a polynomial with exactly two terms.
- **Trinomial**: The sum of three monomials, or a polynomial with exactly three terms.

Review

(A) **Combining algebraic expressions.**

Just as we can add, subtract, or multiply numbers, we can add, subtract or multiply algebraic expressions. (We will get to division later.) Just as we can say that $3(2 + 6) = 24$, we can say $x(3x + 2x) = 5x^2$.

- **Adding and subtracting polynomials.**

Terms with the same variables raised to the same powers are called **like terms**. Polynomials are added and subtracted using a process called combining like terms which utilizes the Distributive Law.

Exercise: Find
$$\left(5x^3 + 3x^2 - 2x + 1\right) + \left(x^3 - 6x^2 + 5x - 1\right).$$

Answer: First group like terms, then combine like terms.
$$\left(5x^3 + 3x^2 - 2x + 1\right) + \left(x^3 - 6x^2 + 5x - 1\right)$$
$$= \left(5x^3 + x^3\right) + \left(3x^2 - 6x^2\right) + \left(-2x + 5x\right) + (1 - 1)$$
$$= (5 + 1)x^3 + (3 - 6)x^2 + (-2 + 5)x$$
$$= 6x^3 - 3x^2 + 3x$$

Exercise: Find
$$\left(7x^3 - x^2 + 5\right) - \left(2x^3 - 6x^2 + 9x - 6\right).$$

Answer: First group like terms, then combine like terms.
$$\left(7x^3 - x^2 + 5\right) - \left(2x^3 - 6x^2 + 9x - 6\right)$$
$$= \left(7x^3 - 2x^3\right) - \left(x^2 - \left(-6x^2\right)\right) - 9x + (5 - (-6))$$
$$= 5x^3 + 5x^2 - 9x + 11$$

• **Multiplying algebraic expressions using the distributive property.**

The product of two polynomials is found by repeated use of the Distributive Law.

Exercise: Find
$(x^2 - 4xy + 2y^2)(x + 4y)$.

Answer: Start by treating the first expression as a single term and distribute each term over $(x + 4y)$.

$(x^2 - 4xy + 2y^2)(x + 4y)$
$$= (x^2 - 4xy + 2y^2)x + (x^2 - 4xy + 2y^2)4y$$

Now distribute x over $(x^2 - 4xy + 2y^2)$ and distribute $4y$ over $(x^2 - 4xy + 2y^2)$.

$(x^2 - 4xy + 2y^2)x + (x^2 - 4xy + 2y^2)4y$
$$= x^3 - 4x^2y + 2xy^2 + 4x^2y - 16xy^2 + 8y^3$$
$$= x^3 - 14xy^2 + 8y^3$$

Exercise: Find $\left(\sqrt{x} - \dfrac{4}{\sqrt{x}}\right)(x + \sqrt{x})$.

Answer: Use the same methods used for multiplying polynomials.

$\left(\sqrt{x} - \dfrac{4}{\sqrt{x}}\right)(x + \sqrt{x})$

$$= \left(\sqrt{x} - \frac{4}{\sqrt{x}}\right)x + \left(\sqrt{x} - \frac{4}{\sqrt{x}}\right)\sqrt{x}$$
$$= x\sqrt{x} - \frac{4x}{\sqrt{x}} + \sqrt{x}\sqrt{x} - \frac{4\sqrt{x}}{\sqrt{x}}$$
$$= x\sqrt{x} - \frac{4x}{\sqrt{x}} + x - 4$$
$$= x\sqrt{x} - 4\sqrt{x} + x - 4$$

• **Multiplying algebraic expressions using special product formulas.**

The text recommends you memorize the following table of products. They come up so often in precalculus, and especially in calculus, that they are crucial to know. If you learn their names, you will find it easier to refer to them in this and in future classes.

Formula	Name
$(a - b)(a + b) = a^2 - b^2$	Difference of Squares
$(a + b)^2 = a^2 + 2ab + b^2$	Perfect Square
$(a - b)^2 = a^2 - 2ab + b^2$	Perfect Square
$(a + b)^3 = a^3 + 3a^2b + 3ab^2 + b^3$	Perfect Cube
$(a - b)^3 = a^3 - 3a^2b + 3ab^2 - b^3$	Perfect Cube

The operation of multiplying algebraic expressions is referred to as *expanding*.

Exercise: Use the special product formulas to find the following products.

Answer:

(a) $\left(3x - 2y^3\right)^2$

Perfect square:
$$\left(3x - 2y^3\right)^2 = (3x)^2 - 2\,(3x)\,(2y^3) + (2y^3)^2$$
$$= 9x^2 - 12xy^3 + 4y^6$$

(b) $\left(w^2 + 2w\right)^3$

Perfect cube:
$$\left(w^2 + 2w\right)^3$$
$$= \left(w^2\right)^3 + 3\left(w^2\right)^2 (2w) + 3\left(w^2\right)(2w)^2 + (2w)^3$$
$$= w^6 + 6w^5 + 12w^4 + 8w^3$$

(c) $\left(\sqrt{x} + \dfrac{1}{\sqrt{x}}\right)\left(\sqrt{x} - \dfrac{1}{\sqrt{x}}\right)$

Difference of squares:
$$\left(\sqrt{x} + \frac{1}{\sqrt{x}}\right)\left(\sqrt{x} - \frac{1}{\sqrt{x}}\right) = (\sqrt{x})^2 - \left(\frac{1}{\sqrt{x}}\right)^2 = x - \frac{1}{x}$$

(d) $\left(x + 3 + x^4\right)\left(x + 3 - x^4\right)$

Difference of squares:
$$\left(x + 3 + x^4\right)\left(x + 3 - x^4\right) = \left[(x + 3) + x^4\right]\left[(x + 3) - x^4\right]$$
$$= (x + 3)^2 - \left(x^4\right)^2$$
$$= (x + 3)^2 - x^8$$

Now we have the square of a binomial:
$$(x + 3)^2 - x^8 = x^2 + 6x + 9 - x^8 = -x^8 + x^2 + 6x + 9$$

Core Exercises

5, 11, 35, 57, 61

P.6   Factoring

Concepts

(B) **Factoring.**

- Factoring a general algebraic expression using common factors.
- Factoring $ax^2 + bx + c$ using trial and error.
- Factoring using the factoring formulas.
- Factoring expressions with fractional exponents.
- Factoring using regrouping.

◤ Definitions

- **Factoring:** Rewriting an algebraic expression as a product of simpler expressions.

◤ Questions to Ask Your Teacher

If you are factoring an expression and wind up with something like $(1 + x)^{-2/3}(1 + 2x)$, is that an acceptable form of the answer, or should you always rewrite it $\dfrac{1 + 2x}{(1 + x)^{2/3}}$ or $\dfrac{1 + 2x}{\sqrt[3]{(1 + x)^2}}$?

◤ Hints and Tips

When I was learning factoring, the hardest part for me was knowing when it was okay to stop. I would have something like $3x^2 + 6x$ and write it as $x(3x + 6)$ and think I was done, only to be informed that I could factor it further by writing $3x(x + 2)$. When you think you are done with a factoring problem, look at every term in your answer, thinking, "can I squeeze another factor out of this term?" The most important thing for you to do is practice — it does get easier.

(B) Factoring.

Factoring is the process of writing an algebraic expression (a sum) as a product of simpler ones. Sometimes a calculus or precalculus problem will be *much* easier if you factor an expression first, and sometimes it will be easier if you do the opposite — multiplying out the expression. Many errors students make in calculus turn out to be factoring errors.

- **Factoring a general algebraic expression using common factors.**

The first step in factoring an expression is to see if each term has a common factor.

Exercise: Factor.

(a) $6x^4 - 2x^7 + 8x^8$

Answer:

The greatest common factor (GCF) of 6, -2 and 8 is 2. The GCF of x^4, x^7 and x^8 is x^4. We now have
$6x^4 - 2x^7 + 8x^8 = 2x^4 \left(3 - x^3 + 4x^4\right)$.
We can check our work by multiplying out our answer using the distributive property (expanding):
$2x^4 \left(3 - x^3 + 4x^4\right) = 6x^4 - 2x^7 + 8x^8$.

(b) $4x^2 w^3 - 16x^5 w^4 + 28x^6 w^6$

The GCF of 4, -16, and 28 is 4; the GCF of x^2, x^5, and x^6 is x^2, and the GCF of w^3, w^4, and w^6 is w^3.
$4x^2 w^3 - 16x^5 w^4 + 28x^6 w^6 = 4x^2 w^3 \left(1 - 4x^3 w + 7x^4 w^3\right)$
We can check our work by multiplying out our answer using the distributive property (expanding):
$4x^2 w^3 \left(1 - 4x^3 w + 7x^4 w^3\right) = 4x^2 w^3 - 16x^5 w^4 + 28x^6 w^6$.

(c) $x^2 - 12x + 20$

Note that x^2, $12x$, and 20 have no common factor, so we move to trying to factor the trinomial into two binomials. Since c is positive and b is negative, r and s are both negative factors of 20 and their sum is 12.

Factors	$1 \cdot 20$	$2 \cdot 10$	$4 \cdot 5$
Corresponding sum	21	12	9

Therefore, taking $r = -2$ and $s = -10$, we get the factorization $x^2 - 12x + 20 = (x - 2)(x - 10)$.

(d) $x^2 + 6x - 16$

Since c is negative, r and s have different signs and we look for a difference of 6.

Factors	$1 \cdot 16$	$2 \cdot 8$	$4 \cdot 4$
Corresponding sum	15	6	0

Since b is positive, the larger factor is positive, so taking $r = 8$ and $s = -2$, we get $x^2 + 6x - 16 = (x + 8)(x - 2)$.

- **Factoring $ax^2 + bx + c$ using trial and error.**
  An algebraic expression of the form $ax^2 + bx + c$ is called a *quadratic expression*. If $a = 1$, the expression will be in the form $x^2 + bx + c$ and the factored form will be $(x + r)(x + s)$. Your job will be to find the r and s by trial and error, given that $rs = c$.

Exercise: Factor.

Answer:

(a) $x^2 + x - 12$

We are looking for r and s such that $(x + r)(x + s) = x^2 + x - 12$, and we are going to use trial and error. We know that $rs = -12$.

Guess	$(x + r)(x + s)$
$r = -1, s = 12$	$(x - 1)(x + 12) = x^2 + 11x - 12$
$r = -2, s = 6$	$(x - 2)(x + 6) = x^2 + 4x - 12$
$r = -3, s = 4$	$(x - 3)(x + 4) = x^2 + x - 12$
$r = 1, s = -12$	$(x + 1)(x - 12) = x^2 - 11x - 12$
$r = 2, s = -6$	$(x + 2)(x - 6) = x^2 - 4x - 12$
$r = 3, s = -4$	$(x + 3)(x - 4) = x^2 - x - 12$

So our final answer is $x^2 + x - 12 = (x - 3)(x + 4)$.

You do not have to try every possibility; you can stop as soon as you get the correct r and s values. Notice that, again, you can check your work by multiplying your answer out.

(b) $x^2 - 12x + 20$

We are looking for r and s such that

$(x + r)(x + s) = x^2 - 12x + 20$, and we are going to use trial and error. We know that $rs = 20$ and that r and s are both negative. [Otherwise, they would both be positive, and the middle term would also have to be positive. Try multiplying $(x + 5)(x + 4)$ if you don't understand this idea.]

Guess	$(x + r)(x + s)$
$r = -1, s = -20$	$(x - 1)(x - 20) = x^2 - 21x + 20$
$r = -2, s = -10$	$(x - 2)(x - 10) = x^2 - 12x + 20$

So our final answer is $x^2 - 12x + 20 = (x - 2)(x - 10)$.

The general case $ax^2 + bx + c$ where $a \neq 1$ is more tricky. Your answer will be of the form $(px + r)(qx + s)$ where $rs = c$ (as before) and $pq = a$.

Exercise: Factor the polynomial

$-2x^2 + 3x + 9$

Answer: We are looking for p, q, r, and s such that

$(px + r)(qx + s) = -2x^2 + 3x + 9$. We know that $rs = 9$ and $pq = -2$. Again, *you don't have to try all the cases; you can stop when you've found a quartet that works.*

p	q	r	s	$(px + r)(qx + s)$
2	-1	1	9	$(2x + 1)(-x + 9) = -2x^2 + 17x + 9$
-2	1	1	9	$(-2x + 1)(x + 9) = -2x^2 - 17x + 9$
2	-1	3	3	$(2x + 3)(-x + 3) = -2x^2 + 3x + 9$
-2	1	3	3	$(-2x + 3)(x + 3) = -2x^2 - 3x + 9$
2	-1	-3	-3	$(2x - 3)(-x - 3) = -2x^2 - 3x + 9$
-2	1	-3	-3	$(-2x - 3)(x - 3) = -2x^2 + 3x + 9$
2	-1	1	-9	$(2x - 1)(-x - 9) = -2x^2 - 17x + 9$
-2	1	1	-9	$(-2x - 1)(x - 9) = -2x^2 + 17x + 9$

So there are two correct answers:

$-2x^2 + 3x + 9 = (2x + 3)(-x + 3)$ and

$-2x^2 + 3x + 9 = (-2x - 3)(x - 3)$

Notes:

1. There are easier ways to factor quadratics. We will see some of them later in this section, and more in future sections.

2. Some quadratics cannot be factored into binomials with integer coefficients, as we've done above. Trial and error will *not* work to factor those quadratics — we will use other techniques.

• **Factoring using the factoring formulas.**

We have seen how tedious factoring can be. Certain special types of expressions turn up frequently. If you can memorize these special forms, your factoring work will be much easier. Formulas 1–3 are used to factor special quadratics, while Formulas 4 and 5 are used for sums or differences of cubes (third-degree polynomials).

	Formula	Name
1.	$a^2 - b^2 = (a - b)(a + b)$	Difference of Squares
2.	$a^2 + 2ab + b^2 = (a + b)^2$	Perfect Square
3.	$a^2 - 2ab + b^2 = (a - b)^2$	Perfect Square
4.	$a^3 - b^3 = (a - b)(a^2 + ab + b^2)$	Difference of Cubes
5.	$a^3 + b^3 = (a + b)(a^2 - ab + b^2)$	Sum of Cubes

Notice that Formulas 1, 4, and 5 have only *two terms* while Formulas 2 and 3 have *three terms*. When factoring a quadratic with three terms and the first and third terms are perfect squares, check the middle term to see if it is $2ab$; if so, then it is a perfect square. Also notice this is almost the same table that appeared earlier in the section. In this version of the table the sums are listed first followed by the product equivalent. In many cases, factoring techniques are combined with the special factoring formulas.

Exercise: Factor.

Answer:

(a) $x^2 - 8x + 16$

Since x^2 and 16 are perfect squares, we check the middle term to see if it fits the form $\pm 2ab$, that is, $-8x = -2(x)(4)$. It does, so we use the perfect square formula:
$x^2 - 8x + 16 = (x - 4)^2$.

(b) $25x^2 - 64$

Since this quadratic is the difference of two terms which are both perfect squares, we use the difference of squares formula:
$25x^2 - 64 = (5x - 8)(5x + 8)$.

(c) $9x^2 + 12x + 4$

Again, since $9x^2$ and 4 are perfect squares, we check the middle term and find: $12x = 2(3x)(2)$. Using the perfect square formula, we get $9x^2 + 12x + 4 = (3x + 2)^2$.

(d) $8x^3 - 64$

Here we have the difference of two terms which are both perfect cubes. Using the difference of cubes formula we get
$$8x^3 - 64 = (2x)^3 - (4)^3$$
$$= (2x - 4)\left[(2x)^2 + (2x)(4) + (4)^2\right]$$
$$= (2x - 4)(4x^2 + 8x + 16)$$
Note that $4x^2 + 8x + 16$ does not factor any further.

Exercise: Factor.

(a) $8x^4 - 24x^3 + 18x^2$

Answer:

First factor out the common factor of $2x^2$:
$8x^4 - 24x^3 + 18x^2 = 2x^2 \left(4x^2 - 12x + 9\right)$. Now factor the second factor (which is a perfect square):
$8x^4 - 24x^3 + 18x^2 = 2x^2 \left(2x - 3\right)^2$.

(b) $3x^2 - 27$

First factor out the common factor 3, then notice that the result is the difference of two squares:
$3x^2 - 27 = 3 \left(x^2 - 9\right) = 3 \left(x - 3\right) \left(x + 3\right)$.

- **Factoring expressions with fractional exponents.**

In cases of fractional exponents, it is best to factor out the power of a common term with the smallest exponent.

Exercise: Factor.

(a) $3x^{-5/3} + 6x^{1/3} + 6x^{4/3}$

Answer:

We first factor out $x^{-5/3}$ because $-\frac{5}{3}$ is the smallest exponent:
$3x^{-5/3} + 6x^{1/3} + 6x^{4/3} = x^{-5/3} \left(3 + 6x^{6/3} + 6x^{9/3}\right)$.

Then we do the arithmetic:
$x^{-5/3} \left(3 + 6x^{6/3} + 6x^{9/3}\right) = x^{-5/3} \left(3 + 6x^2 + 6x^3\right)$.

Finally, we can factor out a 3 from each term (3 is the GCD of 3, 6, and 6):
$x^{-5/3} \left(3 + 6x^2 + 6x^3\right) = 3x^{-5/3} \left(1 + 2x^2 + 2x^3\right)$.

(b) $\left(a^2 + 1\right)^{1/2} + \left(a^2 + 1\right)^{3/2}$

We factor out $\left(a^2 + 1\right)^{1/2}$ because $\frac{1}{2}$ is the smallest exponent:
$\left(a^2 + 1\right)^{1/2} + \left(a^2 + 1\right)^{3/2} = \left(a^2 + 1\right)^{1/2} \left(1 + \left(a^2 + 1\right)^{2/2}\right)$.

Now we simplify:
$\left(a^2 + 1\right)^{1/2} \left(1 + \left(a^2 + 1\right)^{2/2}\right) = \left(a^2 + 1\right)^{1/2} \left(a^2 + 2\right)$. In theory, we could write the answer as $\left(a^2 + 2\right) \sqrt{a^2 + 1}$. Some prefer that form.

- **Factoring using regrouping.**

Polynomials with four terms can sometimes be factored by grouping the terms into groups of 1 and 3 terms, or 2 and 2 terms, or 3 and 1 terms.

Exercise: Factor $x^6 + 2x^4 + 2x^2 + 4$.

Answer: First group into groups of two, then factor out the common factor:
$$x^6 + 2x^4 + 2x^2 + 4 = \left(x^6 + 2x^4\right) + \left(2x^2 + 4\right)$$
$$= x^4 \left(x^2 + 2\right) + 2 \left(x^2 + 2\right) = \left(x^4 + 2\right) \left(x^2 + 2\right)$$
Remember, the sum of squares does not factor.

Exercise: Factor $x^4 - x^2 + 6x - 9$.

Answer: Grouping into groups of two terms each gives no common factor. However, if we group $-x^2 + 6x - 9$ together and factor out -1, we get

$$
\begin{aligned}
x^4 - x^2 + 6x - 9 &= x^4 + \left(-x^2 + 6x - 9\right) \\
&= x^4 - \left(x^2 - 6x + 9\right) = x^4 - (x - 3)^2 \\
&= \left[x^2 - (x - 3)\right]\left[x^2 + (x - 3)\right] \\
&= \left(x^2 - x + 3\right)\left(x^2 + x - 3\right)
\end{aligned}
$$

▉ Core Exercises

19, 33, 59, 93, 95

▉ P.7   Rational Expressions

▉ Concepts

(A) **The domain of an algebraic expression.**

- Computing the domain of an algebraic expression.

(B) **Fractional expressions.**

- Simplifying fractional expressions.
- Multiplying and dividing fractional expressions.
- Adding and subtracting fractional expressions.
- Rationalizing the numerator or denominator of a fractional expression.

(C) **Compound fractions.**

- Simplifying a compound fraction.

(D) **Avoiding common errors.**

▉ Definitions

- **Domain of an algebraic expression**: The set of real values that the variable is allowed to have.
- **Fractional expression**: A quotient of two algebraic expressions.
- **Rational expression**: A fractional expression where both numerator and denominator are polynomials.
- **Compound fraction**: A fraction in which the numerator or the denominator is itself a fractional expression.

▉ Hints and Tips

- The following are two very different expressions: $\dfrac{x^3 + 3}{x}$ and $\dfrac{x}{x^3 + 3}$. The first can be written another way:

$\dfrac{x^3 + 3}{x} = \dfrac{x^3}{x} + \dfrac{3}{x} = x^2 + 3x^{-1}$. The second cannot; there is no way to simplify $\dfrac{x}{x^3 + 3}$.

Review

(A) **The domain of an algebraic expression.**

The domain of an algebraic expression is all of the permitted values of the variable. So what exactly is forbidden? At this stage, you have to look for zeros in the denominator, and negative square roots. (Actually, you have to look for negative even roots). As you go on in precalculus, you will be introduced to other functions that have their own restrictions.

• **Computing the domain of an algebraic expression.**

Exercise: Compute the domain.

Answer:

(a) $\dfrac{1}{x+5}$

We look for values of x that make the denominator 0. $x = -5$ is the only such value. So the domain of $\dfrac{1}{x+5}$ is $\{x \mid x \neq -5\}$.

(b) $\dfrac{\sqrt{x+6}}{x^2 - 3x + 2}$

We check for a zero denominator. It is easiest to do this if we factor the denominator: $\dfrac{\sqrt{x+6}}{x^2 - 3x + 2} = \dfrac{\sqrt{x+6}}{(x-2)(x-1)}$. So we forbid $x = 1$ and $x = 2$. We also have to check for values of x that would make the argument (the "inside") of the square root function negative. Clearly we need $x \geq -6$. So the domain of $\dfrac{\sqrt{x+6}}{x^2 - 3x + 2}$ is $\{x \mid x \geq -6, x \neq 1, x \neq 2\}$.

(c) $x^2 - 2x + 2$

There are no denominators or even roots. Therefore no value of x is forbidden, and the domain is all the real numbers, or $\mathbb{R}$.

(B) **Fractional expressions.**

We simplify a fractional expression by factoring its numerator and denominator and then cancelling out common factors. This should remind you of the way we simplify common, numerical fractions. Remember that rational expressions are just special kinds of fractional expressions, and so get simplified the same way.

• **Simplifying fractional expressions.**

Exercise: Simplify.

Answer:

(a) $\dfrac{x^{1/2} + 3x^{3/2} + x^{5/2}}{x^2 + 3x^3 + x^4}$

We first factor the numerator and denominator: $\dfrac{x^{1/2} + 3x^{3/2} + x^{5/2}}{x^2 + 3x^3 + x^4} = \dfrac{x^{1/2}\left(1 + 3x + x^2\right)}{x^2\left(1 + 3x + x^2\right)}$.

Then we cancel common factors:
$\dfrac{x^{1/2}\left(1 + 3x + x^2\right)}{x^2\left(1 + 3x + x^2\right)} = \dfrac{x^{1/2}}{x^2}$. In this case we can apply a simple law of exponents: $\dfrac{1}{x^{3/2}}$ or $x^{-3/2}$.

(b) $\dfrac{x^2 + 2x - 3}{x^2 - 1}$

$\dfrac{x^2 + 2x - 3}{x^2 - 1} = \dfrac{(x + 3)(x - 1)}{(x + 1)(x - 1)}$ Factor

$= \dfrac{x + 3}{x + 1}$ Cancel

- **Multiplying and dividing fractional expressions.**

The rule $\dfrac{a}{b} \cdot \dfrac{c}{d} = \dfrac{a \cdot c}{b \cdot d}$ is used to multiply two fractional expressions. However, before multiplying the numerator and denominator, first factor each expression, simplify by canceling common factors, and then multiply the remaining factors.

Exercise: Multiply

$\dfrac{x^2 - 2x + 1}{x^2 - 4} \cdot \dfrac{x^2 + 4x + 4}{x^2 + x - 2}$.

Answer: Factor, cancel common factors.

$\dfrac{x^2 - 2x + 1}{x^2 - 4} \cdot \dfrac{x^2 + 4x + 4}{x^2 + x - 2}$

$= \dfrac{(x - 1)(x - 1)}{(x + 2)(x - 2)} \cdot \dfrac{(x + 2)(x + 2)}{(x + 2)(x - 1)}$ Factor

$= \dfrac{x - 1}{x - 2}$ Cancel

The rule $\dfrac{a}{b} \div \dfrac{c}{d} = \dfrac{a}{b} \cdot \dfrac{d}{c} = \dfrac{a \cdot d}{b \cdot c}$ is used to divide two fractional expressions. This rule is commonly referred to as "*invert and multiply.*"

Exercise: Divide

$\dfrac{x^2 - x - 6}{x + 4} \div \dfrac{x^2 - 9}{x^2 + 3x - 4}$.

Answer:

$\dfrac{x^2 - x - 6}{x + 4} \div \dfrac{x^2 - 9}{x^2 + 3x - 4}$

$= \dfrac{x^2 - x - 6}{x + 4} \cdot \dfrac{x^2 + 3x - 4}{x^2 - 9}$ Invert & multiply

$= \dfrac{(x - 3)(x + 2)}{(x + 4)} \cdot \dfrac{(x + 4)(x - 1)}{(x - 3)(x + 3)}$ Factor

$= \dfrac{(x + 2)(x - 1)}{x + 3}$ Cancel

$= \dfrac{x^2 + x - 2}{x + 3}$ Simplify

Exercise: Divide

$\dfrac{x^3 - 7x^2 - 8x}{x^2 - 2x - 15} \div \dfrac{x^2 - 9x + 8}{x^3 - 9x^2 + 20x}$.

Answer:

$\dfrac{x^3 - 7x^2 - 8x}{x^2 - 2x - 15} \div \dfrac{x^2 - 9x + 8}{x^3 - 9x^2 + 20x}$

$= \dfrac{x^3 - 7x^2 - 8x}{x^2 - 2x - 15} \cdot \dfrac{x^3 - 9x^2 + 20x}{x^2 - 9x + 8}$ Invert & multiply

$= \dfrac{x(x + 1)(x - 8)}{(x - 5)(x + 3)} \cdot \dfrac{x(x - 4)(x - 5)}{(x - 1)(x - 8)}$ Factor

$= \dfrac{x^2(x + 1)(x - 4)}{(x + 3)(x - 1)}$ Cancel

$= \dfrac{x^4 - 3x^3 - 4x^2}{x^2 + 2x - 3}$ Simplify

- **Adding and subtracting fractional expressions.**

Just as in the addition of two rational numbers, two fractional expressions must first have a common denominator before they can be added. Then use $\dfrac{a}{b} + \dfrac{c}{b} = \dfrac{a+c}{b}$.

Exercise: Combine and simplify.

Answer:

(a) $\dfrac{2}{x-4} + \dfrac{3x}{4-x}$

Since $4 - x = -(x - 4)$ and $\dfrac{a}{-b} = -\dfrac{a}{b}$,

$$\frac{2}{x-4} + \frac{3x}{4-x} = \frac{2}{x-4} + \frac{3x}{-(x-4)}$$

$$= \frac{2}{x-4} - \frac{3x}{x-4} = \frac{2-3x}{x-4}$$

(b) $\dfrac{2}{x+3} + \dfrac{x}{x+5}$

The LCD (least common denominator) of $x + 3$ and $x + 5$ is $(x+3)(x+5)$.

$$\frac{2}{x+3} + \frac{x}{x+5} = \frac{2}{x+3} \cdot \frac{x+5}{x+5} + \frac{x}{x+5} \cdot \frac{x+3}{x+3}$$

$$= \frac{2x+10}{(x+3)(x+5)} + \frac{x^2+3x}{(x+3)(x+5)}$$

$$= \frac{x^2+5x+10}{(x+3)(x+5)}$$

Notice that the numerator is multiplied and simplified while the denominator is left in factored form.

(c) $\dfrac{2}{(x-3)(x+2)} + \dfrac{3}{(x+3)(x+2)}$

The LCD of $(x-3)(x+2)$ and $(x+3)(x+2)$ is $(x-3)(x+3)(x+2)$.

$$\frac{2}{(x-3)(x+2)} + \frac{3}{(x+3)(x+2)}$$

$$= \frac{2}{(x-3)(x+2)} \cdot \frac{x+3}{x+3} + \frac{3}{(x+3)(x+2)} \cdot \frac{x-3}{x-3}$$

$$= \frac{2x+6}{(x-3)(x+3)(x+2)} + \frac{3x-9}{(x-3)(x+3)(x+2)}$$

$$= \frac{5x-3}{(x-3)(x+3)(x+2)}$$

(d) $\dfrac{-2}{x^2 + 4x + 4} + \dfrac{1}{x^2 + 5x + 6}$

Start by factoring each denominator, then find the LCD.

$x^2 + 4x + 4 = (x + 2)(x + 2)$ $x^2 + 5x + 6 = (x + 2)(x + 3)$,

so the LCD is $(x + 2)(x + 2)(x + 3)$.

$$\dfrac{-2}{x^2 + 4x + 4} + \dfrac{1}{x^2 + 5x + 6}$$

$$= \dfrac{-2}{(x + 2)(x + 2)} + \dfrac{1}{(x + 2)(x + 3)}$$

$$= \dfrac{-2}{(x + 2)(x + 2)} \cdot \dfrac{x + 3}{x + 3} + \dfrac{1}{(x + 2)(x + 3)} \cdot \dfrac{x + 2}{x + 2}$$

$$= \dfrac{-2x - 6}{(x + 2)(x + 2)(x + 3)} + \dfrac{x + 2}{(x + 2)(x + 2)(x + 3)}$$

$$= \dfrac{-x - 4}{(x + 2)(x + 2)(x + 3)}$$

- **Rationalizing the numerator or denominator of a fractional expression.**

In general, it is good style to present fractional expressions without radicals in their denominators, if possible. It allows them to be manipulated more easily in multi-step problems. There are also occasions in calculus when problems become easy to solve once radicals are removed from the numerator. The previous section discussed rationalizing the denominator of a fraction; now we learn how to rationalize the denominator (or numerator) of a fractional expression.

Radicals are eliminated from the denominators by multiplying both numerator and denominator by an appropriate expression. This procedure is called **rationalizing the denominator**. If a fraction has a denominator of the form $A + B\sqrt{C}$, we may rationalize the denominator by multiplying numerator and denominator by the **conjugate radical** $A - B\sqrt{C}$. Then by the difference of squares formula, we have $\left(A + B\sqrt{C}\right)\left(A - B\sqrt{C}\right) = A^2 - B^2 C$.

Exercise: Rationalize the denominator.

Answer:

(a) $\dfrac{3}{1 - \sqrt{2}}$

We use the difference of squares formula to rationalize the denominator:

$$\dfrac{3}{1 - \sqrt{2}} = \dfrac{3}{1 - \sqrt{2}} \cdot \dfrac{1 + \sqrt{2}}{1 + \sqrt{2}}$$

$$= \dfrac{3 + 3\sqrt{2}}{1 - 2} = \dfrac{3 + 3\sqrt{2}}{-1} = -3 - 3\sqrt{2}$$

(b) $\dfrac{\sqrt{3}}{\sqrt{5} + \sqrt{3}}$

$\dfrac{\sqrt{3}}{\sqrt{5} + \sqrt{3}} = \dfrac{\sqrt{3}}{\sqrt{5} + \sqrt{3}} \cdot \dfrac{\sqrt{5} - \sqrt{3}}{\sqrt{5} - \sqrt{3}} = \dfrac{\sqrt{15} - 3}{5 - 3} = \dfrac{\sqrt{15} - 3}{2}$

Exercise: Rationalize the numerator of $\dfrac{\sqrt{3} - \sqrt{x}}{7}$.

Answer:

$\dfrac{\sqrt{3} - \sqrt{x}}{7} = \dfrac{\sqrt{3} - \sqrt{x}}{7} \cdot \dfrac{\sqrt{3} + \sqrt{x}}{\sqrt{3} + \sqrt{x}} = \dfrac{3 - x}{7\left(\sqrt{3} + \sqrt{x}\right)}$

(C) Compound fractions.

• **Simplifying a compound fraction.**

To simplify a compound fractional expression, first write each of the numerator and the denominator as a single term using a common denominator.

Exercise: Simplify $\dfrac{\dfrac{y}{x}+1}{\dfrac{x}{y}-\dfrac{y}{x}}$.

Answer:

Method 1: Use common denominators.

$$\frac{\dfrac{y}{x}+1}{\dfrac{x}{y}-\dfrac{y}{x}}=\frac{\dfrac{y}{x}+1\left(\dfrac{x}{x}\right)}{\left(\dfrac{x}{y}\right)\left(\dfrac{x}{x}\right)-\left(\dfrac{y}{x}\right)\left(\dfrac{y}{y}\right)}$$

$$=\frac{\dfrac{y}{x}+\dfrac{x}{x}}{\dfrac{x^2}{xy}-\dfrac{y^2}{xy}}=\frac{\dfrac{y+x}{x}}{\dfrac{x^2-y^2}{xy}}$$

$$=\frac{y+x}{x}\div\frac{x^2-y^2}{xy}=\frac{y+x}{x}\cdot\frac{xy}{x^2-y^2}$$

$$=\frac{y+x}{x}\cdot\frac{xy}{(x-y)(x+y)}=\frac{y}{x-y}$$

Method 2: Multiply by the LCD.

$$\frac{\dfrac{y}{x}+1}{\dfrac{x}{y}-\dfrac{y}{x}}=\frac{\left(\dfrac{y}{x}+1\right)\cdot xy}{\left(\dfrac{x}{y}-\dfrac{y}{x}\right)\cdot xy}=\frac{y+xy}{x^2-y^2}$$

$$=\frac{y(x+y)}{(x-y)(x+y)}=\frac{y}{x-y}$$

Exercise: Simplify $\dfrac{\dfrac{1}{x}+\dfrac{x}{x+1}}{x-\dfrac{2}{x-1}}$.

Answer:

$$\frac{\dfrac{1}{x}+\dfrac{x}{x+1}}{x-\dfrac{2}{x-1}}=\frac{\dfrac{1}{x}\cdot\dfrac{x+1}{x+1}+\dfrac{x}{x+1}\cdot\dfrac{x}{x}}{x\cdot\dfrac{x-1}{x-1}-\dfrac{2}{x-1}}$$

$$=\frac{\dfrac{x+1}{x(x+1)}+\dfrac{x^2}{x(x+1)}}{\dfrac{x^2-x}{x-1}-\dfrac{2}{x-1}}=\frac{\dfrac{x^2+x+1}{x(x+1)}}{\dfrac{x^2-x-2}{x-1}}$$

$$=\frac{x^2+x+1}{x(x+1)}\div\frac{x^2-x-2}{x-1}$$

$$=\frac{x^2+x+1}{x(x+1)}\cdot\frac{x-1}{(x^2-x-2)}$$

$$=\frac{x^2+x+1}{x(x+1)}\cdot\frac{x-1}{(x+1)(x-2)}$$

$$=\frac{x^3-1}{x(x+1)^2(x-2)}$$

Multiplying by the LCD would be much harder in this example.

(D) **Avoiding common errors.**

The text provides a list of common errors that students make:

Correct multiplication property	Common error with addition
$(a \cdot b)^2 = a^2 \cdot b^2$	$(a + b)^2 \neq a^2 + b^2$
$\sqrt{a \cdot b} = \sqrt{a}\sqrt{b}$ $(a, b \geq 0)$	$\sqrt{a + b} \neq \sqrt{a} + \sqrt{b}$
$\sqrt{a^2 \cdot b^2} = a \cdot b$ $(a, b \geq 0)$	$\sqrt{a^2 + b^2} \neq a + b$
$\dfrac{1}{a} \cdot \dfrac{1}{b} = \dfrac{1}{a \cdot b}$	$\dfrac{1}{a} + \dfrac{1}{b} \neq \dfrac{1}{a + b}$
$\dfrac{ab}{a} = b$	$\dfrac{a + b}{a} \neq b$
$a^{-1} \cdot b^{-1} = (a \cdot b)^{-1}$	$a^{-1} + b^{-1} \neq (a + b)^{-1}$

When doing algebra, it is not only important to remember what you *can* do, it is also important to remember what you *cannot* do. For example, let's compute $\sqrt{9 + 16}$. We add 9 to 16, get 25, and take the square root of 25 to get 5. In other words, it is clear that $\sqrt{9 + 16} = 5$. Notice that if we do the following addition: $\sqrt{9} + \sqrt{16}$ we get the number 7. We have just shown that $\sqrt{9 + 16}$ is not equal to $\sqrt{9} + \sqrt{16}$. Similarly, $\sqrt{x^2 + 3x + 8}$ is not the same thing as $\sqrt{x^2} + \sqrt{3x} + 8$.

◤ Core Exercises

15, 45, 61, 97

1 Equations and Inequalities

1.1 Basic Equations

Concepts

(A) Linear equations.

- Solving linear equations.
- Solving linear equations for one variable in terms of others.

(B) Other equations.

- Equations involving fractional expressions.
- Equations involving radicals.
- Equations involving absolute value.

Definitions

- **Equation:** A statement that two mathematical expressions are equal.
- **Solving an equation:** Finding values of the unknowns of an equation that make it true. These are the **solutions** of the equation.
- **Equivalent equations:** Two equations with the same solutions.
- **Solving an equation for one variable:** Expressing one variable of an equation in terms of the others.
- **Linear equation:** An equation in which each term is either a constant multiple of the variable or a constant.
- **The Zero-Product Property:** $AB = 0$ if and only if $A = 0$ or $B = 0$.
- **Extraneous solution:** When solving an equation, occasionally you will obtain a value that does not actually solve the given equation. This is called an **extraneous solution**. (Note that it is not actually a solution!)

Hints and Tips

- Think of an equation as a balance: the weight on the left hand side (LHS) is the same as the weight on the right hand side (RHS) of the balance. Now if you add the same amount of weight to both sides of the equation, it stays balanced, and the equation remains true. If you double or triple the weight on each side, it again stays balanced. But you have to always make sure that everything you do to the LHS is also done to the RHS.

◤ Review

Ⓐ Linear equations.

A linear equation is also called a **first-degree equation**.   The solution is be the value of the variable that makes the equation true. You solve a linear equation by finding an equivalent equation of the form $x = a$, where a is the solution of the equation. You derive equivalent equations by using the two Properties of Equality:

> 1. You can add or subtract the same quantity to *both sides* of the equation.
> 2. You can multiply or divide *both sides* of the equation by the same nonzero quantity.

Be sure to always check your solution in the original equation.

• Solving linear equations.

Exercise: Solve.   ‖ **Answer:**

(a) $3x - 9 = 0$

$$3x - 9 = 0$$
$$3x = 9 \qquad \text{Add 9}$$
$$x = 3 \qquad \text{Multiply by } \tfrac{1}{3}$$

Check: $3(3) - 9 \overset{?}{=} 0$
$$9 - 9 = 0 \checkmark$$

(b) $5x + 7 = x - 1$

$$5x + 7 = x - 1$$
$$4x = -8 \qquad \text{Subtract } x \text{ and 7}$$
$$x = -2 \qquad \text{Multiply by } \tfrac{1}{4}$$

Check: $5(-2) + 7 \overset{?}{=} (-2) - 1$
$$-10 + 7 \overset{?}{=} -3$$
$$-3 = -3 \checkmark$$

(c) $x - \dfrac{2}{3} = \dfrac{x}{6}$

Start by multiplying both sides of the equation by the LCD of both sides, in this case 6.

$$\left(x - \tfrac{2}{3}\right)6 = \left(\tfrac{x}{6}\right)6 \qquad \text{Multiply by LCD}$$
$$(x)6 - \left(\tfrac{2}{3}\right)6 = x \qquad \text{Distributive property}$$
$$6x - 4 = x \qquad \text{Subtract } 6x$$
$$-4 = -5x \qquad \text{Multiply by } -\tfrac{1}{5}$$
$$\tfrac{4}{5} = x$$

Check: $\left(\tfrac{4}{5}\right) - \tfrac{2}{3} \overset{?}{=} \tfrac{4/5}{6}$
$$\tfrac{12}{15} - \tfrac{10}{15} \overset{?}{=} \tfrac{4}{30}$$
$$\tfrac{2}{15} = \tfrac{2}{15} \checkmark$$

• **Solving linear equations for one variable in terms of others.**

In many formulas, it is often necessary to express one of the variables in terms of the others. As in other equations in this section, we first isolate the terms involving this variable on one side of the equation and then use one of the techniques discussed already.

Exercise: Solve for b_1 in terms of the other variables in the equation $A = \frac{1}{2}\left(b_1 + b_2\right)h$.

Answer:

$$A = \tfrac{1}{2}\left(b_1 + b_2\right)h$$

$$\frac{2A}{h} = b_1 + b_2 \qquad \text{Isolate the term containing } b_1$$

$$b_1 + b_2 = \frac{2A}{h}$$

$$b_1 = \frac{2A}{h} - b_2$$

Exercise: Solve for h in terms of the other variables in the equation $A = 2\pi rh + 2\pi r^2$.

Answer:

$$A = 2\pi rh + 2\pi r^2$$

$$A - 2\pi r^2 = 2\pi rh \qquad \text{Isolate the term containing } h$$

$$2\pi rh = A - 2\pi r^2$$

$$h = \frac{A - 2\pi r^2}{2\pi r}$$

(B) **Other equations.**

There is no magic formula that will enable you to solve all equations. The following are some guidelines to solving various types of equations, but the general rule is always the same. Start with what you have, and focus on getting to either something of the form $x = a$ or the product of several terms that equal zero.

• **Equations involving fractional expressions.**

Equations involving fractional expressions can be simplified by multiplying both sides of the equation by the least common denominator of all fractions involved. This method can introduce **extraneous solutions** — values that look like solutions but don't actually work when you substitute them into the original equation. Because of the existence of extraneous solutions, it is *vital* to check your answers.

Exercise: Solve.

(a) $\dfrac{2x + 6}{3x} + \dfrac{1}{2} = \dfrac{1}{6}$

Answer:

$$\left(\frac{2x + 6}{3x} + \frac{1}{2}\right)6x = \left(\tfrac{1}{6}\right)6x \qquad \text{Multiply by the LCD}$$

$$\left(\frac{2x + 6}{3x}\right)6x + \left(\frac{1}{2}\right)6x = x \qquad \text{Simplify}$$

$$4x + 12 + 3x = x$$

$$7x + 12 = x$$

$$6x = -12$$

$$x = -2 \qquad \text{Potential solution}$$

$$\text{Check: } \frac{2\left(-2\right) + 6}{3\left(-2\right)} + \frac{1}{2} \overset{?}{=} \frac{1}{6}$$

$$\frac{2}{-6} + \frac{1}{2} = \frac{1}{6} \ \checkmark$$

(b) $\dfrac{x}{x-1} = \dfrac{x+3}{x+1}$

The LCD is $(x-1)(x+1)$.

$$\frac{x}{x-1} = \frac{x+3}{x+1}$$

$$\left(\frac{x}{x-1}\right)(x-1)(x+1) = \left(\frac{x+3}{x+1}\right)(x-1)(x+1)$$

$$x(x+1) = (x+3)(x-1) \qquad \text{Expand}$$

$$x^2 + x = x^2 + 2x - 3 \qquad \text{Subtract}$$

$$-x = -3$$

$$x = 3 \qquad \text{Potential solution}$$

Check: $\dfrac{(3)}{(3)-1} \overset{?}{=} \dfrac{(3)+3}{(3)+1}$

$$\frac{3}{2} = \frac{6}{4} \;\checkmark$$

(c) $\dfrac{x+2}{x-3} + 2 = \dfrac{10}{2x-6}$

Since $2x - 6 = 2(x-3)$, the LCD is $2x - 6$.

$$\frac{x+2}{x-3} + 2 = \frac{10}{2x-6}$$

$$\left(\frac{x+2}{x-3} + 2\right)(2x-6) = \left(\frac{10}{2x-6}\right)(2x-6)$$

$$\left(\frac{x+2}{x-3}\right)2(x-3) + 2(2x-6) = \frac{10}{2x-6}(2x-6)$$

$$2x + 4 + 4x - 12 = 10$$

$$6x - 8 = 10$$

$$6x = 18$$

$$x = 3 \qquad \text{Potential solution}$$

Check: $\dfrac{3+2}{3-3} + 2 \overset{?}{=} \dfrac{10}{2(3)-6}$

$$\frac{6}{0} + 2 \overset{?}{=} \frac{10}{0}$$

Since the fractions in the last equation are undefined, there is no solution.

- **Equations involving radicals.**

If an equation involves radicals, we isolate the radical on the LHS, and then raise both sides to a power. Again, because of the existence of extraneous solutions, it vital to check our answers.

Exercise: Solve the equation.

Answer:

(a) $\sqrt{x-4}+3=6$

$$\sqrt{x-4}+3=6$$
$$\sqrt{x-4}=3 \qquad \text{Subtract } 3$$
$$x-4=9 \qquad \text{Square both sides}$$
$$x=13 \qquad \text{Potential solution}$$

Check: $\sqrt{13-4}+3 \overset{?}{=} 6$
$$6=6 \checkmark$$

(b) $3\sqrt{x-5}-\sqrt{x-5}+2=0$

$$3\sqrt{x-5}-\sqrt{x-5}+2=0$$
$$3\sqrt{x-5}-\sqrt{x-5}=-2 \qquad \text{Subtract 2}$$
$$2\sqrt{x-5}=-2 \qquad \text{Simplify}$$
$$\sqrt{x-5}=-1 \qquad \text{Multiply by } \tfrac{1}{2}$$
$$x-5=1 \qquad \text{Square both sides}$$
$$x=6 \qquad \text{Potential solution}$$

Check: $3\sqrt{6-5}-\sqrt{6-5}+2 \overset{?}{=} 0$
$$2+2 \neq 0$$

The solution $x=6$ is extraneous. There is no real solution to this equation. (We could have seen this earlier, when we wrote $\sqrt{x-5}=-1$. There is no real number whose square root is -1.)

- **Equations involving absolute value.**

Notice that if we write $|x|=3$ we have two possible values for x: 3 and -3. In general, an equation involving an absolute value can have multiple solutions. The key is to isolate it on the LHS, and use the definition of absolute value.

Exercise: Solve $|3x+2|=3$.

Answer:

$$|3x+2|=3$$
$$3x+2=3 \quad \text{or} \quad 3x+2=-3 \qquad \text{Definition of absolute value}$$
$$3x=1 \quad \text{or} \quad 3x=-5 \qquad \text{Subtract 2}$$
$$x=\tfrac{1}{3} \quad \text{or} \quad x=-\tfrac{5}{3} \qquad \text{Multiply by } \tfrac{1}{3}$$

We check these solutions, and both are correct.

▸ Core Exercises

15, 49, 79, 85, 91, 93, 95, 97

1.2   Modeling with Equations

Concepts

(A) **Guidelines for modeling with equations.**

- Solving applied problems using mathematical models.

Hints and Tips

- Many students approach modeling by trying to memorize how to do each "type" of problem. This is not a good approach, because it will leave you helpless when you see a problem that does not fit into one of your "types." The purpose of taking math courses is not to learn how to solve only a narrow range of problems, but to have the mental tools necessary to see any form of problem and solve it. For example, doing a bunch of mixture problems until you master them is good mechanical practice, but your larger goal should be to gain the confidence and flexibility to solve an ever-increasing number of types of real-world applications. Your text has a variety of worked-out examples to go through in addition to the ones you see below.

Review

(A) **Guidelines for modeling with equations.**

- **Solving applied problems using mathematical models.**

  The steps involved in solving a word problem are shown below.

  > **1.** Read the entire question. Identify the variable.
  >
  > **2.** Express all unknown quantities in terms of the variable.
  >
  > **3.** Set up an equation or model. Relate the quantities.
  >
  > **4.** Solve the equation and check your solutions.

  Notice the first part of the first step. *Always make sure you have read and understood the problem before trying to solve it.* There is nothing worse than spending a half hour on a problem only to realize that you are working on the wrong problem!

Exercise: A train leaves Station A at 7 A.M. and travels north to Chicago at 42 miles per hour. Two hours later, an express train leaves Station A and travels north to Chicago at 70 miles per hour. When will the express train overtake the slower train?

Answer: After reading through the problem, we find that we are looking for the number of hours the first train travels.

Let t be the number of hours the first train travels. Then $t - 2$ is equal to the number of hours the second train travels, since the second train leaves two hours after the first. The trains will have traveled the same distance when the express train overtakes the other train, so compare the distance traveled by each train using the formula $D = RT$.

First train: $D_1 = 42t$. Second train: $D_2 = 70(t - 2)$. So $42t = 70(t - 2) \Leftrightarrow 42t = 70t - 140 \Leftrightarrow -28t = -140 \Leftrightarrow t = 5$ hours. Thus, the express train overtakes the slower train at 12 noon.

Check: The first train travels $42(5) = 210$ miles and the second train travels $70(5 - 2) = 70(3) = 210$ miles. ✓

Exercise: A bicyclist leaves home and pedals to school at 20 miles per hour. Fifteen minutes later, her roommate leaves home and drives her car to school at 35 miles per hour. When will the car overtake the bicyclist?

Answer: In this problem we seek the time when the distance traveled by the bicyclist is equal to the distance traveled by the car. Let t be the time the car drives, in hours. Then $t + \frac{1}{4}$ is the time the bicyclist rides, since the bicyclist leaves 15 minutes before the car. Using the formula $D = RT$,

$$20\left(t + \tfrac{1}{4}\right) = 35t$$
$$20t + 5 = 35t$$
$$5 = 15t$$
$$t = \tfrac{5}{15} = \tfrac{1}{3} \text{ hr or 20 minutes}$$

Check: The bicyclist travels $20\left(\tfrac{1}{3} + \tfrac{1}{4}\right) = 20\left(\tfrac{7}{12}\right) = \tfrac{35}{3}$ miles. The car travels $35\left(\tfrac{1}{3}\right) = \tfrac{35}{3}$ miles. ✓

Exercise: Wanda has 75 coins. Some of the coins are quarters and the rest are dimes. Altogether, she has $12.90. How many of each type of coin does she have?

Answer: Here we are after the number of quarters and the number of dimes. We let the variable equal one of the quantities. Let x be the number of quarters, so $75 - x$ is the number of dimes (because there are 75 coins altogether). Now translate the fact that the value of the coins adds up to $12.90:

$$\left(\begin{array}{c}\text{value of}\\\text{the quarters}\end{array}\right) + \left(\begin{array}{c}\text{value of}\\\text{the dimes}\end{array}\right) = 12.90$$

$$0.25x + 0.10\left(75 - x\right) = 12.90$$

Solve for x: $0.25x + 7.5 - 0.10x = 12.90 \iff$
$0.15x + 7.5 = 12.90 \iff 0.15x = 5.40 \iff$
$x = \frac{5.40}{0.15} = 36$.

So Wanda has 36 quarters and $75 - 36 = 39$ dimes.

Check: $0.25\left(36\right) + 0.10\left(39\right) = 9.00 + 3.90 = 12.90$ ✓

Exercise: A tug tows a barge 24 miles up a river at 10 miles per hour and returns down the river at 12 miles per hour. The entire trip take $5\frac{1}{2}$ hours. What is the rate of the river's current?

Answer: Let r be the rate of the river's current. Then $10 - r$ is the true rate of the barge upriver and $12 + r$ is the true rate of the barge downriver. Use the distance formula $D = RT$ and solve for time: $T = \dfrac{D}{R}$.

The time it takes the barge to go up the river is $\dfrac{24}{10 - r}$ hours.

The time it takes the barge to go down the river is $\dfrac{24}{12 + r}$ hours.

$$(\text{time upriver}) + (\text{time downriver}) = (\text{total time})$$

$$\frac{24}{10 - r} + \frac{24}{12 + r} = 5\frac{1}{2} = \frac{11}{2}$$

Multiply by the LCD, $2(10 - r)(12 + r)$:

$$2(10 - r)(12 + r)\left(\frac{24}{10 - r} + \frac{24}{12 + r}\right)$$

$$= \left(\tfrac{11}{2}\right)2(10 - r)(12 + r)$$

$$48(12 + r) + 48(10 - r) = 11(10 - r)(12 + r)$$

$$576 + 48r + 480 - 48r = 1320 - 22r - 11r^2$$

$$1056 = 1320 - 22r - 11r^2$$

$$0 = -11r^2 - 22r + 264$$

$$0 = -11\left(r^2 + 2r - 24\right)$$

$$0 = -11(r - 4)(r + 6)$$

$$r - 4 = 0 \qquad r + 6 = 0$$

$$r = 4 \qquad r = -6$$

$r = -6$ does not make sense, since it requires the river to flow backward.

Check $r = 4$: when $r = 4$, the rate upriver is $10 - 4 = 6$ miles per hour, so the trip upriver takes $\frac{24}{6}$ or 4 hours. The rate downriver is $12 + 4 = 16$ miles per hour, so the trip downriver takes $\frac{24}{16}$ hours or $1\frac{1}{2}$ hours. ✓

Mixture and concentration problems can be confusing at first. Often the best approach is to use the various percentages of this and that to compute the absolute quantity (in gallons, ounces or whatever) of the various ingredients. Making tables of the various quantities discussed can help:

	Value in	Value out
Type or name % Amount Value		

Exercise: A butcher at a supermarket has 50 pounds of ground meat that contains 36% fat. How many pounds of ground meat containing 15% fat must be added to obtain a mixture that contains 22% fat?

Answer: Let x be the number of pounds of 15% fat ground meat added to make the mixture. In this case, the amount of fat put into the mixture must equal the amount of fat taken out.

	Value in		Value out
Type	36% fat	15% fat	22% fat
%	36%	15%	22%
Amount	50	x	$50 + x$
Value	$0.36\,(50)$	$0.15x$	$0.22\,(50 + x)$

$0.36\,(50) + 0.15x = 0.22\,(50 + x)$ $\Leftrightarrow$
$1800 + 15x = 1100 + 22x$ $\Leftrightarrow$ $700 = 7x$ $\Leftrightarrow$
$x = 100$ pounds.
Check: The amount of fat put into the mixture is
$0.36\,(50) + 0.15\,(100) = 18 + 15 = 33$ pounds. The amount of fat in the final product is:
$0.22\,[50 + (100)] = 0.22\,(150) = 33$ pounds. $\checkmark$

Exercise: A winery makes a variety wine that, according to the label, contains 95% Cabernet grape juice. Due to a valve error, too much Zinfandel juice is added to the blend. As a result, the winery ends up with 1500 gallons of wine that now contains 93% Cabernet juice. The winery has only 500 gallons of pure Cabernet juice, so they remove a portion of the 93% blend and add 500 gallons of pure Cabernet to reach 95% Cabernet concentration. How many gallons do they need to remove and replace with pure Cabernet juice?

Answer: Let x be the quantity (in gallons) of 93% Cabernet blend that is removed. Then $1500 - x$ is the quantity of the 93% blend that remains. When 500 gallons are added, the resulting blend will have a volume of $2000 - x$ gallons.

	Value in		Value out
Type	Old blend	Pure Cabernet	Final blend
%	93%	100%	95%
Amount	$1500 - x$	500	$2000 - x$
Value	$0.93\,(1500 - x)$	$1\,(500)$	$0.95\,(2000 - x)$

$$0.93\,(1500 - x) + 500 = 0.95\,(2000 - x)$$
$$93\,(1500 - x) + 50{,}000 = 95\,(2000 - x)$$
$$139{,}500 - 93x + 50{,}000 = 190{,}000 - 95x$$
$$189{,}500 - 93x = 190{,}000 - 95x$$
$$2x = 500$$
$$x = 250 \text{ gallons}$$

Check: When 250 gallons of the 93% blend is removed, there are 1250 gallons left, to which 500 gallons of 100% Cabernet juice is added. The result is 1750 gallons of wine. The quantity of Cabernet juice in the blend is
$0.93\,(1250) + 1.00\,(500) = 1162.5 + 500 = 1662.5$ gallons.
Calculated directly, the quantity of Cabernet juice in the blend is $0.95\,(1750) = 1662.5$ gallons. $\checkmark$

▰ Core Exercises

31, 47, 49, 75, 79, 85

1.3   Quadratic Equations

Concepts

(A) **Quadratic equations.**

- Solving quadratic equations by factoring.
- Solving quadratic equations by completing the square.
- Solving quadratic equations using the quadratic formula.
- Using the discriminant.

Definitions

- **Quadratic equation:** An equation equivalent to an equation of the form $ax^2 + bx + c = 0$, where $a \neq 0$.
- **The Zero-Product Property:** $AB = 0$ if and only if $A = 0$ or $B = 0$.
- **Discriminant:** The quantity $b^2 - 4ac$ in a quadratic equation.

(B) **Quadratic equations.**

A quadratic equation is also called a **second-degree equation**. The solutions are the values of the variable that make the equation true. Every quadratic equation has zero, one, or two real solutions. There are several methods for solving quadratic equations, each with its own advantages.

- **Solving quadratic equations by factoring.**

We take an equation of the form $ax^2 + bx + c = 0$, and factor the left hand side. Then we use the zero factor property. (If $b = 0$ then there is a shortcut, which we will see below.)

Exercise: Solve the equation

$x^2 + 3x + 2 = 0$.

Answer:

$$x^2 + 3x + 2 = 0$$
$$(x + 1)(x + 2) = 0 \quad \text{Factor}$$
$$x + 1 = 0 \qquad x + 2 = 0 \quad \text{Set each factor equal to } 0$$
$$x = -1 \qquad x = -2 \quad \text{Solve}$$

Check:
$$(-1)^2 + 3(-1) + 2 = 1 - 3 + 2 = 0 \checkmark$$
$$(-2)^2 + 3(-2) + 2 = 4 - 6 + 2 = 0 \checkmark$$

Exercise: Solve the equation

$x^2 + 12 = 7x$.

Answer:

$$x^2 + 12 = 7x$$
$$x^2 - 7x + 12 = 0 \quad \text{Set equal to } 0$$
$$(x - 3)(x - 4) = 0 \quad \text{Factor}$$
$$x - 3 = 0 \qquad x - 4 = 0 \quad \text{Set each factor equal to } 0$$
$$x = 3 \qquad x = 4 \quad \text{Solve}$$

Check: $(3)^2 + 12 \overset{?}{=} 7(3) \qquad (4)^2 + 12 \overset{?}{=} 7(4)$
$$9 + 12 = 21 \checkmark \qquad 16 + 12 = 28 \checkmark$$

Exercise: Solve the equation $x^2 = 2x + 3$.

Answer:

$$x^2 = 2x + 3$$

$x^2 - 2x - 3 = 0$ Set equal to 0

$(x - 3)(x + 1) = 0$ Factor

$x - 3 = 0 \quad x + 1 = 0$ Set each factor equal to 0

$x = 3 \qquad x = -1$ Solve

Check: $(3)^2 \overset{?}{=} 2(3) + 3 \quad (-1)^2 \overset{?}{=} 2(-1) + 3$

$\qquad\qquad 9 \quad = 6 + 3 \checkmark \qquad\quad 1 = -2 + 3 \checkmark$

Exercise: Solve the equation $x^2 = 9$.

Answer:

$x^2 = 9$

$x = \pm\sqrt{9}$

$x = \pm 3$

The solutions are $x = 3$ and $x = -3$. Check each solution.

Exercise: Solve the equation $x^2 - 8 = 0$.

Answer:

$x^2 - 8 = 0$

$x^2 = 8$ Isolate the variable.

$x = \pm\sqrt{8} = \pm 2\sqrt{2}$

Check each solution.

Exercise: Solve the equation $4x^2 = 24$.

Answer:

$4x^2 = 24$

$x^2 = 6$ Divide by 4

$x = \pm\sqrt{6}$

Check each solution.

Remember: The zero factor property works *only* if the right-hand side of the equation is equal to zero. It does *not* apply otherwise.

- **Solving quadratic equations by completing the square.**

The goal in finding the solution to a quadratic equation by **completing the square** is to add the appropriate constant to make a perfect square $(x + m)^2$ or $(x - m)^2$. Step one is to isolate the terms involving the variables, x^2 and bx, on one side of the equation. Step two is to determine the constant needed to add to both sides to complete the square. Since $(x \pm m)^2 = x^2 \pm 2mx + m^2$, the coefficient of the middle term, b, must equal $2m$, that is, $m = \frac{1}{2}b$; so add $\left(\dfrac{b}{2}\right)^2$ to both sides.

Exercise: Solve $x^2 - 6x + 4 = 0$ by completing the square.

Answer:

$$x^2 - 6x + 4 = 0$$

$$x^2 - 6x \quad\quad = -4 \quad\quad \text{Isolate the variable}$$

$$x^2 - 6x + \mathbf{9} = -4 + \mathbf{9} \quad\quad \text{Add } \left(\tfrac{6}{2}\right)^2 = 9 \text{ to both sides}$$

$$(x - 3)^2 = 5 \quad\quad \text{Simplify}$$

$$x - 3 = \pm\sqrt{5} \quad\quad \text{Take square roots of both sides.}$$

$$x = 3 \pm \sqrt{5}$$

Check each solution.

Exercise: Solve $4x^2 = 12x + 1$ by completing the square.

Answer:

$$4x^2 = 12x + 1$$

$$4x^2 - 12x \quad\quad = 1 \quad\quad \text{Isolate the variable}$$

$$x^2 - 3x \quad\quad = \frac{1}{4} \quad\quad \text{Divide by 4}$$

$$x^2 - 3x + \frac{9}{4} = \frac{1}{4} + \frac{9}{4} \quad\quad \text{Add } \left(\frac{3}{2}\right)^2 = \frac{9}{4} \text{ to both sides}$$

$$\left(x - \frac{3}{2}\right)^2 = \frac{10}{4} \quad\quad \text{Simplify}$$

$$x - \frac{3}{2} = \frac{\pm\sqrt{10}}{2} \quad\quad \text{Take square roots of both sides.}$$

$$x = \frac{3}{2} \pm \frac{\sqrt{10}}{2}$$

$$x = \frac{3 \pm \sqrt{10}}{2}$$

Check each solution.

- **Solving quadratic equations using the quadratic formula.**

The **quadratic formula** $x = \dfrac{-b \pm \sqrt{b^2 - 4ac}}{2a}$ gives the solutions to the quadratic equation $ax^2 + bx + c = 0$.

Exercise: Solve using the quadratic formula.

Answer:

(a) $4x^2 - 4x - 3 = 0$

We substitute $a = 4$, $b = -4$, $c = -3$ into the quadratic formula:

$$x = \frac{-b \pm \sqrt{b^2 - 4ac}}{2a}$$

$$= \frac{-(-4) \pm \sqrt{(-4)^2 - 4(4)(-3)}}{2(4)}$$

$$= \frac{4 \pm \sqrt{16 + 48}}{8} = \frac{4 \pm \sqrt{64}}{8}$$

$$= \frac{4 \pm 8}{8}$$

So the solutions are $x = \dfrac{4 + 8}{8} = \dfrac{3}{2}$ and $x = \dfrac{4 - 8}{8} = -\dfrac{1}{2}$.

Note that the left-hand side factors: $(2x - 3)(2x + 1) = 0$.

(b) $3x^2 - 7x + 3 = 0$

We substitute $a = 3$, $b = -7$, $c = 3$ into the quadratic formula:

$$x = \frac{-b \pm \sqrt{b^2 - 4ac}}{2a}$$

$$= \frac{-(-7) \pm \sqrt{(-7)^2 - 4(3)(3)}}{2(3)}$$

$$= \frac{7 \pm \sqrt{49 - 36}}{6}$$

$$= \frac{7 \pm \sqrt{13}}{6}$$

So the solutions are $x = \dfrac{7 + \sqrt{13}}{6}$ and $x = \dfrac{7 - \sqrt{13}}{6}$.

(c) $4x^2 + 20x + 25 = 0$

We substitute $a = 4$, $b = 20$, $c = 25$ into the quadratic formula:

$$x = \frac{-b \pm \sqrt{b^2 - 4ac}}{2a}$$

$$= \frac{-(20) \pm \sqrt{(20)^2 - 4(4)(25)}}{2(4)}$$

$$= \frac{-20 \pm \sqrt{400 - 400}}{8} = \frac{-20 \pm \sqrt{0}}{8}$$

$$= \frac{-20 \pm 0}{8}$$

So the only solution is $x = \frac{-20}{8} = -\frac{5}{2}$. Since $4x^2 + 20x + 25 = (2x + 5)^2$, the solution checks.

- **Using the discriminant.**

The number of solutions to a quadratic equation will depend on the term $\pm\sqrt{b^2 - 4ac}$ in the quadratic formula. If $b^2 - 4ac$ is negative, then its square root does not exist. If $b^2 - 4ac = 0$, then $\pm\sqrt{b^2 - 4ac} = 0$ and the quadratic formula gives one solution. If $b^2 - 4ac$ is greater than zero, then $\pm\sqrt{b^2 - 4ac}$ takes on two values, and the quadratic formula gives two solutions. We call the quantity $D = b^2 - 4ac$ the **discriminant** of a quadratic equation.

Exercise: Without solving the equation, determine how many real solutions it has.	**Answer:**
(a) $2x^2 + 5x + 3 = 0$	The discriminant D is $5^2 - 4\,(2)\,(3) = 1$. Since $D > 0$, there are two solutions.
(b) $x^2 + 10x + 25 = 0$	$D = 10^2 - 4\,(1)\,(25)$. Since $D = 0$, there is one solution.
(c) $2x^2 + 4x + 3$	$D = 4^2 - 4\,(2)\,(3)$. Since $D < 0$, there is no solution.

1.4   Complex Numbers

Concepts

(A) Complex numbers: the basics.

- Adding, subtracting, and multiplying complex numbers.
- Dividing complex numbers.
- Taking square roots of negative numbers.

(B) Solving quadratic equations.

- Finding complex roots of quadratic equations.

Definitions

- **Complex number:** An expression of the form $a + bi$, where i is a quantity defined by the property $i^2 = -1$.
- **Imaginary number:** An expression of the form bi where i is defined as above.

Questions to Ask Your Teacher

- If your teacher starts writing js on the board, you may want to ask if j is the same thing as i in the textbook. Some people use j as the imaginary constant instead of i.

Hints and Tips

- You will note that almost every mathematician you talk to seems a little defensive about the "reality" of complex numbers. That will include your textbook authors, your teacher, and me, if you catch me on an ordinary (non-imaginary) day. We tend to say that complex numbers are just as "real" as the irrational numbers, but in your heart you know that doesn't seem right. You can picture holding $\sqrt{2}$ inches of chalk in a way that you cannot picture holding $3 + 2i$ inches. I never really believed that complex numbers were useful until I was getting my engineering degree. There were certain quantities that had both a magnitude and an angle associated with them. One example you may have seen is vectors. Another example is alternating current — there is a magnitude, and a "phase angle". Time and time again, I saw these quantities represented as complex numbers, and then all sorts of operations done with them, and then, at the end of the problem, the answer would be brought back to a real-world quantity. The identity $i^2 = -1$ was a consequence of the way we used complex numbers, not a definition.

Think of it this way. We start with a real-world problem. When we make our mathematical model, complex numbers turn out to be useful. We use complex arithmetic in the model. Then, when we get the real-world solution, it is something that can be pictured as a real answer.

Review

(A) **Complex numbers: the basics.**

Complex numbers are expressions of the form $a + bi$, where a and b are real numbers and i is the **imaginary number** defined by $i^2 = -1$. The term a is called the **real part** and the term b is called the **imaginary part**. Two complex numbers are equal if their real parts are equal and their imaginary parts are equal. The number bi, $b \neq 0$, is called a **pure imaginary number**. You add two complex numbers by adding the real part and adding the imaginary part. The distributive property or the FOIL mnemonic (First, Outer, Inner, Last) can be used along with the identity $i^2 = -1$ to multiply two complex numbers.

- **Adding, subtracting, and multiplying complex numbers.**

Exercise: Evaluate the expression.

(a) $(3 + 4i) - (4 - 5i)$

(b) $\left(\sqrt{3} - i\right) + \left(1 + \sqrt{3}i\right)$

Answer:

$$(3 + 4i) - (4 - 5i) = 3 + 4i - 4 + 5i$$
$$= (3 - 4) + (4 + 5)\,i = -1 + 9i$$

$$\left(\sqrt{3} - i\right) + \left(1 + \sqrt{3}i\right) = \sqrt{3} - i + 1 + \sqrt{3}i$$
$$= \sqrt{3} + 1 + \left(-1 + \sqrt{3}\right)i$$

This is as far as this expression can be simplified.

Exercise: Evaluate the expression.

(a) $(2 - 3i)(7 - 5i)$

Answer:

We use the distributive property:

$$(2 - 3i)(7 - 5i) = 2(7 - 5i) + (-3i)(7 - 5i)$$
$$= 14 - 10i - 21i + 15i^2$$
$$= 14 - 31i + 15(-1) = -1 - 31i$$

(b) $(3 + 2i)(4 + 9i)$

We use FOIL:

$$(3 + 2i)(4 + 9i) = (3)(4) + (3)(9i) + (2i)(4) + (2i)(9i)$$
$$= 12 + 27i + 8i + 18i^2$$
$$= 12 + 35i + 18(-1) = -6 + 35i$$

We used different methods in parts (a) and (b) to show how each method works.

Exercise: Show that $1 + 2i$ is a solution to $x^2 - 2x + 5 = 0$.

Answer: Substitute $1 + 2i$ for x and simplify:

$$(1 + 2i)^2 - 2(1 + 2i) + 5$$
$$= \left[(1)^2 + 2(1)(2i) + (2i)^2\right] - 2 - 4i + 5$$
$$= 1 + 4i + 4i^2 - 4i + 3$$
$$= 1 - 4 + 3 = 0$$

So $1 + 2i$ is a solution to $x^2 - 2x + 5 = 0$.

• **Dividing complex numbers.**

The difference of squares product $(a - bi)(a + bi) = a^2 - (bi)^2 = a^2 + b^2$ is used to simplify fractional expressions or to divide complex numbers. The numbers $a + bi$ and $a - bi$ are called **complex conjugates**.

Exercise: Simplify $\dfrac{2}{3 - i}$.

Answer:

$$\frac{2}{3 - i} = \left(\frac{2}{3 - i}\right)\left(\frac{3 + i}{3 + i}\right)$$
$$= \frac{2(3 + i)}{9 - i^2} = \frac{2(3 + i)}{9 + 1}$$
$$= \frac{2(3 + i)}{10} = \frac{3 + i}{5} \text{ or } \frac{3}{5} + \frac{1}{5}i$$

Exercise: Divide $4 - 3i$ by $3 + 2i$.

Answer:

$$(4 - 3i) \div (3 + 2i) = \frac{4 - 3i}{3 + 2i} = \left(\frac{4 - 3i}{3 + 2i}\right)\left(\frac{3 - 2i}{3 - 2i}\right)$$
$$= \frac{12 - 8i - 9i + 6i^2}{9 - 4i^2} = \frac{12 - 17i - 6}{9 + 4}$$
$$= \frac{6 - 17i}{13} \text{ or } \frac{6}{13} - \frac{17}{13}i$$

- **Taking square roots of negative numbers.**

If $-r < 0$, then the square roots of $-r$, are $i\sqrt{r}$ and $-i\sqrt{r}$, where $i\sqrt{r}$ is called the **principal square root** of $-r$. Remember $\sqrt{a}\sqrt{b} = \sqrt{ab}$ only when a and b are *not both negative*. For example: if a and b are both negative then $\sqrt{-3}\sqrt{-6} = i\sqrt{3} \cdot i\sqrt{6} = 3\sqrt{2}i^2 = -3\sqrt{2}$, while $\sqrt{(-3)(-6)} = \sqrt{18} = 3\sqrt{2}$.

Exercise: Evaluate the expression.

Answer:

(a) $\sqrt{-1}$

$\sqrt{-1} = i$. This is the definition of i.

(b) $\sqrt{-9}$

$\sqrt{-9} = i\sqrt{9} = 3i$

(c) $\sqrt{-12}$

$\sqrt{-12} = i\sqrt{12} = 2\sqrt{3}i$

(d) $-\sqrt{-25}$

$-\sqrt{-25} = -i\sqrt{25} = -5i$.

Note that $-\sqrt{-25} \neq \sqrt{25} = 5$; we must evaluate the expression inside the radical first.

(B) Solving Quadratic Equations.

In the quadratic formula, if the discriminant $D = b^2 - 4ac$ is less than 0, then there is no real solution to $ax^2 + bx + c = 0$. Both solutions are complex, and they are complex conjugates.

- **Solving quadratic equations with complex solutions.**

Exercise: Solve $x^2 + 4x + 6 = 0$.

Answer: We substitute $a = 1$, $b = 4$, and $c = 6$ into the quadratic formula:

$$x = \frac{-b \pm \sqrt{b^2 - 4ac}}{2a} = \frac{-(4) \pm \sqrt{(4)^2 - 4(1)(6)}}{2(1)}$$

$$= \frac{-4 \pm \sqrt{16 - 24}}{2} = \frac{-4 \pm \sqrt{-8}}{2}$$

$$= \frac{-4 \pm 2\sqrt{2}i}{2} = \frac{2(-2 \pm \sqrt{2}i)}{2}$$

$$= -2 \pm \sqrt{2}i$$

Exercise: Solve $5x^2 - 6x + 5 = 0$.

Answer: We substitute $a = 5$, $b = -6$, and $c = 5$ into the quadratic formula:

$$x = \frac{-b \pm \sqrt{b^2 - 4ac}}{2a} = \frac{-(-6) \pm \sqrt{(-6)^2 - 4(5)(5)}}{2(5)}$$

$$= \frac{6 \pm \sqrt{36 - 100}}{10} = \frac{6 \pm \sqrt{-64}}{10} = \frac{6 \pm 8i}{10}$$

$$= \frac{2(3 \pm 4i)}{10} = \frac{3 \pm 4i}{5} \text{ or } \tfrac{3}{5} \pm \tfrac{4}{5}i$$

▚ Core Exercises

19, 55, 59, 71, 79

1.5   Other Types of Equations

▲ Concepts

(A) **Polynomial equations.**

(B) **Equations with radicals.**

(C) **Equations of quadratic type.**

▲ Hints and Tips

- You know that a quadratic equation looks like this: $5x^2 - 3x + 2 = 0$. Informally: a number times x squared, plus a number times x, plus a number. Well, we can replace the x by any expression: $5(3x + 8)^2 - 3(3x + 8) + 2 = 0$. Now we have an **equation of quadratic type**. Once you understand what these equations look like, you will be able to recognize them.

  - **Higher-degree polynomial equations.**

    Higher degree equations are solved by getting one side equal to zero, factoring, and using the Zero-Product Property.

Exercise: Find all solutions of the equation $x^3 - 4x^2 = 5x$.

Answer:

$$x^3 - 4x^2 = 5x$$
$$x^3 - 4x^2 - 5x = 0 \qquad \text{Collect terms}$$
$$x(x^2 - 4x - 5) = 0 \qquad \text{Factor}$$
$$x(x - 5)(x + 1) = 0 \qquad \text{Set each factor equal to } 0$$

$$x = 0 \qquad x - 5 = 0 \qquad x + 1 = 0$$
$$x = 5 \qquad x = -1$$

The three solutions are $x = 0$, $x = 5$, and $x = -1$.

Note: Many students forget the lone x factor and lose the solution $x = 0$. Another common mistake is to divide both sides by x at the first step. This also loses the solution $x = 0$.

Exercise: Find all solutions of the equation $x^4 - 16 = 0$

Answer:

$$x^4 - 16 = 0$$
$$(x^2 - 4)(x^2 + 4) = 0 \qquad \text{Factor}$$
$$(x - 2)(x + 2)(x^2 + 4) = 0 \qquad \text{Set each factor equal to } 0$$

$$x - 2 = 0 \qquad x + 2 = 0$$
$$x = 2 \qquad x = -2$$

There is no real solution to $x^2 + 4 = 0$, so the only real solutions are $x = 2$ and $x = -2$.

(B) **Equations with radicals.**

Equations involving one radical are solved by isolating the radical on one side and then *squaring both sides*. When there is more than one radical, put a radical on either side and then *square both sides*. It is very important to check your solutions to these problems. **Extraneous solutions** are incorrect solutions that can be introduced into a problem by squaring, for example, $3 \neq -3$ but $(3)^2 = (-3)^2$.

Exercise: Find all real solutions to $\sqrt{2x-7} - 5 = x + 4$.

Answer:

$$\sqrt{2x-7} - 5 = x + 4$$
$$\sqrt{2x-7} = x + 9 \qquad \text{Isolate the square root}$$
$$2x - 7 = x^2 + 18x + 81 \qquad \text{Square both sides}$$
$$0 = x^2 + 16x + 88 \qquad \text{Set equal to } 0$$

Now using the quadratic formula,

$$x = \frac{-(16) \pm \sqrt{(16)^2 - 4(1)(88)}}{2(1)} = \frac{-16 \pm \sqrt{256 - 352}}{2} = \frac{-16 \pm \sqrt{-96}}{2}.$$

But since $\sqrt{-96}$ is not a real number, there is no solution.

Exercise: Find all real solutions to $\sqrt{2x} + \sqrt{x+1} = 7$.

Answer:

$$\sqrt{2x} + \sqrt{x+1} = 7$$
$$\sqrt{x+1} = 7 - \sqrt{2x} \qquad \text{Move one root}$$
$$\left(\sqrt{x+1}\right)^2 = \left(7 - \sqrt{2x}\right)^2 \qquad \text{Square both sides}$$
$$x + 1 = 49 - 14\sqrt{2x} + 2x$$
$$14\sqrt{2x} = x + 48 \qquad \text{Isolate the root}$$
$$392x = 196(2x) = x^2 + 96x + 2304 \qquad \text{Square both sides}$$
$$0 = x^2 - 296x + 2304$$
$$0 = (x - 8)(x - 288) \qquad \text{Factor}$$

So either $x - 8 = 0 \Leftrightarrow x = 8$ or $x - 288 = 0 \Leftrightarrow x = 288$. Check $x = 288$ in the *original equation*:

$$\sqrt{2(288)} + \sqrt{(288) + 1} = \sqrt{576} + \sqrt{289}$$
$$= 24 + 17 = 41 \neq 7$$

Therefore, $x = 288$ is not a solution.

Check $x = 8$ in the *original equation*.

$$\sqrt{2(8)} + \sqrt{(8) + 1} = \sqrt{16} + \sqrt{9} = 4 + 3 = 7.$$ Thus, $x = 8$ is the only solution.

Exercise: Find all real solutions to
$\sqrt{2x+9} = x+4$.

Answer:

$$\sqrt{2x+9} = x+4$$
$$\left(\sqrt{2x+9}\right)^2 = (x+4)^2 \qquad \text{Square both sides}$$
$$2x+9 = x^2+8x+16$$
$$0 = x^2+6x+7 \qquad \text{Simplify}$$

Thus,

$$x = \frac{-(6)\pm\sqrt{(6)^2-4(1)(7)}}{2(1)} = \frac{-6\pm\sqrt{36-28}}{2}$$
$$= \frac{-6\pm\sqrt{8}}{2} = \frac{-6\pm2\sqrt{2}}{2} = \frac{2(-3\pm\sqrt{2})}{2} = -3\pm\sqrt{2}$$

Check $x = -3+\sqrt{2}$ in the *original equation*:

$$\sqrt{2\left(-3+\sqrt{2}\right)+9} \overset{?}{=} \left(-3+\sqrt{2}\right)+4$$
$$\sqrt{3+2\sqrt{2}} \overset{?}{=} 1+\sqrt{2}$$

This is hard to check directly. We use the fact if $b \geq 0$ and $b^2 = a$, then $b = \sqrt{a}$:

$\left(1+\sqrt{2}\right)^2 = 1+2\sqrt{2}+2 = 3+2\sqrt{2}$, so this solution checks.

Check $x = -3-\sqrt{2}$ in the *original equation*:

$\sqrt{2\left(-3-\sqrt{2}\right)+9} \overset{?}{=} -3-\sqrt{2}+4$. The right-hand side is

$-3-\sqrt{2}+4 = 1-\sqrt{2} < 0$ and the left-hand side

is $\sqrt{2\left(-3-\sqrt{2}\right)+9} \geq 0$, so this cannot be a solution. The

only solution is $x = -3+\sqrt{2}$.

(C) **Equations of quadratic type.**

An equation of the form $aw^2 + bw + c = 0$, where w is an algebraic expression, is an **equation of quadratic type**. We solve equations of quadratic type by substituting for the algebraic expression.

Exercise: Find all solutions of the equation $x^4 - 3x^2 + 2 = 0$.

Answer: Since all variables are raised to even powers, we use the substitution $w = x^2$. The equation then becomes

$$x^4 - 3x^2 + 2 = \left(x^2\right)^2 - 3\left(x^2\right) + 2 = w^2 - 3w + 2.$$

$$w^2 - 3w + 2 = 0 \qquad \text{Factor}$$
$$(w-2)(w-1) = 0 \qquad \text{Set each factor equal to } 0$$

$w - 2 = 0$	$w - 1 = 0$
$x^2 - 2 = 0$	$x^2 - 1 = 0$
$x^2 = 2$	$x^2 = 1$
$x = \pm\sqrt{2}$	$x = \pm1$

The four real solutions are $x = 1$, $x = -1$, $x = \sqrt{2}$, and $x = -\sqrt{2}$.

Exercise: Find all solutions of the equation

$$\left(x + \frac{1}{x}\right)^2 + 4\left(x + \frac{1}{x}\right) - 5 = 0$$

Answer: We substitute w for the algebraic expression $x + \frac{1}{x}$:

$$\left(x + \frac{1}{x}\right)^2 + 4\left(x + \frac{1}{x}\right) - 5 = w^2 + 4w - 5. \text{ Now}$$

$$w^2 + 4w - 5 = 0 \Leftrightarrow (w - 1)(w + 5) = 0.$$

$w - 1 = 0$	$w + 5 = 0$	
$x + \dfrac{1}{x} - 1 = 0$	$x + \dfrac{1}{x} + 5 = 0$	Multiply by x
$x^2 + 1 - x = 0$	$x^2 + 1 + 5x = 0$	
$x^2 - x + 1 = 0$	$x^2 + 5x + 1 = 0$	

The discriminant of the first equation, $x^2 - x + 1 = 0$, is $D = 1 - 4(1)(1) = -3 < 0$, so it has no solution.

Applying the quadratic formula to $x^2 + 5x + 1 = 0$, we get

$$x = \frac{-(5) \pm \sqrt{(5)^2 - 4(1)(1)}}{2(1)}$$

$$= \frac{-5 \pm \sqrt{25 - 4}}{2} = \frac{-5 \pm \sqrt{21}}{2} = -\frac{5}{2} \pm \frac{\sqrt{21}}{2}$$

So the two real solutions are $x = -\dfrac{5}{2} + \dfrac{\sqrt{21}}{2}$ and

$x = -\dfrac{5}{2} - \dfrac{\sqrt{21}}{2}$.

Core Exercises

13, 17, 37, 53, 59, 69

1.6   Inequalities

Concepts

(A) **Rules for inequalities.**

- Solving linear inequalities.
- Solving simultaneous inequalities.
- Solving nonlinear inequalities.

(B) **Modeling with inequalities.**

Definitions

- **Solving an inequality:** To solve an inequality that contains a variable means to find all values of the variable that will make the inequality true. Usually (but not always) the solution will be an interval or set of intervals.

◤ Questions to Ask Your Teacher

- If the solution to an equality is, say, $x \leq 5$, what is the preferred form of the answer? Possibilities are $x \leq 5$, $\{x \mid x \leq 5\}$, $(-\infty, 5]$, or the following picture:

5

◤ Review

(A) **Rules for inequalities.**

Inequalities are statements involving the symbols $<$, $>$, $\geq$, and $\leq$. The following rules should be mastered to the point that you know what rule to use and how to use it.

Rules	Description
If $a \leq b$, then $a + c \leq b + c$.	Add or subtract the same value from each side.
If $a \leq b$ and $c > 0$, then $ac \leq bc$.	Multiplying each side by a positive number maintains the direction of the inequality sign.
If $a \leq b$ and $c < 0$, then $ac \geq bc$.	Multiplying each side by a negative number reverses the direction of the inequality sign.
If $0 < a \leq b$, then $\dfrac{1}{a} \geq \dfrac{1}{b}$.	Taking reciprocals of both sides of an inequality of positive numbers reverses the direction of the inequality sign.
If $a \leq b$ and $c \leq d$, then $a + c \leq b + d$.	Inequalities can be added.

- **Solving linear inequalities.**

An inequality is **linear** if each term is either a constant multiple of the variable or a constant.

Exercise: Solve $4x - 5 < 3$ and graph the solution on the number line below.

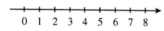

Answer:

$$4x - 5 < 3$$
$$4x < 8 \qquad \text{Add 5 to each side}$$
$$x < 2 \qquad \text{Divide each side by 4}$$

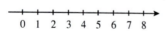

Exercise: Solve $-\frac{1}{3}x + 7 \geq 2x$ and graph the solution on the number line below.

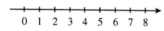

Answer:

$$-\tfrac{1}{3}x + 7 \geq 2x$$
$$7 \geq 2x + \tfrac{1}{3}x \qquad \text{Add } \tfrac{1}{3}x \text{ to each side}$$
$$7 \geq \tfrac{7}{3}x$$
$$3 \geq x \qquad \text{Multiply both sides by } \tfrac{3}{7}$$

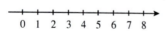

Exercise: Solve $-8x < 2x - 15$ and graph the solution on the number line below.

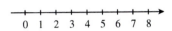

Answer:

$$-8x < 2x - 15$$

$$-10x < -15 \qquad \text{Subtract } 2x \text{ from each side.}$$

$$x > \tfrac{3}{2} \qquad \text{Divide each side by } -10; \text{ reverse sign}$$

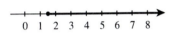

- **Solving simultaneous inequalities.**

Simultaneous inequalities use the same rules as simple inequalities. Whichever inequality rule you apply to one of the inequality expressions, you *must* apply to each of the other expressions. These inequalities are also usually expressed in increasing order.

Exercise: Solve the inequality $3 \le 2x - 7 < 8$.

Answer:

$$3 \le 2x - 7 < 8$$

$$10 \le 2x < 15 \qquad \text{Add 7 to each quantity}$$

$$5 \le x < \tfrac{15}{2} \qquad \text{Divide by 2}$$

Exercise: Solve the inequality $3 < 8 + 4x \le 8$.

Answer:

$$3 < 8 + 4x \le 8$$

$$-5 < 4x \le 0 \qquad \text{Subtract 8 from each quantity}$$

$$-\tfrac{5}{4} < x \le 0 \qquad \text{Divide by 4}$$

Exercise: Solve the inequality $-7 < 2 - 3x < 11$.

Answer:

$$-7 < 2 - 3x < 11$$

$$-9 < -3x < 9 \qquad \text{Subtract 2 from each quantity}$$

$$3 > x > -3 \qquad \text{Divide by } -3; \text{ reverse signs}$$

$$-3 < x < 3 \qquad \text{Rewrite in ascending order}$$

- **Solving nonlinear inequalities.**

These steps are used to solve inequalities that are not linear:

1. Move all terms to one side of the inequality sign. If the nonzero side of the inequality involves quotients, bring them to a common denominator.

2. Factor the nonzero side of the inequality.

3. List the intervals determined by the factorization.

4. Make a table or diagram of the signs of each factor on each interval. In the last row of the table determine the sign of the product (or quotient) of these factors.

5. Determine the solution set from the last row of the table. Be sure to check whether the inequality is satisfied by some or all of the endpoints of the intervals (only necessary when $\le$ and $\ge$ are involved).

When doing the fourth step, you will need to remember that a negative number times a negative number is positive, and a positive number times a negative number is negative. (Also, a negative number divided by a negative number is positive, and a positive number divided by a negative number is negative.)

Exercise: Solve the inequality $(x-2)(x+1) < 0$.

Answer: $(x-2)(x+1) < 0$

$$x - 2 = 0 \qquad x + 1 = 0$$
$$x = 2 \qquad x = -1$$

Interval	$(-\infty, -1)$	$(-1, 2)$	$(2, \infty)$
Test value	-2	0	3
Sign of $x - 2$	$-$	$-$	$+$
Sign of $x + 1$	$-$	$+$	$+$
Sign of $(x-2)(x+1)$	$+$	$-$	$+$

We want the interval where the product is negative, so the solution is $(-1, 2)$.

Exercise: Solve the inequality $\dfrac{x-5}{x+7} > 0$.

Answer: $\dfrac{x-5}{x+7} > 0$

$$x - 5 = 0 \qquad x + 7 = 0$$
$$x = 5 \qquad x = -7$$

Interval	$(-\infty, -7)$	$(-7, 5)$	$(5, \infty)$
Test value	-10	0	10
Sign of $x - 5$	$-$	$-$	$+$
Sign of $x + 7$	$-$	$+$	$+$
Sign of $\dfrac{x-5}{x+7}$	$+$	$-$	$+$

Since we want the interval where the quotient is positive, the solution is $(-\infty, -7) \cup (5, \infty)$.

Exercise: Solve the inequality $x^2 + 6 > 5x$.

Answer:

$$x^2 + 6 > 5x$$
$$x^2 - 5x + 6 > 0 \quad \text{Move all terms to one side.}$$
$$(x-2)(x-3) > 0 \quad \text{Factor.}$$

$$x - 2 = 0 \qquad x - 3 = 0$$
$$x = 2 \qquad x = 3$$

Interval	$(-\infty, 2)$	$(2, 3)$	$(3, \infty)$
Test value	0	2.5	4
Sign of $x - 2$	$-$	$+$	$+$
Sign of $x - 3$	$-$	$-$	$+$
Sign of $(x-2)(x-3)$	$+$	$-$	$+$

We want the interval where the product is positive, so the solution is $(-\infty, 2) \cup (3, \infty)$.

Exercise: Solve the inequality $\dfrac{2x}{x^2+1} > 0$.

Answer: $\dfrac{2x}{x^2+1} > 0$. $2x = 0 \Leftrightarrow x = 0$. Since $x^2+1 \neq 0$, this is the only boundary.

Interval	$(-\infty, 0)$	$(0, \infty)$
Test value	-1	1
Sign of $2x$	$-$	$+$
Sign of x^2+1	$+$	$+$
Sign of $\dfrac{2x}{x^2+1}$	$-$	$+$

We want the interval where the quotient is positive, so the solution is $(0, \infty)$.

Exercise: Solve the inequality $\dfrac{9}{x} \leq x$.

Answer: Move all terms to one side: $\dfrac{9}{x} \leq x \Leftrightarrow \dfrac{9}{x} - x \leq 0$.

Now find a common denominator and factor: $\dfrac{9 - x^2}{x} \leq 0 \Leftrightarrow$

$\dfrac{(3-x)(3+x)}{x} \leq 0$.

$$3 - x = 0 \qquad 3 + x = 0 \qquad x = 0$$
$$3 = x \qquad\quad x = -3$$

Interval	$(-\infty, -3)$	$(-3, 0)$	$(0, 3)$	$(3, \infty)$
Test value	-5	-1	1	5
Sign of $3 - x$	$+$	$+$	$+$	$-$
Sign of $3 + x$	$-$	$+$	$+$	$+$
Sign of x	$-$	$-$	$+$	$+$
Sign of $\dfrac{9 - x^2}{x}$	$+$	$-$	$+$	$-$

Since the inequality is not strict, we check the endpoints. The endpoint $x = 0$ is excluded from the solution since it makes a denominator 0. We want the interval where the quotient is nonpositive, so the solution is $[-3, 0) \cup [3, \infty)$.

Exercise: Solve the inequality

$$\frac{4}{x-1} \geq \frac{3}{x-4}.$$

Answer: $\dfrac{4}{x-1} \geq \dfrac{3}{x-4} \Leftrightarrow \dfrac{4}{x-1} - \dfrac{3}{x-4} \geq 0 \Leftrightarrow$

$$\frac{4(x-4)}{(x-1)(x-4)} - \frac{3(x-1)}{(x-4)(x-1)} \geq 0 \Leftrightarrow$$

$$\frac{4x-16-3x+3}{(x-1)(x-4)} \geq 0 \Leftrightarrow \frac{x-13}{(x-1)(x-4)} \geq 0.$$

$$x - 13 = 0 \qquad x - 1 = 0 \qquad x - 4 = 0$$
$$x = 13 \qquad\quad x = 1 \qquad\quad x = 4$$

Interval	$(-\infty, 1)$	$(1, 4)$	$(4, 13)$	$(13, \infty)$
Test value	0	2	10	15
Sign of $x - 13$	$-$	$-$	$-$	$+$
Sign of $x - 1$	$-$	$+$	$+$	$+$
Sign of $x - 4$	$-$	$-$	$+$	$+$
Sign of $\dfrac{x-13}{(x-1)(x-4)}$	$-$	$+$	$-$	$+$

Since the inequality is not strict, we check the endpoints. The endpoints $x = 1$ and $x = 4$ are excluded from the solution since each makes a denominator 0. We want the interval where the quotient is nonnegative, so the solution is $(1, 4) \cup [13, \infty)$.

Exercise: Solve the inequality

$$\frac{x^2 - 11}{x+1} \leq 1.$$

Answer: Move all terms to one side: $\dfrac{x^2 - 11}{x+1} \leq 1 \Leftrightarrow$

$\dfrac{x^2 - 11}{x+1} - 1 \leq 0$. Find a common denominator, simplify, and

factor: $\dfrac{x^2 - 11}{x+1} - \dfrac{x+1}{x+1} \leq 0 \Leftrightarrow \dfrac{x^2 - x - 12}{x+1} \leq 0 \Leftrightarrow$

$\dfrac{(x-4)(x+3)}{x+1} \leq 0.$

$$x - 4 = 0 \qquad x + 3 = 0 \qquad x + 1 = 0$$
$$x = 4 \qquad\quad x = -3 \qquad\quad x = -1$$

Interval	$(-\infty, -3)$	$(-3, -1)$	$(-1, 4)$	$(4, \infty)$
Test value	-5	-2	0	5
Sign of $x - 4$	$-$	$-$	$-$	$+$
Sign of $x + 3$	$-$	$+$	$+$	$+$
Sign of $x + 1$	$-$	$-$	$+$	$+$
Sign of $\dfrac{x^2 - x - 12}{x+1}$	$-$	$+$	$-$	$+$

Since the inequality is not strict, we first check the endpoints. The endpoint $x = 1$ is excluded from the solution since it makes a denominator 0. We want the interval where the quotient is negative, so the solution is $(-\infty, -3] \cup (-1, 4]$.

(B) **Modeling with inequalities.**

The basic ideas of modeling applied problems with inequalities are the same as modeling applied problems with equalities. The only difference is that you wind up with an inequality to solve instead of an equation.

Exercise: After an evening of playing video games in an arcade, some friends find they have only 20 coins left. If they have only quarters and dimes, how much money might they have left?

Answer: The least they could have left is 20×0.10 (all dimes) and the most they could have is 20×0.25 (all quarters). So if they have x dollars left, then

$$20 \times 0.10 \ \leq \ x \leq 20 \times 0.25$$
$$2 \ \leq \ x \leq 5$$

So they have between $2 and $5 left.

▰ Core Exercises

7, 9, 29, 35, 43, 63, 75, 83, 101, 105

▰ 1.7   Absolute Value Equations and Inequalities

▰ Concepts

(A) **Equations and inequalities involving absolute value.**

- Solving absolute value equations.
- Properties of absolute value inequalities.

▰ Definitions

- **Absolute value equation:** An equation involving an absolute value, such as $|x + 3| = 5$.
- **Absolute value inequality:** An inequality involving an absolute value, such as $|x + 3| \geq 5$.

▰ Question to Ask Your Teacher

- If the solution to an equation is of a form $-5 \leq x \leq 6$, is it permissible to leave it in that form, or is interval notation or a number line required?

▰ Review

(A) **Equations and inequalities involving absolute value.**

- **Solving absolute value equations.**

  Equations that involve absolute value are solved by the following property:

  For $c \geq 0$, $|x| = c$ is equivalent to the *two* equations $x = c$ or $x = -c$.

  Exercise: Solve $|x + 6| = 8$.

  Answer: This is equivalent to $x + 6 = 8$ (in which case $x = 2$) or $x + 6 = -8$ (in which case $x = -14$). Check: $|2 + 6| = 8$ and $|-14 + 6| = |-8| = 8$, so both $x = 2$ and $x = -14$ are solutions.

Exercise: Solve $|3x + 7| = 5$.

Answer: Either $3x + 7 = 5 \Leftrightarrow 3x = -2 \Leftrightarrow x = -\frac{2}{3}$
or $3x + 7 = -5 \Leftrightarrow 3x = -12 \Leftrightarrow x = -4$.
Check: $\left|3\left(-\frac{2}{3}\right) + 7\right| = |-2 + 7| = 5$ and
$|3(-4) + 7| = |-12 + 7| = |-5| = 5$, so both are solutions.

- **Properties of Absolute Value Inequalities.**

Equations involving absolute values are solved using the fact that $|x| = c$ is equivalent to $x = \pm c$. To solve inequalities that involve absolute value, we use the following properties:

Inequality	Equivalent Form	Graph
$|x| < c$	$-c < x < c$	
$|x| \le c$	$-c \le x \le c$	
$|x| > c$	$x < -c \text{ or } c < x$	
$|x| \ge c$	$x \le -c \text{ or } c \le x$	

Exercise: Solve $|3x + 5| < 7$ and graph the solution.

Answer: $|3x + 5| < 7 \Leftrightarrow -7 < 3x + 5 < 7 \Leftrightarrow$
$-12 < 3x < 2 \Leftrightarrow -4 < x < \frac{2}{3}$

Check by substituting values of x that are inside the interval and outside the interval.

Exercise: Solve $|5 - 2x| \ge 9$ and graph the solution.

Answer: $|5 - 2x| \ge 9$, so either $5 - 2x \ge 9 \Leftrightarrow -2 \ge 4 \Leftrightarrow$
$x \le -2$, or $5 - 2x \le -9 \Leftrightarrow -2x \le -14 \Leftrightarrow x \ge 7$.

Exercise: Solve $\left|\dfrac{3}{2x + 1}\right| \ge 2$ and graph the solution.

Answer: $\left|\dfrac{3}{2x + 1}\right| \ge 2 \Leftrightarrow \dfrac{|3|}{|2x + 1|} \ge 2 \Leftrightarrow 3 \ge 2|2x + 1|$
$\Leftrightarrow \frac{3}{2} \ge |2x + 1|$. Express in the form $|Q| \le c$: $|2x + 1| \le \frac{3}{2}$
$\Leftrightarrow -\frac{3}{2} \le 2x + 1 \le \frac{3}{2} \Leftrightarrow -\frac{5}{2} \le 2x \le \frac{1}{2} \Leftrightarrow -\frac{5}{4} \le x \le \frac{1}{4}$.
However, $\left|\dfrac{3}{2x + 1}\right|$ is not defined for $x = -\frac{1}{2}$. Therefore, the
solution is $-\frac{5}{4} \le x \le \frac{1}{4}$, $x \ne -\frac{1}{2}$ or $\left[-\frac{5}{4}, -\frac{1}{2}\right) \cup \left(-\frac{1}{2}, \frac{1}{4}\right]$.

▰ Core Exercises

23, 49, 51, 53

2   Coordinates and Graphs

2.1   The Coordinate Plane

▸ Concepts

(A) The coordinate plane.

- Plotting points and regions.
- Finding the distance between two points.
- Finding the midpoint between two points.

▸ Definitions

- **Coordinate Plane:** The plane defined by the x- and y-axes. Also called the xy-**plane**.
- x-**axis**: The horizontal axis.
- y-**axis**: The vertical axis.
- **Origin**: The point $(0, 0)$ where the x- and y-axes meet.
- **Quadrants**: The coordinate axes divide the plane into four pieces. These pieces are called the four **quadrants** and are numbered I, II, III, IV:

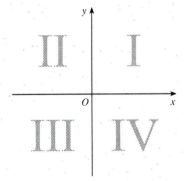

▸ Hints and Tips

- Don't ignore Figure 4 in the text. If you can truly understand that figure, and what every part of the figure means, and how the Pythagorean Theorem applies, you will never forget the distance formula. If you ask your teacher to explain that figure to you, he or she will probably react like you offered him or her a piece of chocolate. Teachers love talking about the Pythagorean Theorem.

Review

(A) The coordinate plane.

The (x, y) coordinate system is called the **rectangular coordinate system** or the **Cartesian coordinate system**. The plane used with this coordinate system is called the **coordinate plane** or the **Cartesian plane**. Points are specified in the form $(x\text{-coordinate}, y\text{-coordinate})$.

• Plotting points and regions.

Exercise: Plot the points $(1, 2)$, $(-1, 4)$, $(0, 0)$, and $\left(\frac{1}{2}, -2\right)$.

Answer:

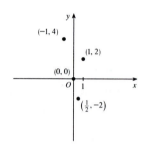

Exercise: Sketch the region $\{(x, y) \mid |x| \geq 2\}$.

Answer:

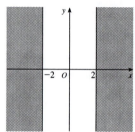

Note that the lines $x = \pm 2$ satisfy the inequality, because it is not strict. We show this by using solid lines for $x = \pm 2$ on the graph.

• Finding the distance between two points

The **distance** between the points $P_1(x_1, y_1)$ and $P_2(x_2, y_2)$ is given by the formula $d(P_1, P_2) = \sqrt{(x_2 - x_1)^2 + (y_2 - y_1)^2}$.

Exercise: Find the distance between the following pairs of points.

(a) $(4, 7)$ and $(5, 5)$

Answer:

$$d = \sqrt{(x_2 - x_1)^2 + (y_2 - y_1)^2}$$
$$= \sqrt{(5 - 4)^2 + (5 - 7)^2}$$
$$= \sqrt{(1)^2 + (-2)^2} = \sqrt{1 + 4}$$
$$= \sqrt{5}$$

(b) $(-2, 3)$ and $(1, -4)$

$$d = \sqrt{(x_2 - x_1)^2 + (y_2 - y_1)^2}$$
$$= \sqrt{(1 - (-2))^2 + (-4 - 3)^2}$$
$$= \sqrt{(3)^2 + (-7)^2} = \sqrt{9 + 49}$$
$$= \sqrt{58}$$

(c) $(-5, -1)$ and $(7, -6)$

$$d = \sqrt{(x_2 - x_1)^2 + (y_2 - y_1)^2}$$
$$= \sqrt{(7 - (-5))^2 + ((-6) - (-1))^2}$$
$$= \sqrt{(12)^2 + (-5)^2} = \sqrt{144 + 25}$$
$$= \sqrt{169} = 13$$

Exercise: Which point is closer to the origin?

Answer:

(a) $(5, 4)$ or $(-3, -3)$

The distance from $(5, 4)$ to the origin is
$$d = \sqrt{(5 - 0)^2 + (4 - 0)^2} = \sqrt{25 + 16} = \sqrt{41}.$$
The distance from $(-3, -3)$ to the origin is
$$d = \sqrt{(-3 - 0)^2 + (-3 - 0)^2} = \sqrt{9 + 9} = \sqrt{18} = 3\sqrt{2}.$$
Since $\sqrt{18} < \sqrt{41}$, $(-3, -3)$ is closer to the origin.

(b) $(7, 3)$ or $(-6, 5)$

The distance from $(7, 3)$ to the origin is
$$d = \sqrt{(7 - 0)^2 + (3 - 0)^2} = \sqrt{49 + 9} = \sqrt{56}.$$
The distance from $(-6, 5)$ to the origin is
$$d = \sqrt{(-6 - 0)^2 + (5 - 0)^2} = \sqrt{36 + 25} = \sqrt{61}.$$
Since $\sqrt{56} < \sqrt{61}$, $(7, 3)$ is closer to the origin.

- **Finding the midpoint between two points.**

  The **midpoint** of the line segment joining $P_1(x_1, y_1)$ and $P_2(x_2, y_2)$ is $\left(\dfrac{x_1 + x_2}{2}, \dfrac{y_1 + y_2}{2}\right)$.

Exercise: Find the midpoint of the segment joining P_1 and P_2.

Answer:

(a) $P_1 = (1, 5)$ and $P_2 = (-3, -3)$

The midpoint is $\left(\frac{1+(-3)}{2}, \frac{5+(-3)}{2}\right) = \left(\frac{-2}{2}, \frac{2}{2}\right) = (-1, 1)$.

(b) $P_1 = (7, -2)$ and $P_2 = (5, 8)$

The midpoint is $\left(\frac{7+5}{2}, \frac{(-2)+8}{2}\right) = \left(\frac{12}{2}, \frac{6}{2}\right) = (6, 3)$.

(c) $P_1 = (-5, 4)$ and $P_2 = (-2, 1)$

The midpoint is $\left(\frac{(-5)+(-2)}{2}, \frac{4+1}{2}\right) = \left(-\frac{7}{2}, \frac{5}{2}\right)$.

▜ Core Exercises

7, 17, 35, 37, 43, 51, 55

2.2   Graphs of Equations in Two Variables

Concepts

(A) **Graphs of equations in two variables.**

- Sketching a graph by plotting points.
- Finding x- and y-intercepts.
- Graphing a circle from its equation.

(B) **Symmetry.**

- Testing an equation for various types of symmetry.
- Using symmetry to help graph an equation.

Definitions

- **x-intercepts:** The x-coordinates of the points where a graph intersects the x-axis. These occur when the y-coordinate is zero.
- **y-intercepts:** The y-coordinates of the points where a graph intersects the y-axis. These occur when the x-coordinate is zero.
- **Symmetry with respect to the x-axis:** A graph is symmetric with respect to the x-axis if the graph looks like a mirror has been placed along the x-axis.
- **Symmetry with respect to the y-axis:** A graph is symmetric with respect to the y-axis if the graph looks like a mirror has been placed along the y-axis.
- **Symmetry with respect to the origin:** A graph is symmetric with respect to the origin if the graph looks the same when the whole page is turned $180°$.

Review

(A) **Graphs of equations in two variables.**

- **Sketching a graph by plotting points.**

  The graph of an equation in x and y is the set of all points (x, y) that satisfy the equation. The point (x, y) lies on the graph of the equation if and only if its coordinates satisfy the equation.

Exercise: Make a table and sketch the graph of the equation $2x + y = 5$.

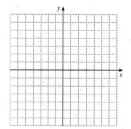

Answer: Start by solving $2x + y = 5$ for y: $y = -2x + 5$. Now put in different values for x and find the corresponding y-values. Make a list and plot these points.

x	y	(x, y)
-2	9	$(-2, 9)$
-1	7	$(-1, 7)$
0	5	$(0, 5)$
1	3	$(1, 3)$
2	1	$(2, 1)$
3	-1	$(3, -1)$

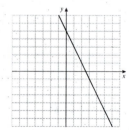

Exercise: Make a table and sketch the graph of the equation $x = 2y^2 - 1$.

Answer: Since the equation is already solved for x, substitute various values for y and find the corresponding x-values. Make a list and plot these points.

y	x	(x, y)
-2	7	$(7, -2)$
-1	1	$(1, -1)$
0	-1	$(-1, 0)$
1	1	$(1, 1)$
2	7	$(7, 2)$
3	17	$(17, 3)$

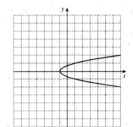

• **Finding x- and y-intercepts.**

The x-coordinates of points where the graph of an equation intersects the x-axis are called x-**intercepts**. The y-coordinates of points where the graph of an equation intersects the y-axis are called y-**intercepts**.

Intercepts	How to Find Them	Where They Are on the Graph
x-**intercepts:** The x-coordinates of the points where the graph intersects the x-axis.	Set $y = 0$ and solve for x.	
x-**intercepts:** The y-coordinates of the points where the graph intersects the y-axis.	Set $x = 0$ and solve for y.	

Exercise: Find the x- and y-intercepts of the graph of the given equation.

Answer:

(a) $y = 2x - 3$

To find the x-intercept(s), we set $y = 0$ and solve for x:
$0 = 2x - 3 \iff 2x = 3 \iff x = \frac{3}{2}$, so the x-intercept is $\frac{3}{2}$.

To find the y-intercept(s), we set $x = 0$ and solve for y:
$y = 2(0) - 3 = -3$, so the y-intercept is -3.

(b) $x^2 - 4x - 2xy = y + 5$

To find the x-intercept(s), we set $y = 0$ and solve for x:

$x^2 - 4x - 2x\,(0) = (0) + 5 \Leftrightarrow x^2 - 4x = 5 \Leftrightarrow$

$x^2 - 4x - 5 = (x - 5)(x + 1) = 0.$

$$x - 5 = 0 \qquad x + 1 = 0$$
$$x = 5 \qquad x = -1$$

So the x-intercepts are 5 and -1.

To find the y-intercept(s), we set $x = 0$ and solve for y:

$(0)^2 - 4\,(0) - 2\,(0)\,y = y + 5 \Leftrightarrow 0 = y + 5 \Leftrightarrow y = -5$, so the y-intercept is -5.

(c) $x = 2y^2 - 1$

To find the x-intercept(s), we set $y = 0$ and solve for x:

$x = 2\,(0)^2 - 1 \Leftrightarrow x = -1$, so the x-intercept is -1.

To find the y-intercept(s), we set $x = 0$ and solve for y:

$0 = 2y^2 - 1 \Leftrightarrow 2y^2 = 1 \Rightarrow y = \pm\sqrt{\frac{1}{2}}$, so the y-intercepts are

$\pm\sqrt{\frac{1}{2}}.$

- **Graphing circles from their equation.**

The **general equation of a circle** with radius r and centered at (h, k) is $(x - h)^2 + (y - k)^2 = r^2$. When the center is the origin $(0, 0)$, this equation becomes $x^2 + y^2 = r^2$.

Exercise: Describe the graph of $x^2 + y^2 = 49$.

Answer: We write the equation as $x^2 + y^2 = 7^2$, and see that this is the equation of a circle of radius 7, centered at the origin.

Exercise: Find an equation of the circle with radius 8 units centered at $(2, -4)$.

Answer: Substituting $h = 2$, $k = -4$, and $r = 8$ into the general equation of a circle, we have

$(x - 2)^2 + (y - (-4))^2 = 8^2$ or $(x - 2)^2 + (y + 4)^2 = 64.$

Exercise: Determine the center and radius of the circle given by

$x^2 + y^2 + 6x - 4y - 68 = 0.$

Answer: We group the terms by variable and complete the square in each variable. $x^2 + y^2 + 6x - 4y - 68 = 0 \Leftrightarrow$

$x^2 + 6x + \mathbf{9} + y^2 - 4y + \mathbf{4} = 68 + \mathbf{9} + \mathbf{4} \Leftrightarrow$

$(x + 3)^2 + (y - 2)^2 = 81 = 9^2$. The center of this circle is $(-3, 2)$ and its radius is 9 units.

B Symmetry.

Symmetry is an important tool that can be used in graphing equations.

> 1. A curve is **symmetric with respect to the x-axis** if its equation is unchanged when y is replaced by $-y$.
> 2. A curve is **symmetric with respect to the y-axis** if its equation is unchanged when x is replaced by $-x$.
> 3. A curve is **symmetric with respect to the origin** if its equation is unchanged when x is replaced by $-x$ and y is replaced by $-y$.

• **Testing an equation for various types of symmetry.**

Exercise: Determine the symmetry (if any) of the graph of $x - y^2 = 6$.

Answer: Check each type of symmetry:

With respect to the x-axis? Yes, since $x - (-y)^2 = 6 \Leftrightarrow$ $x - y^2 = 6$ is the same as the original equation.

With respect to the y-axis? No:
$(-x) - y^2 = -x - y^2 \neq x - y^2$.

With respect to the origin? No:
$(-x) - (-y)^2 = -x - y^2 \neq x - y^2$.

Exercise: Determine the symmetry (if any) of the graph of $x^2 + y^2 - 6y = 2$.

Answer: Check each type of symmetry:

With respect to the x-axis? No, since
$x^2 + (-y)^2 - 6(-y) = x^2 + y^2 + 6y$ and
$x^2 + y^2 + 6y \neq x^2 + y^2 - 6y$.

With respect to the y-axis? Yes, since
$(-x)^2 + y^2 - 6y = x^2 + y^2 - 6y = 2$.

With respect to the origin? No:
$(-x)^2 + (-y)^2 - 6(-y) = x^2 + y^2 + 6y$ and
$x^2 + y^2 + 6y \neq x^2 + y^2 - 6y$.

Exercise: Determine the symmetry (if any) of the graph of $x^4 + 3y^2 = 6$.

Answer: Check each type of symmetry:

With respect to the x-axis? Yes, since $x^4 + 3(-y)^2 = 6 \Leftrightarrow$ $x^4 + 3y^2 = 6$ is the same as the original equation.

With respect to the y-axis? Yes, since
$(-x)^4 + 3y^2 = x^4 + 3y^2 = 6$.

With respect to the origin? Yes, since
$(-x)^4 + 3(-y)^2 = x^4 + 3y^2 = 6$.

• **Using symmetry to help graph an equation.**

Exercise: Graph $y = \sqrt[3]{x}$.

Answer: A quick check shows that this graph is symmetric about the origin. We find some points on the graph:

x	$y = \sqrt[3]{x}$	(x, y)
0	0	$(0, 0)$
1	1	$(1, 1)$
8	2	$(8, 2)$

Now, because this graph is symmetric about the origin, we know that $(-1, -1)$ and $(-8, -2)$ are on the graph. We get these points automatically, by symmetry.

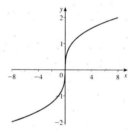

Core Exercises

13, 27, 39, 49, 53, 57, 65, 75, 79, 81, 83, 87

2.3   Graphing Calculators; Solving Equations and Inequalities Graphically

Concepts

(A) **Using a graphing calculator.**

• Sketching graphs of equations with the use of a graphing calculator, including choosing an appropriate viewing rectangle.

• Solving equations graphically.

• Solving inequalities graphically.

(B) **Applications.**

Definitions

- **Viewing rectangle:** When graphing an equation, we don't have an infinite amount of paper to use. The viewing rectangle is the combination of the range of x-values and y-values that we are going to draw. It is important to choose a viewing rectangle that displays the interesting features of a graph.

- **Algebraic method of solving an equation:** Solving an equation by using algebra to get a sequence of equivalent equations, ending with something like $x = a$.

- **Graphical method of solving an equation:** Solving an equation by moving all terms to one side, setting that side equal to y, and using the resulting graph to approximate where $y = 0$.

Questions to Ask Your Teacher

- What are the circumstances under which we are allowed to use a calculator?

Hints and Tips

- Some students find it difficult to get a good viewing rectangle. The first thing to try is using your head — if you see a function like $y = \sqrt{\text{a big mess}}$, you should think, "I *know* y is always positive, because of the radical sign, so I'm not going to bother including negative values of y in my viewing rectangle." Be liberal with the use of the "Zoom $\rightarrow$ box" key on your calculator. It is your friend.

- An unpleasant fact about math courses is that time often becomes a factor on exams. Even if your teacher's exams are not too long, you don't know what next semester will bring. It is important to practice to the point where you can graph a function efficiently, without spending a lot of time messing with parenthesis errors and viewing rectangles. Because it is portable, you can practice with your calculator on long car rides (assuming you are not driving) and while waiting in line. You don't want to spend fifteen minutes on a fifty-minute exam finding out that your calculator was graphing $3x + 2^2$ when you wanted it to graph $(3x + 2)^2$.

- A good rule of thumb for calculator use in general is this: Always try to know more about how to use your calculator than your teacher does. If you spend a good four or five hours with your calculator manual, you can achieve this goal. This up-front time will reward you for years to come.

Review

(A) **Using a graphing calculator.**

- **Sketching graphs of equations with the use of a graphing calculator, including choosing an appropriate viewing rectangle.**
  Calculators and computers graph by plotting points. Sometimes the points are connected by straight lines, other times just the points are displayed. The viewing area of a calculator or computer is referred to as the **viewing rectangle**. This is an $[a, b] \times [c, d]$ portion of the coordinate plane. Choosing the *appropriate* viewing rectangle is one of the most important aspects of using a graphing calculator. Choose a relatively large viewing rectangle to obtain a global view of the graphs. Then, to investigate details, zoom to a smaller viewing rectangle that shows just the details of interest.

Exercise: Use a graphing calculator to draw the graph of the equation $f(x) = \sqrt{x^2 + 9}$ in the following viewing rectangles.

Answer: The solutions shown below are representations of what you should see on your graphing calculator.

(a) $[-3, 3]$ by $[-3, 3]$

Only the point $(0, 3)$ should appear, and it will be lost in the y-axis.

(b) $[-5, 5]$ by $[-5, 5]$

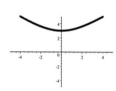

(c) $[-5, 5]$ by $[0, 10]$

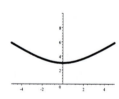

Exercise: Graph the equation $y = |x^2 - 4|$ on a graphing calculator.

Answer: Make sure you have chosen a viewing window that shows the "W" shape of the graph, the sharp corners on the x-axis, and how every part curves. One possible good viewing rectangle is $[-5, 5]$ by $[-5, 5]$.

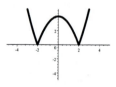

Exercise: In parts (a)–(c), graph the equation $9x^2 + 4y^2 = 36$ using a graphing calculator with the given viewing rectangle. In part (d), sketch the graph by hand.

Answer:

(a) $[-2, 2]$ by $[-3, 3]$

We solve for y: $9x^2 + 4y^2 = 36 \Leftrightarrow 4y^2 = 36 - 9x^2 \Leftrightarrow y^2 = 9 - \frac{9}{4}x^2 \Rightarrow y = \pm\sqrt{9 - \frac{9}{4}x^2}$. The top half of the graph is given by $y = \sqrt{9 - \frac{9}{4}x^2}$, and the bottom half is given by $y = -\sqrt{9 - \frac{9}{4}x^2}$.

This example shows how the choice of viewing rectangle can give different impressions of the same graph.

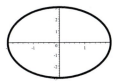

(b) $[-3, 3]$ by $[-6, 6]$

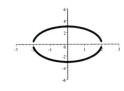

(c) $[-6, 6]$ by $[-4, 4]$

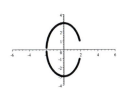

(d)

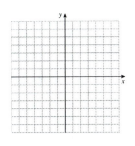

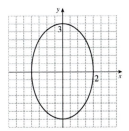

Exercise: Graph $y = x^2$ and $y = 4 - x^2$ on the same screen.

Answer: Your viewing window should show what both graphs look like, and where they intersect. One good viewing rectangle is $[-3, 3]$ by $[-2, 5]$.

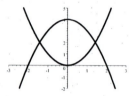

• **Solving equations graphically.**

In the **algebraic method**, we use the rules of algebra to isolate x on one side of the equation. In the **graphical method**, we move all terms to one side and set the expression equal to y. We then sketch the graph of the equation and find the values of x for which $y = 0$. Another method is to set each side of the equation equal to y, then sketch the graph of the equations and determine where the graphs intersect. Solutions found using a graphing calculator are only approximations. Exact solutions can be confirmed only by checking.

Exercise: Solve $4x^2 - 3 = x^2 - 6x + 9$ both algebraically and graphically.

Answer: *Algebraically:*

$$4x^2 - 3 = x^2 - 6x + 9$$
$$3x^2 + 6x - 12 = 0 \qquad \text{Set equal to zero.}$$
$$x^2 + 2x - 4 = 0$$

Now use the quadratic formula:

$$x = \frac{-2 \pm \sqrt{2^2 - (4)(1)(4)}}{2(1)} = \frac{-2 \pm \sqrt{20}}{2} = -1 \pm \sqrt{5}.$$

Thus, $x = -1 + \sqrt{5}$ or $x = -1 - \sqrt{5}$.

Graphically:

Start by setting each side of the equation equal to y to get $y = 4x^2 - 3$ and $y = x^2 - 6x + 9$. We graph the two equations in the viewing rectangle $[-10, 10]$ by $[-5, 50]$. It appears that these graphs have two points of intersection.

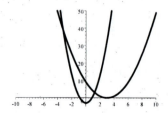

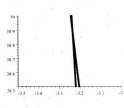

In the viewing rectangle $[-3.5, -3]$ by $[38.5, 39]$, we find the solution $x \approx -3.236$.

In the viewing rectangle $[1, 1.5]$ by $[2.75, 3.25]$ we find the solution $x \approx 1.236$.

Note: $x = -1 + \sqrt{5} \approx 1.236$ and $x = -1 - \sqrt{5} \approx -3.236$.

Exercise: Solve the equation $x^3 + 3x^2 - 3 = 0$ graphically.

Answer: We graph the equation $y = x^3 + 3x^2 - 3$ in the viewing rectangle $[-5, 5]$ by $[-5, 5]$. The solutions occur at the x-intercepts. Using zoom and trace functions, we can approximate the solutions as $x = -2.53$, $x = -1.35$, and $x = 0.88$.

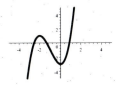

Exercise: Solve $x^3 + 3 = \dfrac{2}{x^2 + 2}$ by moving all terms to one side and graphically locating the x-intercept.

Answer:
$$x^3 + 3 = \frac{2}{x^2 + 2}$$
$$x^3 + 3 - \frac{2}{x^2 + 2} = 0 \qquad \text{Isolate terms}$$

Now we set the result equal to y: $y = x^3 + 3 - \dfrac{2}{x^2 + 2}$. We graph this expression in the viewing rectangle $[-5, 5]$ by $[-5, 5]$.

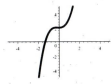

The solutions occur at the x-intercepts of the equation. The only solution appears to be $x \approx -1.35$.

Exercise: Solve $x^3 + 3 = \dfrac{2}{x^2 + 2}$ by graphing each side of the equation.

Answer: We graph $y = x^3 + 3$ and $y = \dfrac{2}{x^2 + 2}$ in the viewing rectangle $[-5, 5]$ by $[-5, 5]$. The solutions occur at the intercepts of the two graphs.

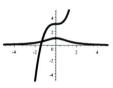

Again the solution is $x \approx -1.35$.

- **Solving inequalities graphically.**

We solve inequalities graphically by graphing each side of the inequality in the same viewing rectangle. We determine which graph goes with which equation by evaluating each equation at an x value. Then determine where the graph of one equation lies above the graph of the other.

Exercise: Solve the inequality $\sqrt{x^2 + x + 2} > 3 - x$ graphically.

Answer: We set $y_1 = \sqrt{x^2 + x + 2}$ (solid) and $y_2 = 3 - x$ (dashed) in the viewing rectangle $[-5, 5]$ by $[-10, 10]$.

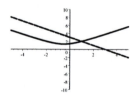

From the graph, we see that $y_1 < y_2$ when $x > 1$. So the solution is $(1, \infty)$.

Exercise: Solve the inequality $\sqrt{x^2 + 2x + 4} \geq \sqrt{2x^2 + 1}$ graphically.

Answer: We set $y_1 = \sqrt{x^2 + 2x + 4}$ (solid) and $y_2 = \sqrt{2x^2 + 1}$ (dashed) in the viewing rectangle $[-5, 5]$ by $[-5, 10]$.

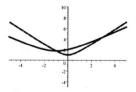

From the graph, we see that $y_1 \geq y_2$ when $-1 \leq x \leq 3$, so the solution is $[-1, 3]$.

(B) Applications

The process of setting up applications has not changed; you just have a new tool to solve the resultant equations.

Exercise: An umbrella manufacturer estimates that the cost per week of producing x umbrellas is $0.01x^2 + 4x + 800$. How many umbrellas do they need to manufacture each week so that the *average* cost per umbrella is less than \$12?

Answer: The average cost per umbrella manufactured during the week is $\dfrac{0.01x^2 + 4x + 800}{x}$. So we need to solve

$\dfrac{0.01x^2 + 4x + 800}{x} < 12$. We set $y_1 = \dfrac{0.01x^2 + 4x + 800}{x}$

and $y_2 = 12$ and graph both equations in the viewing rectangle $[0, 800]$ by $[0, 50]$.

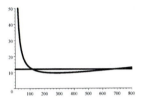

Note that x represents the number of umbrellas manufactured during the week, so we must have $x > 0$. From the viewing rectangle, the average cost is less than \$12 when $117.2 < x < 682.8$. Thus, they must manufacture between 118 and 682 umbrellas.

▚ Core Exercises

53, 59, 61

▚ 2.4   Lines

▚ Concepts

(A) The slope of a line.

- Estimating the slope of a line.
- Finding the slope of a line through two points.

(B) Equations of lines.

- Finding an equation of a line given a point and a slope.
- Finding an equation of a line given two points.
- Finding equations of vertical and horizontal lines.

(C) Parallel and perpendicular lines.

- Determining if given lines are parallel or perpendicular.
- Finding an equation of a line parallel or perpendicular to a given line.

(D) Applications.

- Using the slope as a rate of change.
- Economic applications.

Definitions

- **Slope:** The slope of a line is a measure of its steepness. Positive slope means the line is ascending as we go from left to right. Negative slope means the line is descending as we go from left to right. Zero slope means the line is horizontal.
- **Parallel:** Two lines are parallel if they have the same slope.
- **Perpendicular:** Two lines are perpendicular if they intersect at a right angle. This occurs if and only if their slopes are negative reciprocals of each other.
- **Equilibrium point:** Economists call the equilibrium point the price where the amount produced of a good is equal to the amount sold.

Questions to Ask Your Teacher

- Is there a preferred form of an equation of a line?

Hints and Tips

- In this course, in calculus, and in every other math course you will ever take, there is one thing that will not change. If you want to find an equation of a line, you are going to need to find a point and a slope. Sometimes, if you are lucky, both will be given to you. Sometimes, you will have to find one or both. But, at some point, you are going to need to find a point and a slope. In this section, for example, you are asked to find an equation of a line given two points. The first thing you do is find the slope.

Review

(A) **The slope of a line.**

Slope is an extremely important concept in mathematics. Slope is a central focus in the study of calculus. The slope of a nonvertical line that passes through the points $P_1\,(x_1, y_1)$ and $P_2\,(x_2, y_2)$ is $m = \dfrac{y_2 - y_1}{x_2 - x_1} = \dfrac{\text{rise}}{\text{run}} = \dfrac{\text{change in } y}{\text{change in } x}$.
The slope of a vertical line is not defined.

- **Estimating the slope of a line.**

Given the graph of a line, you should be able to roughly estimate its slope, provided the x- and y-axes are drawn to the same scale.

Exercise: For each of the following lines, *estimate* the slope. Your answer need not be exact.

Answer: (a) 1 (b) −1 (c) 2 (d) 0 (e) −$\frac{1}{2}$

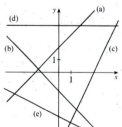

Notice that if the x and y axes are *not* drawn to the same scale, then it is not possible to casually estimate the slope. For example, the slope of the line $y = x + 1$ is equal to 1. We can estimate the slope in the first picture below. But if the scales on the axes are not the same, then the slope of the line looks different.

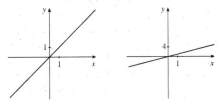

- **Finding the slope of a line through two points.**

Exercise: Find the slope of the line that passes through the points P and Q.

Answer:

(a) $P(1, 3)$ and $Q(-2, 9)$

$m = \dfrac{9 - 3}{-2 - 1} = \dfrac{6}{-3} = -2$ or $m = \dfrac{3 - 9}{1 - (-2)} = \dfrac{-6}{3} = -2$.

The order in which the points are taken is not important, but you must be consistent throughout the calculation.

(b) $P(2, -5)$ and $Q(-3, 4)$

$m = \dfrac{4 - (-5)}{-3 - 2} = \dfrac{9}{-5} = -\dfrac{9}{5}$

(c) $P(-7, -4)$ and $Q(-4, -3)$

$m = \dfrac{-3 - (-4)}{-4 - (-7)} = \dfrac{1}{3}$

(d) $P(5, 2)$ and $Q(-3, 2)$

$m = \dfrac{2 - 2}{-3 - 5} = \dfrac{0}{-8} = 0$. This is a horizontal line.

Note: $\dfrac{0}{-8}$ is defined and is equal to 0, but $\dfrac{-8}{0}$ *is not defined*; many students get these two concepts confused.

(e) $P(4, 2)$ and $Q(4, -1)$

$m = \dfrac{-1 - 2}{4 - 4} = \dfrac{-3}{0}$ which is undefined. So this line has no slope (it is a vertical line).

(B) Equations of lines.

There are three forms of the equation of a line that you should know. All of the equations are equivalent, as they should be, since they are all describing the same set of points. Some of them are easier to work with than others, and some are easier to derive than others. Assume that we know a line goes through the point (x_1, y_1) and has slope m.

Equation Type	Equation	Notes
Point-slope	$y - y_1 = m(x - x_1)$	Given a point and a slope, this is the easiest form to derive.
Slope-intercept	$y = mx + b$	b is the y-intercept; that is, the line goes through $(0, b)$.
General equation	$ax + by = c$	It is usually easiest to find one of the other forms, and then convert to this form.

● **Finding an equation of a line given a point and a slope.**

Given a point and a slope, it is usually easiest to derive the point-slope form of the line's equation.

Exercise: Find an equation of the line that passes through the point $(2, -1)$ with slope $\frac{1}{3}$. Sketch the line.

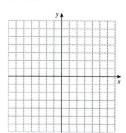

Answer:

Point-slope: Substitute into the equation $y - y_1 = m(x - x_1)$ with $(x_1, y_1) = (2, -1)$ and $m = \frac{1}{3}$. $y - (-1) = \frac{1}{3}(x - 2) \Leftrightarrow y + 1 = \frac{1}{3}(x - 2)$.

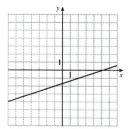

Slope-intercept: $y + 1 = \frac{1}{3}(x - 2) \Leftrightarrow y + 1 = \frac{1}{3}x - \frac{2}{3} \Leftrightarrow y = \frac{1}{3}x - \frac{5}{3}$. Notice that the slope-intercept form of the equation tells us that the y-intercept is $-\frac{5}{3}$.

General equation: $y = \frac{1}{3}x - \frac{5}{3} \Leftrightarrow -\frac{1}{3}x + y = -\frac{5}{3}$

Exercise: Find an equation of the line that passes through the point $(1, 3)$ with slope $-\frac{2}{5}$. Sketch the line.

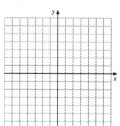

Answer:

Point-slope: Substitute $(x_1, y_1) = (1, 3)$ and $m = -\frac{2}{5}$ into the equation $y - y_1 = m(x - x_1)$: $y - 3 = -\frac{2}{5}(x - 1)$.

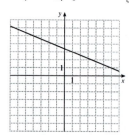

Slope-intercept: $y - 3 = -\frac{2}{5}(x - 1) \Leftrightarrow y - 3 = -\frac{2}{5}x + \frac{2}{5} \Leftrightarrow y = -\frac{2}{5}x + \frac{17}{5}$. Notice that the slope-intercept form of the equation tells us that the y-intercept is $\frac{17}{5}$.

General equation: $y = -\frac{2}{5}x + \frac{17}{5} \Leftrightarrow y + \frac{2}{5}x = \frac{17}{5}$

• **Finding an equation of a line given two points.**

To find the equation of a line, any line, we need a point and a slope. So if we are given two points, we have to find the slope of the line joining them.

Exercise: Find an equation of the line between $(2, 3)$ and $(4, 9)$. Express your answer in both point-slope form and slope-intercept form.

Answer: We first find the slope of the line: $m = \dfrac{9 - 3}{4 - 2} = 3.$
We choose the point $(2, 3)$. [The answer would also be correct if we choose $(4, 9)$.] The point-slope form is $y - 2 = 3\,(x - 3)$ and the slope-intercept form is $y = 3x - 7.$

Exercise: Find an equation of the line going through $(-2, 3)$ with y-intercept 1.

Answer: To determine the slope we use two points on the line, $(-2, 3)$ and $(0, 1)$ (by definition of the y-intercept). The slope is $m = \dfrac{1 - 3}{0 - (-2)} = -1.$ So the point-slope form is
$y - 3 = -1\,(x + 2)$ and the slope-intercept form is
$y = -x + 1.$
Note: We could have used a shortcut in this problem. The fact that we know that the y-intercept is 1 allows us to write
$y = -x + 1$ as soon as we know that $m = -1.$

Note that if you are given an equation of a line, and want to find the slope or the y-intercept, you can put it in the form $y = mx + b.$

• **Finding equations of vertical and horizontal lines.**

Horizontal lines have slope 0, and vertical lines have no slope (the slope is undefined, which is not the same thing as zero). Horizontal lines have the form $y = k$ and vertical lines have the form $x = k$ for some constant k.

Exercise: Find the equation of the horizontal line that passes through the point $(3, -2)$.

Answer: $y = -2$

Exercise: Find an equation of the vertical line that passes through the point $(2, -1)$.

Answer: $x = 2$

Exercise: Sketch the line with equation $y = \frac{3}{2}.$

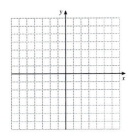

Answer:

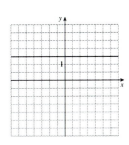

(C) **Parallel and perpendicular lines.**

Two lines are **parallel** if and only if they have the same slope (or both have no slope). Two lines with slopes m_1 and m_2 are **perpendicular** if $m_1 m_2 = -1$, that is, their slopes are negative reciprocals, so $m_2 = -\dfrac{1}{m_1}$. Also, horizontal lines (0 slope) are perpendicular to vertical lines (no slope).

● **Determining if given lines are parallel or perpendicular.**

Exercise: Determine if each pair of lines is parallel, perpendicular, or neither.

Answer:

(a) $2x + 6y = 7$ and $x + 3y = 9$

Solve each equation for y in order to put each line into slope-intercept form:

$$2x + 6y = 7 \qquad\qquad x + 3y = 9$$
$$6y = -2x + 7 \qquad\qquad 3y = -x + 9$$
$$y = -\tfrac{1}{3}x + \tfrac{7}{6} \qquad\qquad y = -\tfrac{1}{3}x + 3$$

Both lines have the same slope, $-\tfrac{1}{3}$, thus they are parallel.

(b) $5x + 2y = 0$ and $-5x + 2y = 2$

Solve each equation for y in order to put each line into slope-intercept form:

$$5x + 2y = 0 \qquad\qquad -5x + 2y = 2$$
$$2y = -5x \qquad\qquad 2y = 5x + 2$$
$$y = -\tfrac{5}{2}x \qquad\qquad y = \tfrac{5}{2}x + 1$$

Since $\left(-\tfrac{5}{2}\right) \cdot \left(\tfrac{5}{2}\right) \neq -1$ and the slopes are not equal, the lines are neither parallel nor perpendicular.

(c) $3x - y = 10$ and $x + 3y = 9$

Solve each equation for y in order to put each line into slope-intercept form:

$$3x - y = 10 \qquad\qquad x + 3y = 9$$
$$-y = -3x + 10 \qquad\qquad 3y = -x + 9$$
$$y = 3x - 10 \qquad\qquad y = -\tfrac{1}{3}x + 3$$

Since $3 = \dfrac{-1}{-1/3}$, the lines are perpendicular.

Exercise: Are $A\,(1,6)$, $B\,(7,8)$, $C\,(9,3)$, and $D\,(3,1)$ the vertices of a rectangle, or parallelogram, or neither?

Answer: Plot the points to find their relative position.

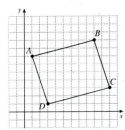

Now find the slopes of the lines passing through each pair of points.

Slope of $AB = \frac{8-6}{7-1} = \frac{1}{3}$

Slope of $BC = \frac{3-8}{9-7} = -\frac{5}{2}$

Slope of $CD = \frac{1-3}{3-9} = \frac{1}{3}$

Slope of $DA = \frac{6-1}{1-3} = -\frac{5}{2}$

Since AB and CD are parallel (denoted by $AB \parallel CD$) and $BC \parallel DA$, $ABCD$ is at least a parallelogram. Since $\frac{1}{3} \cdot \left(-\frac{5}{2}\right) \neq -1$, AB is *not* perpendicular to BC. So $ABCD$ is a parallelogram, but not a square.

• **Finding the equation of a line parallel or perpendicular to a given line.**

Remember that to find the equation of *any* line you are looking for a point and a slope.

Exercise: Find an equation of the line parallel to $y = 3x + 5$ that goes through the origin.

Answer:
Point: $(0,0)$
Slope: Since the line is parallel to $y = 3x + 5$, it has the same slope: $m = 3$. Thus the line has equation $y - 0 = 3\,(x - 0)$ or $y = 3x$.

Exercise: Find an equation of the line perpendicular to $y = 3x + 5$ that goes through the point $(-1, -1)$.

Answer:
Point: $(-1, -1)$
Slope: Since it is perpendicular to $y = 3x + 5$, its slope is the negative reciprocal of 3: $m = -\frac{1}{3}$. Thus the line has equation $y + 1 = -\frac{1}{3}\,(x + 1)$.

(D) Applications.

• **Using the slope as a rate of change.**

In following any line (except for a horizontal or vertical line), we go over a distance x and we go up (or down) a distance y. We can think of the slope as the rate at which this occurs. In calculus, we often use the words "slope" and "rate of change" interchangeably. There are some problems in the text that are designed to get you thinking about that concept.

Exercise: When we put mercury in a glass tube, it will rise or fall depending on the temperature. (This principle allows us to make mercury thermometers.)   Assume that we have a thermometer whose column of mercury is 10 centimeters tall when the temperature is 15° C and 12 centimeters tall when the temperature is 20° C.

‖ **Answer:**

(a) Write an equation of the line that gives the amount of mercury as a function of temperature. Put your answer in slope-intercept form.

We have two points: $(15, 10)$ and $(20, 12)$. Since we are thinking of height as a function of temperature, we let y be the height and x be the temperature.

Slope: $m = \dfrac{12 - 10}{20 - 15} = \dfrac{2}{5}$. The point-slope form of the line is thus $y - 10 = \frac{2}{5}(x - 15)$ and the slope-intercept form is $y = \frac{2}{5}x + 4$.

(b) What do the y-intercept and the slope of this line represent in real-world terms?

The y-intercept 4 represents the height of the mercury when the temperature is 0° C. The slope $\frac{2}{5}$ represents the rate of mercury climb with respect to temperature change.

- **Economic applications.**
In these applications, we have two lines. One gives the supply of an item as a function of its selling price. The more expensive something is, the more people are likely to produce it. The other gives the demand of an item as a function of its selling price. The more expensive something is, the less people are likely to buy. The selling price p will be the price at which the amount produced is equal to the amount sold, that is, the price where the two lines meet.

Exercise: A product's supply curve is $S = \frac{1}{3}p$, where p is given in dollars, and S is given in thousands of units. Its demand curve is $D = -p + 200$. At what price will the product be sold, and how many units will sell?

Answer: We graph the supply and demand curves and estimate the point of intersection:

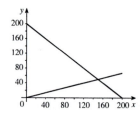

From the graph, we see that the intersection point is at about $(150, 50)$. [It turns out that this is the exact answer, which can be verified by substituting the point $(150, 50)$ into both equations.] Therefore the price will be $150. Substituting into the demand (or supply) function, we find that 50 units will be sold at that price.

◤ Core Exercises

2.5   Modeling Variation

Concepts

(A) **Proportionality.**

- Direct variation.
- Inverse variation.
- Joint variation.

Definitions

- **Direct variation/Directly proportional:** If the quantities x and y are related by $y = kx$ for some constant k, we say that y varies directly as x, or that y is directly proportional to x.

- **Inverse variation/Inversely proportional:** If the quantities x and y are related by $y = \dfrac{k}{x}$ for some constant k, we say that y varies inversely as x, or that y is inversely proportional to x.

- **Joint variation/Jointly proportional:** If the quantities x and y are related by $z = kxy$ for some constant k, we say that z varies jointly as x and y, or that z is jointly proportional to x and y.

Hints and Tips

- When someone says "y is proportional to x" they mean that y is directly proportional to x.

- The most important things to remember are that "y is proportional to x" implies that when x gets bigger, y gets bigger (for example, the size of my bald spot is proportional to my age), and that "y is inversely proportional to x" implies that when x gets bigger, y gets smaller (for example, my appetite at dinnertime is inversely proportional to the size of my lunch that day).

Review

(A) **Proportionality.**

- **Direct variation.**

  A *mathematical model* is a function that describes, at least approximately, the dependence of one physical quantity on another physical quantity. In the direct variation model, y **is directly proportional to** x if x and y are related by the equation $y = kx$ for some constant $k \neq 0$. The constant k is called the **constant of proportionality**.

  Exercise: Federal excise tax on gasoline is 13¢ per gallon. Write an equation that directly relates the tax and the quantity of gasoline.

  Answer: Let g be the quantity of gas (in gallons) and let T be the tax. Federal excise tax on gasoline is 13¢ per gallon. Since tax is 13¢ per gallon we have $T = 0.13g$. Here 0.13 is the constant of proportionality.

Exercise: The volume of liquid in a soda can is directly related to the height of the liquid in the can.

Answer:

(a) If the can is 12.5 cm tall and contains 355 mL when full, determine the constant of proportionality and write the equation for the variation.

Let h be the height of the can, V its volume, and k the constant of proportionality. Then $V = kh$. When $h = 12.5$, $V = 355$. Substituting, we have $355 = 12.5k \Leftrightarrow k = 28.4$. Thus the model is $V = 28.4h$.

(b) How much soda is left in the can when the liquid is 3 cm deep?

Setting $h = 3$ in the equation gives $V = 28.4h = 28.4\,(3) = 85.2$ mL.

(c) How deep is the liquid when there is 300 mL of soda left?

Setting $V = 300$ and solving for h gives $V = 28.4h \Leftrightarrow 300 = 28.4h \Leftrightarrow h \approx 10.56$ cm.

- **Inverse variation.**

In another model, y **is inversely proportional to** x if x and y are related by the equation $y = \dfrac{k}{x}$ for some constant $k \neq 0$.

Exercise: Write an equation that relates the width to the length of all rectangles whose area is 4 square units.

Answer: Let w be the width and l the length. Since $A = wl$, we have $4 = wl \Leftrightarrow w = \dfrac{4}{l}$. Thus the width of such a rectangle is inversely proportional to its length.

- **Joint variation.**

We say that z **is jointly proportional to** x **and** y if x, y, and z are related by the equation $z = kxy$ for some constant $k \neq 0$.

Exercise: Glu-lam beams are long beams used in construction. They are made by laminating 2 inch by 8 inch planks of wood together to create a wide beam.

Answer:

(a) Write an equation that directly relates the volume (in ft^3) of the wood used to the length and width of the glu-lam (in ft).

Let V be the volume of wood in ft^3, l the length (in ft) and w the width (in ft). Since the plank is 8 inches $= \frac{2}{3}$ foot thick, $V = \frac{2}{3}lw$. The volume of wood is jointly proportional to the length and width of the beam.

(b) How much wood is in a 2-foot wide, 50-foot long glu-lam beam?

Substituting $w = 2$ and $l = 50$, we find that $V = \frac{2}{3}lw = \frac{2}{3}\,(50)\,(2) = \frac{200}{3}\ \text{ft}^3 \approx 66.7\ \text{ft}^3$.

Core Exercises

5, 7, 13, 23, 43

3 Functions

3.1 What is a Function?

Concepts

(A) **Working with functions.**
- Evaluating a function given its formula.
- Finding the domain of a function.

(B) **Four ways to represent a function.**

Definitions

- **Function, Domain, Range:** A **function** f is a rule that assigns to each element x in a set A exactly one element, called $f(x)$, in a set B. The set A is called the **domain** of the function, and B is the **range** of the function. In this course, A and B will usually be some subset of the real numbers. In this context, the domain is the set of allowable inputs, and the range is the set of all possible outputs.
- **Independent and dependent variables:** If we write a function as $y = f(x)$, the "dummy variable" x is the independent variable (we are allowed to choose its value) and y is the "dependent variable" (its value depends on what we choose as x).

Review

(A) **Working with functions.**

In this course, we can think of a **function** as a rule that takes a real number and transforms it to another real number. "$f(x)$" is read as "f of x" or "f at x" and represents the **value of f at x**, or the **image of x under f**. Since $f(x)$ depends on the value x, $y = f(x)$ is the called the **dependent variable** and x is called the **independent variable**.

- **Evaluating a function given its formula.**

Exercise: Let $f(x) = \dfrac{3x^2 + x}{2 - x}$.

Answer:

(a) Find $f(-2)$.

Substituting -2 for x, we find that
$$f(-2) = \frac{3(-2)^2 + (-2)}{2 - (-2)} = \frac{12 - 2}{2 + 2} = \frac{10}{4} = \frac{5}{2}.$$

(b) Find $f(4)$.

Substituting 4 for x, we find that
$$f(4) = \frac{3(4)^2 + (4)}{2 - (4)} = \frac{48 + 4}{2 - 4} = \frac{52}{-2} = -26.$$

(c) Find $f(0)$.

Substituting 0 for x, we find that
$$f(0) = \frac{3(0)^2 + (0)}{2 - (0)} = \frac{0 - 0}{2 - 0} = \frac{0}{2} = 0.$$

- **Finding the domain of a function.**

If not explicitly stated, the **domain** of a function is the set of all real numbers for which the function makes sense and defines a real number. The domain is the "input" to the function. The **range** is the set of all possible values $f(x)$ as x varies throughout the domain of f. The range is sometimes called the **image**, because it is the "image of x under the rule f." The range depends on the domain. The range is the "output" of the function.

Exercise: Find the domain of the function
$$f(x) = \frac{4}{x+2}.$$

Answer: The only values of x for which f does not make sense occur when the denominator is 0, that is, when $x + 2 = 0$ or $x = -2$. So the domain of f is $(-\infty, -2) \cup (-2, \infty)$. The domain can also be expressed as $\{x \mid x \neq -2\}$, or simply $x \neq -2$.

Exercise: Find the domain of the function
$$f(x) = \frac{x-9}{x^3 - 25x}.$$

Answer: Again, the only values of x for which f does not make sense occur when the denominator is 0. We factor the denominator: $x^3 - 25x = x(x-5)(x+5)$. This is 0 when $x = 0$, $x = 5$, or $x = -5$, so the domain is $\{x \mid x \neq -5, x \neq 0, x \neq 5\}$. The domain can also be expressed as $(-\infty, -5) \cup (-5, 0) \cup (0, 5) \cup (5, \infty)$.

Exercise: Find the domain of the function
$$f(x) = \sqrt{x^2 - 4}.$$

Answer: We are looking for the values of x for which the function makes sense *and defines a real number*. Since only nonnegative numbers have real square roots, we use the methods from Section 1.7 to determine where $x^2 - 4 \geq 0$. Factoring, we find that $x^2 - 4 = (x-2)(x+2)$.

Interval	$(-\infty, -2)$	$(-2, 2)$	$(2, \infty)$
Test value	-5	0	5
Sign of $x - 2$	$-$	$-$	$+$
Sign of $x + 2$	$-$	$+$	$+$
Sign of $x^2 - 4$	$+$	$-$	$+$

Since the inequality is not strict, we check the endpoints of the intervals and see that both $x = -2$ and $x = 2$ are acceptable as input values. Thus, the domain is $(-\infty, -2] \cup [2, \infty)$ or $\{x \mid x \leq -2 \text{ or } x \geq 2\}$.

B **Four ways to represent a function.**

There are four ways to describe a specific function:

Verbally:	By a description in words
Algebraically:	By an explicit formula
Visually:	By a graph
Numerically:	By a table of values

▼ Core Exercises

3, 15, 23, 29, 51, 71, 73, 75

3.2   Graphs of Functions

Concepts

(A) **The relationship of a function and its graph.**

- Sketching the graph of a function from its formula.

- Determining function values from a graph.

- Finding the domain and range of a function from its graph.

(B) **Piecewise defined functions.**

- Sketching the graph of a piecewise defined function.

- Understanding the greatest integer function.

(C) **The Vertical Line Test.**

- Testing an equation to see if it defines a function.

Definitions

- **Linear function:** A function of the form $f(x) = mx + b$. It is called linear because its graph is a line.

- **Constant function:** A function of the form $f(x) = b$. This is a special case of a linear function with $m = 0$, and its graph is a horizontal line.

- **Power function**: A function of the form $f(x) = ax^n$.

- **Greatest integer function**: Written as $f(x) = [\![x]\!]$, this function is defined as the greatest integer less than or equal to x.

Hints and Tips

- Graphs on a calculator can be misleading, but they usually give you a good idea of the shape of a given function, assuming that the viewing rectangle is set appropriately. It is possible to graph piecewise functions on a calculator, but the process is usually more trouble than it is worth. Instead, use a calculator to graph each piece individually, and then draw the whole graph using pencil and paper.

Review

(A) **The relationship of a function and its graph.**

If f is a function, its graph consists of all ordered pairs $(x, f(x))$. The domain of the function is the set of allowable inputs x, and its range is the set of all possible outputs $f(x)$.

● **Sketching the graph of a function from its formula.**

Exercise: If $f(x) = 3x + 1$, find the domain and range of f and sketch the graph of f.

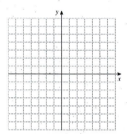

Answer: This is a linear function with slope 3 and y-intercept 1. Domain: All real numbers. Range: All real numbers. Note that the domain of a linear function is $\mathbb{R}$. Generate some points and plot:

x	$f(x)$
-1	-2
0	1
1	4
2	7

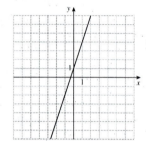

Exercise: If $f(x) = \sqrt{2x + 1}$ find the domain and range of f and sketch the graph of f.

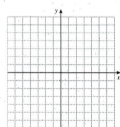

Answer: This is not a linear function. We need to find the domain *before* we can generate points to plot: $2x + 1 \geq 0 \Leftrightarrow 2x \geq -1 \Leftrightarrow x \geq -\frac{1}{2}$. The domain is $\left[-\frac{1}{2}, \infty\right)$ and the range is $[0, \infty)$.

x	$f(x)$
$-\frac{1}{2}$	0
0	1
$\frac{3}{2}$	2
4	3

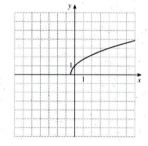

Exercise: If $f(x) = -4$, find the domain and range of f and sketch the graph of f.

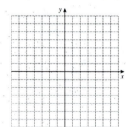

Answer: This is a constant function, so every input value gives the same output value, in this case -4. The domain is all real numbers, and the range is $\{-4\}$. The domain of any constant function is $\mathbb{R}$.

x	$f(x)$
-100	-4
0	-4
5	-4

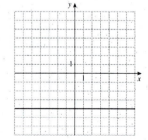

A graphing calculator graphs functions the same way, using its computational power to plot many points.

Exercise: Use a graphing calculator to sketch $y = -x^4 + 2x^2 - x$.

Answer: The hardest part of using a graphing calculator is finding an appropriate range. Both bumps should be visible, and it should be evident that the curve shoots down on both ends.

• **Determining function values from a graph.**

Exercise: The pressure of a fixed amount of gas is a function of its volume, assuming its temperature remains constant. So we can write $P = f(V)$. Volume is given in cubic centimeters, and pressure is given in atmospheres. Here is the graph of f:

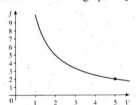

What is the pressure when the volume is 5 cubic centimeters?

Answer: From the graph, $f(5) = 2$, so when the volume is 5 cubic centimeters, the pressure is 2 atmospheres.

• **Finding the domain and range of a function from its graph.**

In the previous section we discussed how to find the domain of a function from its equation. It is easy to determine the domain and range of a function for its graph. For the domain, we see what x-coordinates appear on the graph, and for the range we see what y-coordinates appear.

Exercise: Find the domain and range of the following functions.

(a)

Answer:

Domain: $\{x \mid x \geq 0\}$. Range: $\{y \mid y \geq 0\}$.

(b)

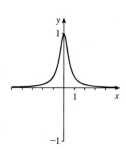

Domain: $\mathbb{R}$. Range: $\{y \mid 0 < y \leq 1\}$ or $(0, 1]$.

(c)

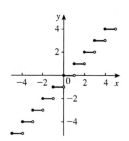

Domain: All real numbers. Range: All integers.

(B) **Piecewise defined functions.**

A function is called **piecewise defined** if it is defined differently for distinct intervals of its domain.

• **Sketching the graph of a piecewise defined function.**

Exercise: Sketch the graph of

$$f(x) = \begin{cases} x + 1 & \text{if } x < 0 \\ 2x^2 + 1 & \text{if } x \geq 0 \end{cases}$$

Answer:

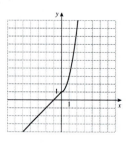

Exercise: Sketch the graph of

$$f(x) = \begin{cases} \sqrt{-5 - 3x} & \text{if } x \leq -2 \\ \sqrt{5 + 2x} & \text{if } -2 < x < 2 \\ -x + 8 & \text{if } 2 \leq x \end{cases}$$

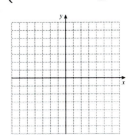

Answer:

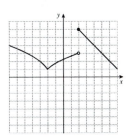

- **Understanding the greatest integer function.**

The function $[\![x]\!]$ is the greatest integer less than or equal to x.   So if $f(x) = [\![x]\!]$, then $f(4.6) = 4$, $f(3.14159) = 3$ and $f(8) = 8$. Negative numbers are a bit trickier: $f(-3.4) = -4$, because $-4 \leq -3.4$ but $-3 > -3.4$.

Exercise: Compute the following.

(a) $[\![422.64]\!]$

(b) $[\![\pi]\!]$

(c) $[\![-6]\!]$

(d) $[\![-6.1]\!]$

Answer:

$[\![422.64]\!] = 422$

$[\![\pi]\!] = 3$

$[\![-6]\!] = -6$

$[\![-6.1]\!] = -7$

Exercise: Sketch the graph of
$f(x) = [\![x]\!] + 2.$

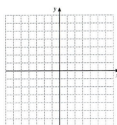

Answer:

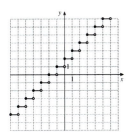

ⓒ The Vertical Line Test.

The **Vertical Line Test** states that a curve in the plane is the graph of a function if and only if no vertical line intersects the curve more than once. It works because a function defines a unique value $f(a)$, so if the vertical line $x = a$ intersects a graph in more than one place, the value of $f(a)$ is not unique. The Vertical Line Test is very easy to apply.

● **Testing an equation to see if it defines a function.**

Exercise: Use the Vertical Line Test to determine which curves are graphs of functions of x.

Answer:

(a)

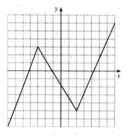

This graph passes the Vertical Line Test. It is the graph of a function.

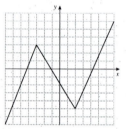

(b)

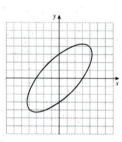

This graph fails the Vertical Line Test. It is not the graph of a function.

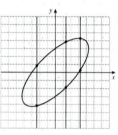

(c)

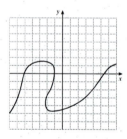

This graph fails the Vertical Line Test. It is not the graph of a function.

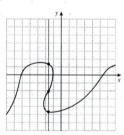

◢ Core Exercises

7, 25, 55, 61, 73, 85

3.3   Increasing and Decreasing Functions; Average Rate of Change

Concepts

(A) Increasing and decreasing functions.

- Using a graph to find intervals where a function increases and where it decreases.

(B) Average rate of change.

- Calculating the average rate of change of a function between two points.

Definitions

- **Increasing:** A function is increasing if its graph rises as we go from left to right.

- **Decreasing:** A function is decreasing if its graph falls as we go from left to right.

- **Secant line:** Given a graph and two x-values a and b, the secant line between $x = a$ and $x = b$ is the line connecting the points $(a, f(a))$ and $(b, f(b))$. See the illustration in the text.

- **Average rate of change:** The average rate of change of the function f between $x = a$ and $x = b$ is
  $$\frac{\text{change in } f(x)}{\text{change in } x} = \frac{f(b) - f(a)}{b - a},$$ that is, the slope of the secant line between $x = a$ and $x = b$.

Questions to Ask Your Teacher

- What emphasis will be placed on the technical definitions of increasing and decreasing as given in the textbook?

Hints and Tips

- The text points out that the average value of a function between two values can be thought of as the slope of the secant line between those values. That fact is a key idea of calculus. If you make sure you understand the concept now, it will make calculus easier next year.

Review

(A) Increasing and decreasing functions.

Graphs of functions are described by what the function does as x increases over an interval in the domain.

A function f is called **increasing** on an interval I if $f(x_1) < f(x_2)$ whenever $x_1 < x_2$ in I.

A function f is called **decreasing** on an interval I if $f(x_1) > f(x_2)$ whenever $x_1 < x_2$ in I.

• **Using a graph to find intervals where a function increases and where it decreases.**

Exercise: State the intervals in which the function whose graph is shown below is increasing or decreasing.

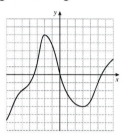

Answer: The function is increasing on $(-\infty, -2]$ and $[3, \infty)$ and decreasing on $[-2, 3]$.

Exercise: Use a graphing device to graph the function $f(x) = x^3 - x^2 - x + 2$. State approximately the intervals on which f is increasing and on which f is decreasing.

Answer:

The function is increasing on approximately $(-\infty, -0.33]$ and $[1, \infty)$. The function is decreasing on approximately $[-0.33, 1]$.

(B) **Average rate of change.**

The **average rate of change** of the function $y = f(x)$ between $x = a$ and $x = b$ is

$$\text{average rate of change} = \frac{\text{change in } y}{\text{change in } x} = \frac{f(b) - f(a)}{b - a}$$

The average rate of change is the slope of the **secant line** that joins the points $(a, f(a))$ and $(b, f(b))$.

• **Calculating the average rate of change of a function between two points.**

Exercise: A function is given. Determine the average rate of change of the function between the given values of the variable.

Answer:

(a) $f(x) = 5x - 7$; $x = 2$, $x = 4$

The average rate of change is

$$\frac{f(4) - f(2)}{4 - 2} = \frac{(5 \cdot 4 - 7) - (5 \cdot 2 - 7)}{2}$$
$$= \frac{13 - 3}{2} = \frac{10}{2} = 5$$

(b) $g(x) = 3x - x^2$; $x = -1$, $x = 3$

The average rate of change is

$$\frac{g(3) - g(-1)}{3 - (-1)} = \frac{\left[3(3) - (3)^2\right] - \left[3(-1) - (-1)^2\right]}{3 + 1}$$

$$= \frac{(9 - 9) - (-3 - 1)}{4} = \frac{0 - (-4)}{4}$$

$$= \frac{4}{4} = 1$$

(c) $h(x) = x^3 - x$; $x = -1$, $x = 1$

The average rate of change is

$$\frac{h(1) - h(-1)}{1 - (-1)} = \frac{\left[(1)^3 - (1)\right] - \left[(-1)^3 - (-1)\right]}{1 + 1}$$

$$= \frac{(1 - 1) - (-1 + 1)}{2} = \frac{0 - 0}{2} = 0$$

Exercise: A graph of a function is given. Determine the average rate of change of the function between $x = -3$ and $x = 0$.

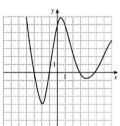

Answer: We have $a = -3$ and $b = 0$. The points in question are $(-3, 0)$ and $(0, 6)$.

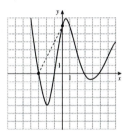

The average rate of change is

$$\frac{f(b) - f(a)}{b - a} = \frac{6 - 0}{0 - (-3)} = \frac{6}{3} = 2.$$

Core Exercises

1, 11, 29, 35, 39

3.4   Transformations of Functions

◤ Concepts

(A) Shifting.

- Recognizing and sketching vertical shifts of graphs.
- Recognizing and sketching horizontal shifts of graphs.

(B) Reflecting.

- Recognizing and sketching reflections of graphs.

(C) Stretching and shrinking.

- Recognizing and sketching vertical stretches of graphs.
- Recognizing and sketching horizontal stretches of graphs.

(D) Even and odd functions.

- Determining if a function is even or odd.

◤ Definitions

- **Vertical shift:** Taking a given graph and moving it straight up or down.
- **Horizontal shift:** Taking a given graph and moving it straight left or right.
- **Vertical stretch:** Extending a graph in the vertical direction.
- **Vertical shrink:** Compressing a graph in the vertical direction.
- **Horizontal stretch:** Extending a graph in the horizontal direction.
- **Horizontal shrink:** Compressing a graph in the horizontal direction.
- **Even function:** A function whose graph is symmetric with respect to the y-axis.
- **Odd function:** A function whose graph is symmetric with respect to the origin.

◤ Hints and Tips

- It is easy to get so wrapped up in the formulas of this section that you miss what is really going on. The idea is that, through experience, we *know* what the graphs of certain functions look like. For example, by the end of this course, if you see $y = x^2$ you will automatically picture this graph:

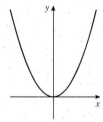

Trust me on this one. The idea of this section is to make the link between familiar graphs like $y = x^2$ and variant graphs, like $y = x^2 + 2$ (shifted up by 2) and $y = -x^2$ (reflected across the x-axis). If you master the concept of transformations, it will be amazing how many functions you will come across in calculus that you can quickly sketch before your classmates have had time to take the covers off their calculators!

◤ Review

Ⓐ Shifting.

When $c > 0$, the equation $y = f(x) + c$ moves the graph of $y = f(x)$ vertically c units up, and the equation $y = f(x) - c$ moves the graph of $y = f(x)$ vertically c units down.

When $c > 0$, the equation $y = f(x - c)$ moves the graph of $y = f(x)$ horizontally c units to the right, and the equation $y = f(x + c)$ moves the graph of $y = f(x)$ horizontally c units to the left.

- **Recognizing and sketching vertical shifts of graphs.**

Exercise: Graph the functions $f(x) = x^3$ and $g(x) = x^3 + 4$ on the same set of axes.

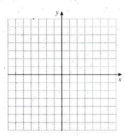

Answer: First graph $f(x) = x^3$. Then graph $g(x) = x^3 + 4$ by translating the graph of f up 4 units.

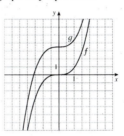

Exercise: Graph the functions $f(x) = |x|$ and $g(x) = |x| - 2$ on the same set of axes.

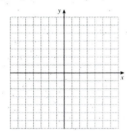

Answer: First graph $f(x) = |x|$. Then graph $g(x) = |x| - 2$ by translating the graph of f down 2 units.

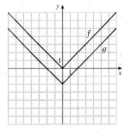

- **Recognizing and sketching horizontal shifts of graphs.**

Exercise: Let $f(x) = x^2$. Evaluate the functions $f(x - 3)$ and $f(x + 2)$. Sketch all three on the same set of axes.

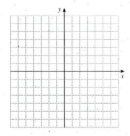

Answer: $f(x - 3) = (x - 3)^2 = x^2 - 6x + 9$ moves the graph of f right 3 units.

$f(x + 2) = (x + 2)^2 = x^2 + 4x + 4$ moves the graph of f left 2 units.

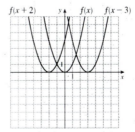

Exercise: Let $f(x) = \dfrac{5}{x^2+1}$. Evaluate the functions $f(x-2)$ and $f(x+1)$. Sketch all three on the same set of axes.

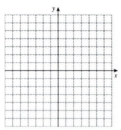

Answer: $f(x-2) = \dfrac{5}{(x-2)^2+1} = \dfrac{5}{x^2-4x+5}$ moves the graph of f right 2 units.

$f(x+1) = \dfrac{5}{(x+1)^2+1} = \dfrac{5}{x^2+2x+2}$ moves the graph of f left 1 unit.

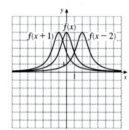

Notice that these functions are much easier to graph using horizontal shifts than by plotting points.

(B) Reflecting.

There are two simple ways we can reflect a graph: about the x-axis and about the y-axis. If we reflect in *both* ways, that turns out to be a "reflection about the origin."

• Recognizing and sketching reflections of graphs.

We reflect about the y axis by replacing x with $-x$ in the function definition. That is, if $f(x)$ is a function, $f(-x)$ will be that function, reflected about the y-axis. We reflect about the x-axis by replacing the y-coordinates of all the points on the graph by their opposites. In other words, if $f(x)$ is a function, $-f(x)$ will be that function, reflected about the y-axis.

Exercise: Let $f(x) = x^3 - x + 2$. Find $-f(x)$ and sketch both graphs on the same set of axes.

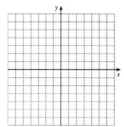

Answer: $-f(x) = -x^3 + x - 2$

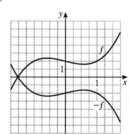

Exercise: Let $f(x) = x^2 + 2x - 2$. Find $f(-x)$ and sketch both graphs on the same set of axes.

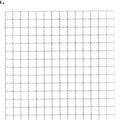

Answer: $f(-x) = x^2 - 2x - 2$.

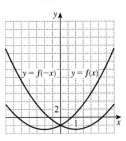

Ⓒ **Stretching and shrinking.**

• **Recognizing and sketching vertical stretches of graphs.**

For $c > 1$, the equation $y = cf(x)$ *stretches* the graph of $y = f(x)$ vertically by a factor of c. For $0 < c < 1$, the equation $y = cf(x)$ *shrinks* the graph of $y = f(x)$ vertically by a factor of c.

Exercise: Let $f(x) = 2x - 1$. Find $3f(x)$ and $\frac{1}{2}f(x)$. Sketch all three graphs on the same set of axes.

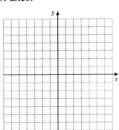

Answer: $3f(x) = 3(2x - 1) = 6x - 3$

$\frac{1}{2}f(x) = \frac{1}{2}(2x - 1) = x - \frac{1}{2}$

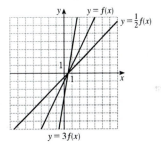

Exercise: Let $f(x) = \sqrt{x} - 1$. Find $2f(x)$ and $-2f(x)$. Sketch all three graphs on the same set of axes.

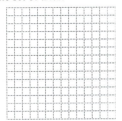

Answer: $2f(x) = 2(\sqrt{x} - 1) = 2\sqrt{x} - 2$

$-2f(x) = -2(\sqrt{x} - 1) = -2\sqrt{x} + 2$

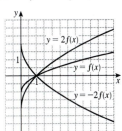

Exercise: Let $f(x) = 3x + 2/x$. Find $-f(x)$ and $-\frac{1}{2}f(x)$. Sketch all three graphs on the same set of axes.

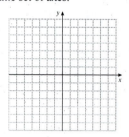

Answer: $-f(x) = -(3x + 2/x) = -3x - 2/x$

$-\frac{1}{2}f(x) = -\frac{1}{2}(3x + 2/x) = -\frac{3}{2}x - 1/x$

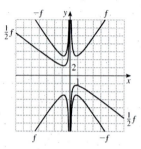

Exercise: Let $f(x) = \dfrac{5}{x^2 + 1}$. Find the functions $\frac{1}{2}f(x)$ and $-2f(x)$. Sketch all three graphs on the same set of axes.

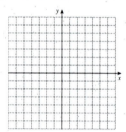

Answer: $\frac{1}{2}f(x) = \frac{1}{2}\left(\dfrac{5}{x^2 + 1}\right) = \dfrac{5}{2(x^2 + 1)}$

$-2f(x) = -2\left(\dfrac{5}{x^2 + 1}\right) = \dfrac{-10}{x^2 + 1}$. Compare these to the transformations of the same function at the end of Part A.

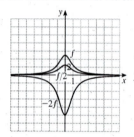

- **Recognizing and sketching horizontal stretches of graphs.**

  For $c > 1$, the equation $y = f(cx)$ *shrinks* the graph of $y = f(x)$ horizontally by a factor of $1/c$. For $0 < c < 1$, the equation $y = f(cx)$ *stretches* the graph of $y = f(x)$ horizontally by a factor of $1/c$.

Exercise: Let $f(x) = x + 1$. Find $f(3x)$ and $f\left(\frac{1}{2}x\right)$. Sketch all three graphs on the same set of axes.

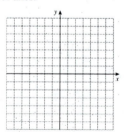

Answer: $f(3x) = (3x) + 1 = 3x + 1$

$f\left(\frac{1}{2}x\right) = \left(\frac{1}{2}x\right) + 1 = \frac{1}{2}x + 1$

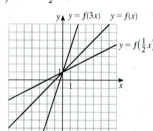

Exercise: Let $f(x) = x^2 + x$. Find $f(2x)$ and $f(-x)$. Sketch all three graphs on the same set of axes.

Answer: $f(2x) = (2x)^2 + 2x = 4x^2 + 2x$

$f(-x) = (-x)^2 + (-x) = x^2 - x$

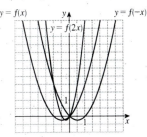

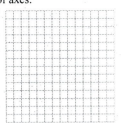

(D) **Even and odd functions.**

Symmetry often helps in graphing a function. An **even function** f is a function for which $f(-x) = f(x)$ for all x in its domain. The graph of an even function is symmetric about the y-axis. An **odd function** f is a function for which $f(-x) = -f(x)$ for all x in its domain. The graph of an odd function is symmetric about the origin. Before graphing, quickly check to see if the function is odd, even, or neither.

• **Determining if a function is even or odd.**

Exercise: Check each function to see if it is odd, even, or neither. Then graph the function.

Answer:

(a) $f(x) = x^2 - 4$

$f(x) = x^2 - 4 \quad \Rightarrow$

$f(-x) = (-x)^2 - 4 = x^2 - 4 = f(x)$. Since $f(-x) = f(x)$, f is even.

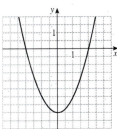

(b) $f(x) = \dfrac{8x}{x^2 + 1}$

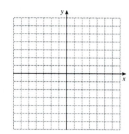

$f(x) = \dfrac{8x}{x^2 + 1}$ $\Rightarrow$

$f(-x) = \dfrac{8(-x)}{(-x)^2 + 1} = \dfrac{-8x}{x^2 + 1} = -\dfrac{8x}{x^2 + 1} = -f(x)$.

Since $f(-x) = -f(x)$, f is odd.

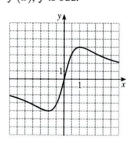

(c) $f(x) = |x|$

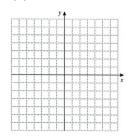

$f(x) = |x|$ $\Rightarrow$ $f(-x) = |-x| = |x| = f(x)$. Since $f(-x) = f(x)$, f is even.

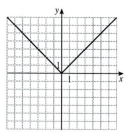

(d) $f(x) = x^3 - x^2 - 8x + 4$

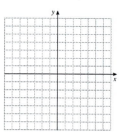

$f(x) = x^3 - x^2 - 8x + 4$ $\Rightarrow$

$f(-x) = (-x)^3 - (-x)^2 - 8(-x) + 4 = -x^3 - x^2 + 8x + 4$

and $-f(x) = -x^3 + x^2 + 8x - 4$. Since $f(-x) \neq -f(x)$

and $f(-x) \neq f(x)$, f is neither even nor odd.

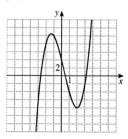

▰ Core Exercises

17, 31, 75

3.5   Quadratic Functions; Maxima and Minima

▚ Concepts

(A) Quadratic functions.

- Putting quadratic functions into standard form.
- Graphing quadratic functions using the standard form.
- Finding the maximum and minimum values of a quadratic function.

(B) Local maxima and minima.

- Using a graphing device to find local maxima and minima.

▚ Definitions

- **Quadratic function:** A function of the form $f(x) = ax^2 + bx + c$.
- **Standard form of a quadratic function:** $f(x) = a(x - h)^2 + k$. Any quadratic function can be put into standard form.
- **Local maximum value of a function:** If there exists a viewing window for which a given point is the highest point on a graph of a function, then the y-coordinate of that point is a local maximum of that function.
- **Local minimum value of a function:** If there exists a viewing window for which a given point is the lowest point on a graph of a function, then the y-coordinate of that point is a local minimum of that function.

▚ Hints and Tips

- The local maximum value of a function is as defined in the text. Some people use the phrase "local maximum" to mean the local maximum value, and some people use it to mean the point $(a, f(a))$, not just $f(a)$. Some people even incorrectly use the phrase "local maximum" to mean the x-value at which f attains its local maximum. Be careful when you speak and write to express exactly what you mean.

▚ Review

(A) Quadratic functions.

A **quadratic function** is a function of the form $f(x) = ax^2 + bx + c$, where a, b, and c are real numbers, with $a \neq 0$. It is often easier to graph and work with quadratic functions when we have completed the square and expressed f in the form $f(x) = a(x - h)^2 + k$.

- **Putting quadratic functions into the standard form.**

Exercise: Let $f(x) = 2x^2 - 8x + 3$. Express f in standard form, and find its x- and y-intercepts.	**Answer:** $f(x) = 2x^2 - 8x + 3 = 2(x^2 - 4x) + 3$ $\qquad = 2(x^2 - 4x + 4) + 3 - 2(4) = 2(x - 2)^2 - 5$ Thus, in standard form $f(x) = 2(x - 2)^2 - 5$. The x-intercepts occur where $2(x - 2)^2 - 5 = 0$ $\Leftrightarrow$ $(x - 2)^2 = \frac{5}{2}$ $\Leftrightarrow$ $x = 2 \pm \sqrt{\frac{5}{2}}$, and the y-intercept is $f(0) = 3$.

- **Graphing quadratic functions using the standard form.**

Exercise: Express f in the form $f(x) = a(x-h)^2 + k$ and sketch the graph of f.

Answer:

(a) $f(x) = x^2 + 6x$

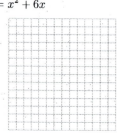

$f(x) = x^2 + 6x = x^2 + 6x + 9 - 9 = (x+3)^2 - 9$

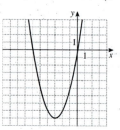

(b) $f(x) = x^2 - 4x - 2$

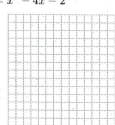

$f(x) = x^2 - 4x - 2 = x^2 - 4x + 4 - 2 - 4 = (x-2)^2 - 6$

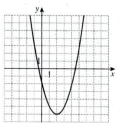

(c) $f(x) = 2x^2 - 6x - \frac{5}{2}$

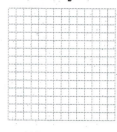

$$f(x) = 2x^2 - 6x - \tfrac{5}{2} = 2\left(x^2 - 3x\right) - \tfrac{5}{2}$$
$$= 2\left(x^2 - 3x + \tfrac{9}{4}\right) - \tfrac{5}{2} - 2\left(\tfrac{9}{4}\right)$$
$$= 2\left(x - \tfrac{3}{2}\right)^2 - 7$$

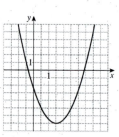

- **Finding the maximum or minimum value of a quadratic function.**

When $a > 0$, the parabola (quadratic function) $f(x) = a(x-h)^2 + k$ opens up. The lowest point is the vertex (h, k), so the **minimum value** of the function occurs when $x = h$, and this minimum value is $f(h) = k$. Also $f(x) = a(x-h)^2 + k \geq k$ for all x.

When $a < 0$, the parabola (quadratic function) $f(x) = a(x-h)^2 + k$ opens down. The highest point is the vertex (h, k), so the **maximum value** of the function occurs when $x = h$ and this maximum value is $f(h) = k$. Also $f(x) = a(x-h)^2 + k \leq k$ for all x.

Exercise: Determine the maximum or minimum values of f by completing the square. Find the intercepts and then sketch the graph.

Answer:

(a) $f(x) = x^2 + 5x + 4$

$f(x) = x^2 + 5x + 4 = (x^2 + 5x) + 4$

$\quad = (x^2 + 5x + \frac{25}{4}) + 4 - \frac{25}{4}$

$\quad = (x + \frac{5}{2})^2 - \frac{9}{4}$

f has a minimum value of $-\frac{9}{4}$ at $x = -\frac{5}{2}$.

To find the x-intercepts, set $f(x) = 0$ and solve:

$f(x) = x^2 + 5x + 4 = 0 \quad \Leftrightarrow \quad (x + 1)(x + 4) = 0 \quad \Leftrightarrow$

$x = -1$ or $x = -4$, so the x-intercepts are -1 and -4. The y-intercept is $f(0) = 4$.

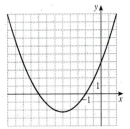

(b) $f(x) = 3x^2 - 6x + 4$

$f(x) = 3x^2 - 6x + 4 = 3(x^2 - 2x) + 4$

$\quad = 3(x^2 - 2x + 1) + 4 - 3(1)$

$\quad = 3(x - 1)^2 + 1$

f has a minimum value of 1 at $x = 1$. Since $f(x) = 3(x - 1)^2 + 1 \geq 1$, there is no x-intercept. The y-intercept is $f(0) = 4$.

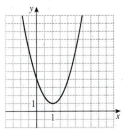

(c) $f(x) = -2x^2 + 5x$

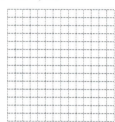

$$f(x) = -2x^2 + 5x = -2\left(x^2 - \tfrac{5}{2}x\right)$$
$$= -2\left(x^2 - \tfrac{5}{2}x + \tfrac{25}{16}\right) + 2\left(\tfrac{25}{16}\right)$$
$$= -2\left(x - \tfrac{5}{4}\right)^2 + \tfrac{25}{8}$$

f has a maximum value of $\tfrac{25}{8}$ at $x = \tfrac{5}{4}$. To find the x-intercepts, set $f(x) = 0$ and solve. $f(x) = -2x^2 + 5x = 0$ $\Leftrightarrow$ $x(-2x + 5) = 0$ $\Leftrightarrow$ $x = 0$ or $x = \tfrac{5}{2}$. The x-intercepts are 0 and $\tfrac{5}{2}$. (Don't forget $x = 0$!) The y-intercept is $f(0) = 0$.

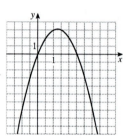

(d) $f(x) = -5x^2 - 10x + 1$

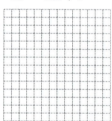

$$f(x) = -5x^2 - 10x + 1 = -5\left(x^2 + 2x\right) + 1$$
$$= -5\left(x^2 - 2x + 1\right) + 1 + 5(1)$$
$$= -5\left(x + 1\right)^2 + 6$$

f has a maximum value of 6 at $x = -1$. To find the x-intercepts, set $f = 0$ and solve.

$f(x) = -5x^2 - 10x + 1$ cannot be factored with integers, so we use the quadratic formula:

$$x = \frac{-(-10) \pm \sqrt{(-10)^2 - 4(-5)(1)}}{2(-5)} = \frac{10 \pm \sqrt{100 + 20}}{-10}$$
$$= \frac{10 \pm \sqrt{120}}{-10} = -1 \pm \frac{\sqrt{30}}{5} \approx -1 \pm 1.10$$

Thus, the x-intercepts are approximately -2.10 and 0.10. The y-intercept is $f(0) = 1$.

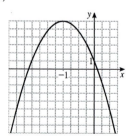

Exercise: A farmer wants to enclose a rectangular field on three sides by a fence and divide it into two smaller rectangular fields by a fence perpendicular to the second side. He has 3000 yards of fencing. Find the dimensions of the field so that the total enclosed area is a maximum.

Answer: First make a drawing and label the sides.

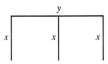

The farmer has 3000 yards of fencing, so we have the equation $3x + y = 3000$. We want to maximize the area given by the equation $A = xy$. So we solve the first equation for y and substitute into the second equation:

$y = 300 - 3xA = x(3000 - 3x) = -3x^2 + 3000x$. Since the leading coefficient is negative, A will have a maximum. Now complete the square:

$$A = -3(x^2 - 1000x)$$
$$= -3(x^2 - 1000x + 250,000) + 750,000$$
$$= -3(x - 500)^2 + 750,000$$

So A is a maximum of 750,000 yd^2 when $x = 500$ yd and $y = 3000 - 3(500) = 1500$ yd.

B **Local maxima and minima.**

A point a is called a **local maximum** of the function f if $f(a) > f(x)$ for all x in some interval containing a. Likewise, a **local minimum** of the function f is a point a where $f(a) < f(x)$ for all x in some interval containing a. When using a graphing device, these local extreme values can be found by looking for the highest (or lowest) point *within* a viewing rectangle.

• **Using a graphing device to find local maxima and minima.**

Exercise: Find the approximate local maximum and minimum values of
$f(x) = x^3 - 2x + 4.$

Answer: Local minimum: $y \approx 2.911$ at $x \approx 0.816$.

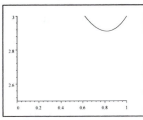

$[0, 1]$ by $[2.5, 3]$

Local maximum: $y \approx 5.089$ at $x \approx -0.816$.

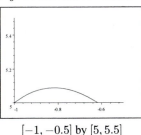

$[-1, -0.5]$ by $[5, 5.5]$

Exercise: Find the approximate local maximum and minimum values of
$f(x) = x^4 - 4x^3 + 16x$.

Answer: Local minimum: $y \approx -11$ at $x \approx -1$.

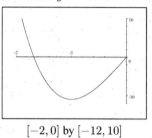

$[-2, 0]$ by $[-12, 10]$

No local maximum.

Exercise: Find the approximate local maximum and minimum values of
$f(x) = x^3 + 3x^2 - 2\sqrt{x^2 + 1}$.

Answer: Local minimum: $y \approx -2$ at $x \approx 0$. Local maximum: $y \approx -0.18$ at $x \approx -1.66$.

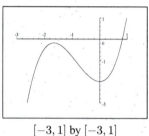

$[-3, 1]$ by $[-3, 1]$

Core Exercises

9, 33, 45

3.6   Combining Functions

Concepts

(A) **The algebra of functions.**

- Adding, subtracting, multiplying, and dividing functions.

(B) **Composition of functions.**

- Finding compositions of functions.
- Recognizing a composition of functions.
- Applying compositions of functions.

Definitions

- **Composite function:** Given functions f and g, the composite function, $f \circ g$ is given by $(f \circ g)(x) = f(g(x))$. The function $f \circ g$ is usually pronounced "f circle g" although some people also call it "fog".

Questions to Ask Your Teacher

- While I don't have a specific question for you to ask your teacher, I suggest you pay very close attention to the notation used in class. Many teachers (myself included) are particular about the notation used for combinations of functions.

Review

(A) The algebra of functions.

Let $f(x)$ be a function with domain A, and let $g(x)$ be a function with domain B. We can combine the two functions to get new functions in some very natural ways. The function $f + g$ is defined this way: $(f + g)(x) = f(x) + g(x)$.

Similarly, $(f - g)(x) = f(x) - g(x)$. We also can get functions fg and $\dfrac{f}{g}$, defined the natural way. The domains

of the new functions are $A \cap B$, with the exception of the function $\dfrac{f}{g}$. We aren't allowed to divide by zero, so the

domain of $\dfrac{f}{g}$ is $\{x \mid x \in A \cap B \text{ and } g(x) \neq 0\}$.

- **Adding, subtracting, multiplying, and dividing functions.**

Exercise: Let $f(x) = 8x + 3$ and $g(x) = x^2 - 4$. Find $f + g$ and its domain.

Answer: $(f + g)(x) = f(x) + g(x) = (8x + 3) + (x^2 - 4) = x^2 + 8x - 1$. Since the domain for both f and g is all real numbers, the domain of $f + g$ is all real numbers.

Exercise: Let $f(x) = \sqrt{x + 3}$ and $g(x) = \sqrt{9 - x^2}$. Find $f + g$ and its domain.

Answer: $(f + g)(x) = f(x) + g(x) = \sqrt{x + 3} + \sqrt{9 - x^2}$. Domain of f: $x + 3 \geq 0 \iff x \geq -3$ or $[-3, \infty)$. Domain of g: $9 - x^2 \geq 0 \iff (3 - x)(3 + x) \geq 0$. Using earlier methods we find that the domain is $[-3, 3]$. The domain of $f + g$ is thus $[-3, \infty) \cap [-3, 3] = [-3, 3]$.

Exercise: Let $f(x) = \dfrac{2}{x - 3}$ and $g(x) = \dfrac{1}{x + 5}$. Find $f - g$ and its domain.

Answer: $(f - g)(x) = f(x) - g(x) = \dfrac{2}{x - 3} - \dfrac{1}{x + 5}$. Domain of f: $x - 3 \neq 0 \iff x \neq 3$. Domain of g: $x + 5 \neq 0 \iff x \neq -5$. The domain of $f - g$ is thus $\{x \mid x \neq 3 \text{ and } x \neq -5\}$.

Exercise: Let $f(x) = \sqrt{x + 12}$ and $g(x) = \sqrt{x - 5}$. Find fg and its domain.

Answer: $(fg)(x) = f(x)g(x) = \left(\sqrt{x + 12}\right)\left(\sqrt{x - 5}\right)$
$= \sqrt{(x + 12)(x - 5)} = \sqrt{x^2 + 7x - 60}$

Domain of f: $x + 12 \geq 0 \iff x \geq -12$ or $[-12, \infty)$. Domain of g: $x - 5 \geq 0 \iff x \geq 5$ or $[5, \infty)$. The domain of fg is $[-12, \infty) \cap [5, \infty) = [5, \infty)$.

Exercise: Let $f(x) = \sqrt{x-1}$ and $g(x) = x^2 - 4$.

Answer:

(a) Find $\dfrac{f}{g}$ and its domain.

$\dfrac{f(x)}{g(x)} = \dfrac{\sqrt{x-1}}{x^2-4}$. Domain of f: $x - 1 \geq 0 \quad \Leftrightarrow \quad x \geq 1$ or $[1, \infty)$. We find the values of x for which g exists and is nonzero: $x^2 - 4 \neq 0 \quad \Leftrightarrow \quad (x-2)(x+2) \neq 0 \quad \Rightarrow$ $x \neq \pm 2$. The domain of $\dfrac{f}{g}$ is thus $[1, \infty) \cap \{x \mid x \neq \pm 2\} = [1, 2) \cup (2, \infty)$.

(b) Find $\dfrac{g}{f}$ and find its domain.

$\dfrac{f(x)}{g(x)} = \dfrac{x^2-4}{\sqrt{x-1}}$. Domain of g: all real numbers. We find the values of x for which f exists and is nonzero: $x - 1 > 0 \quad \Leftrightarrow$ $x > 1$. The domain of $\dfrac{g}{f}$ is thus $(1, \infty)$.

(B) **Composition of Functions.**

Given two functions f and g, the **composition function** $f \circ g$ is defined by $(f \circ g)(x) = f(g(x))$. We sometimes say that $f \circ g$ is "f acting on $g(x)$," in other words, x is the input to g and $g(x)$ is the input to f. The domain of $f \circ g$ is that subset of the domain of g where $g(x)$ is in the domain of f. Another way of saying this is that the domain of $f \circ g$ is those points where both $g(x)$ and $f(g(x))$ are defined.

• **Finding the composition of functions.**

Exercise: Let $f(x) = \sqrt{x-10}$ and $g(x) = x + 9$.

Answer:

(a) Find $f \circ g$ and its domain.

$(f \circ g)(x) = f(g(x)) = f(x+9) = \sqrt{(x+9) - 10}$
$= \sqrt{x-1}$
Domain: $x - 1 \geq 0 \quad \Leftrightarrow \quad x \geq 1$ or $[1, \infty)$.

(b) Find $g \circ f$ and its domain.

$(g \circ f)(x) = g(f(x)) = f\left(\sqrt{x-10}\right) = \sqrt{x-10} + 9$
Domain: $x - 10 \geq 0 \quad \Leftrightarrow \quad x \geq 10$ or $[10, \infty)$. Notice that $f \circ g \neq g \circ f$.

Exercise: Let $f(x) = \dfrac{1}{x} - 4$ and let $g(x) = \dfrac{1}{x+4}$.

Answer:

(a) Find $f \circ g$ and find its domain.

$(f \circ g)(x) = f(g(x)) = f\left(\dfrac{1}{x+4}\right) = \dfrac{1}{\dfrac{1}{x+4}} - 4$

$= x + 4 - 4 = x$

Domain: $x + 4 \neq 0 \quad \Leftrightarrow \quad x \neq -4$ or $\{x \mid x \neq -4\}$. Note that $\dfrac{1}{x+4}$ is never 0, so there is no additional restriction on the domain of $f \circ g$.

(b) Find $g \circ f$ and find its domain.

$(g \circ f)(x) = g(f(x)) = f\left(\dfrac{1}{x} - 4\right) = \dfrac{1}{\left(\dfrac{1}{x} - 4\right) + 4}$

$= \dfrac{1}{\dfrac{1}{x}} = x$

Domain: $x \neq 0$ or $\{x \mid x \neq 0\}$.

Exercise: Let $f(x) = \sqrt{x^2 + x + 1}$ and $g(x) = x^2 - 1$.

Answer:

(a) Find $f \circ g$ and find its domain.

$(f \circ g)(x) = f(g(x)) = f(x^2 - 1)$

$= \sqrt{(x^2 - 1)^2 + (x^2 - 1) + 1} = \sqrt{x^4 - x^2 + 1}$

Domain: Since the domain of g is all real numbers and since $x^4 - x^2 + 1 = \left(x^2 - \frac{1}{2}\right)^2 + \frac{3}{4}$ is always positive, the domain is all real numbers.

(b) Find $g \circ f$ and find its domain.

$(g \circ f)(x) = g(f(x)) = f\left(\sqrt{x^2 + x + 1}\right)$

$= \left(\sqrt{x^2 + x + 1}\right)^2 - 1$

$= x^2 + x + 1 - 1 = x^2 + x$

At the third line above, we can write $\left(\sqrt{x^2 + x + 1}\right)^2 = x^2 + x + 1$ because $x^2 + x + 1 = \left(x + \frac{1}{2}\right)^2 + \frac{3}{4}$. The domain is all real numbers.

- **Recognizing a composition of functions.**

Sometimes, particularly in calculus, it makes things easier to take a complicated function and write it as the composition of two simpler functions. For example, we would think of the function $\sqrt{x^2+2x+32}$ as the function $\sqrt{x}$ composed with the function $x^2+2x+32$.

Exercise: Given $f(x) = (x^2-9)^{37}$, find functions g and h such that $f = g \circ h$.

Answer: A good answer is $g(x) = x^{37}, h(x) = x^2 - 9$. This answer is not unique. We could also have said $g(x) = (x^2-9)^{37}, h(x) = x$, but that answer violates the spirit of trying to find two *simpler* functions.

- **Applying composition of functions.**

Understanding composition of functions gives us a new tool to use when solving applied problems.

Exercise: I tend to buy books in significant quantities at my local used bookstore. They have a "remainder" section where the books cost only fifty cents each. When I'm not reading, I often write student study guides for Mel/Porter, a publishing company. They pay me based on page count. I get $\$100$ plus an extra twenty-five cents per page.

(a) Find a function f that models the number of books I can buy as a function of the amount of money I have x, where x is in dollars.

Answer: $f(x) = 2x$

(b) Find a function g that models how much I am paid as a function of the number of pages I've written, where x is the number of pages.

Answer: $g(x) = 100 + \frac{x}{4}$.

(c) Find $(f \circ g)(x)$. What does the function $f \circ g$ represent in real-world terms?

Answer:
$(f \circ g)(x) = f(g(x)) = f\left(100 + \frac{x}{4}\right) = 2\left(100 + \frac{x}{4}\right)$.
This function represents how many remaindered books I can buy as a function of the number of pages I've written.

Core Exercises

7, 15, 21, 33, 45, 59, 63

3.7   One-to-One Functions and Their Inverses

Concepts

(A) One-to-one functions.

- Deciding whether a function is one-to-one using the definition of one-to-one.
- Deciding whether a function is one-to-one using the Horizontal Line Test.

(B) The inverse of a function.

- Finding f^{-1} for specific values of a function.
- Finding the inverse of a one-to-one function.
- Sketching the inverse of a one-to-one function.
- Verifying that two functions are inverses.

Definitions

- **One-to-one function:** A function is called a one-to-one function if two different inputs always give two different outputs.
- **Horizontal Line Test:** A function is one-to-one if and only if no horizontal line intersects its graph more than once.

Questions to Ask Your Teacher

- Would you consider the Horizontal Line Test sufficient to show that a function does not have an inverse?

Hints and Tips

- When trying to understand inverse functions (as opposed to just pushing the symbols around) I find it useful to think about the units. Let's say that $f(x)$ tells us how many inches of rainfall have occurred after x hours. I think, "For f, the units of input are hours and the units of output are inches." Now for the inverse function, the units always reverse. So I think, "For f^{-1}, the units of input are inches and the units of output are hours." That tells me that $f^{-1}(x)$ is going to tell me the time at which x inches of rain have fallen.

- There is a bit of very unfortunate notation in this section. It is standard notation, and we are stuck with it. You know that $3^{-1} = \frac{1}{3}$ and that $(x+3)^{-1} = \frac{1}{x+3}$. It is tempting to think that f^{-1} follows the pattern, but *it does not!* f^{-1} is the inverse function of f, and has *nothing to do with* $\frac{1}{f}$. When you are under the pressure of a test, try not to make that mistake.

Review

(A) One-to-one functions.

A function with domain A is called a **one-to-one function** if no two elements of A have the same image; that is, $f(x_1) \neq f(x_2)$ whenever $x_1 \neq x_2$. Another way of stating this is, "If $f(x_1) = f(x_2)$, then $x_1 = x_2$." Care must be taken to avoid some common errors in algebraically showing a function is one-to-one. While it is true that $a \neq b$ implies that $a + c \neq b + c$, it is *not* true that $a \neq b$ and $c \neq d$ implies that $a + c \neq b + d$.

• **Deciding whether a function is one-to-one using the definition of one-to-one.**

Exercise: Show that $f(x) = x^2 + 2x - 3$ is not one-to-one by finding two points $x_1 \neq x_2$ where $f(x_1) = f(x_2)$.

Answer: There are many possible answers. An easy pair of points is found by setting $f(x) = 0$ and factoring.

$x^2 + 2x - 3 = (x-1)(x+3) = 0$ if $x = 1$ or $x = -3$. So $f(1) = f(-3) = 0$, and thus f is not one-to-one. In general, since $f(x) = (x+1)^2 - 4$, if $f(x_1) = f(x_2)$ then $|x_1 + 1| = |x_2 + 1|$. Example: $f(1) = f(-3)$ since $|(1) + 1| = |(-3) + 1| = 2$.

Exercise: Show that $f(x) = \sqrt[3]{x - 5}$ is one-to-one, using the definition of one-to-one.

Answer: Suppose there are numbers x_1 and x_2 such that $f(x_1) = f(x_2)$. Then $\sqrt[3]{x_1 - 5} = \sqrt[3]{x_2 - 5}$ $\Leftrightarrow$ $x_1 - 5 = x_2 - 5$ $\Leftrightarrow$ $x_1 = x_2$. Therefore, f is one-to-one. Note that when we raise both sides of the equation to an odd power, we do not introduce extraneous solutions.

• **Deciding whether a function is one-to-one using the Horizontal Line Test.**

One test for determining when a function is one-to-one is called the **Horizontal Line Test**, which states that a function is one-to-one if and only if no horizontal line intersects its graph more than once.

Exercise: Show that $f(x) = \sqrt[3]{x - 5}$ is one-to-one using the Horizontal Line Test.

Answer:

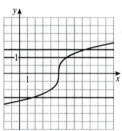

The graph of f passes the Horizontal Line Test. Therefore, $\sqrt[3]{x - 5}$ is a one-to-one function.

Exercise: Use the Horizontal Line Test to determine whether each function is one-to-one.

(a)

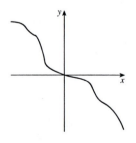

Answer:

This graph passes the Horizontal Line Test, so this is a one-to-one function.

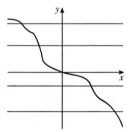

(b)

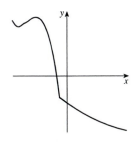

This graph does not pass the Horizontal Line Test, so this function is not one-to-one.

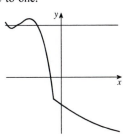

(c)

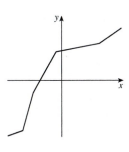

This graph passes the Horizontal Line Test, so this is a one-to-one function.

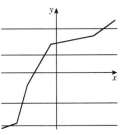

(B) The inverse of a function.

Let f be a one-to-one function with domain A and range B. Then its **inverse function** f^{-1} has domain B and range A, and $f^{-1}(y) = x \iff f(x) = y$ for every y in B.

• Finding f^{-1} for specific values of a function.

Exercise: Suppose f is one-to-one and $f(0) = 3$, $f(1) = 4$, $f(3) = 5$, and $f(4) = 6$.

Answer:

(a) Find $f^{-1}(3)$.

$f^{-1}(3) = 0$ because $f(0) = 3$.

(b) Find $f^{-1}(4)$.

$f^{-1}(4) = 1$ because $f(1) = 4$.

• Finding the inverse of a one-to-one function.

If f is a one-to-one function, then we find the inverse function by the following procedure.

1.	Write $y = f(x)$.
2.	Solve this equation for x in terms of y (if possible).
3.	Interchange x and y. The resulting function is $y = f^{-1}(x)$.

Exercise: Find the inverse function of $f(x) = \sqrt{2x-1}$. What are the domain and range of f^{-1}?

Answer: $y = \sqrt{2x-1}$ $\Rightarrow$ $y^2 = 2x-1$ $\Leftrightarrow$ $y^2+1 = 2x$ $\Leftrightarrow$ $\frac{y^2+1}{2} = x$ $\Leftrightarrow$ $x = \frac{1}{2}y^2 + \frac{1}{2}$. Now interchange x and y: $y = \frac{1}{2}x^2 + \frac{1}{2}$ $\Leftrightarrow$ $f^{-1}(x) = \frac{1}{2}x^2 + \frac{1}{2}$.

Domain of f^{-1} (which is the range of f): $[0, \infty)$. Range of f^{-1} (which is the domain of f): $[\frac{1}{2}, \infty)$. Note that in this case, the easiest way to find the domain of f^{-1} is to find the range of f.

Exercise: Find the inverse function of $f(x) = \dfrac{3x-1}{2+x}$. What is the domain and range of f^{-1}?

Answer: $y = \dfrac{3x-1}{2+x}$ $\Leftrightarrow$ $y(2+x) = 3x-1$ $\Leftrightarrow$ $2y + xy = 3x - 1$ $\Leftrightarrow$ $xy - 3x = -2y - 1$ $\Leftrightarrow$ $x(y-3) = -(2y+1)$ $\Leftrightarrow$ $x = -\dfrac{2y+1}{y-3}$. Interchange x and y: $y = -\dfrac{2x+1}{x-3}$, so $f^{-1}(x) = -\dfrac{2x+1}{x-3}$. The domain of f^{-1} (range of f) is $\{x \mid x \neq 3\}$. The range of f^{-1} (domain of f) is $\{y \mid y \neq -2\}$. Note that in this case it is easier to find the domain of f in order to find the range of f^{-1}.

- **Sketching the inverse of a one-to-one function.**

Even though we might not be able to find the formula for f^{-1}, we can still find the graph of f^{-1}. The graph of f^{-1} is obtained by reflecting the graph of f about the line $y = x$. The important thing here is that (a, b) is a point on the graph of f if and only if (b, a) is a point on the graph of f^{-1}.

Exercise: Find the graph of f^{-1} for each graph of f shown below.

Answer:

(a)

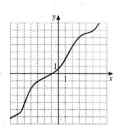

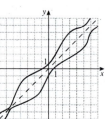

(b)

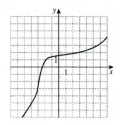

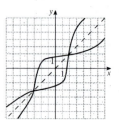

- **Verifying that two functions are inverses.**

To verify that f and g are inverses of each other, we must show that $(f \circ g)(x) = x$ and $(g \circ f)(x) = x$ for all x.

Exercise: Verify that the two given functions are inverses of each other.

(a) $f(x) = 2x + 3$, $g(x) = \dfrac{x - 3}{2}$

Answer:

$(f \circ g)(x)$:

$$f(g(x)) = f\left(\frac{x-3}{2}\right) = 2\left(\frac{x-3}{2}\right) + 3 = x - 3 + 3 = x.$$

$(g \circ f)(x)$: $g(f(x)) = g(2x + 3) = \dfrac{(2x + 3) - 3}{2} = x.$

Since $(f \circ g)(x) = (g \circ f)(x) = x$, f and g are inverses of each other.

(b) $f(x) = \sqrt[3]{x - 5}$, $g(x) = x^3 + 5$.

$(f \circ g)(x)$: $f(g(x)) = f(x^3 + 5) = \sqrt[3]{(x^3 + 5) - 5} = x.$

$(g \circ f)(x)$: $g(f(x)) = g\left(\sqrt[3]{x - 5}\right) = \left(\sqrt[3]{x - 5}\right)^3 + 5 = x.$

Since $(f \circ g)(x) = (g \circ f)(x) = x$, f and g are inverses of each other.

◤ Core Exercises

13, 17, 23, 43, 63, 69, 79

4 Polynomial and Rational Functions

4.1 Polynomial Functions and Their Graphs

Concepts

(A) **Graphs of polynomials.**
- Determining the end behavior of a polynomial.
- Finding the zeros of a polynomial.
- Graphing a polynomial.

(B) **The Intermediate Value Theorem for Polynomials.**

(C) **Local extrema of polynomials.**
- Finding the number of local extrema of a given polynomial from its graph.

Definitions
- **Polynomial function of degree** n: A function of the form $P(x) = a_n x^n + a_{n-1} x^{n-1} + \cdots + a_1 x + a_0$.
- **Coefficients of a polynomial:** The constant terms a_i above.
- **Monomial:** A polynomial with just one term.
- **Binomial:** A polynomial with two terms.
- **Trinomial:** A polynomial with three terms.
- **Zeros of a polynomial:** Solutions of the equation $P(x) = 0$. These are the x-intercepts of the graph of the polynomial.

Questions to Ask Your Teacher
- Do you prefer the term "zeros" or "roots" of a polynomial?

Hints and Tips
- If you are like me, you may be thinking, "Why don't I just graph polynomials on my calculator? Why do I have to mess with the steps it the text?" There is a real answer: It is notoriously tough to find a good graphing window for polynomials, because they can have many little bumps, and they zoom off to infinity and negative infinity very quickly. For example, if you graph $f(x) = x^4 - 5x^2 + 4$, it is easy to get a graph like this:

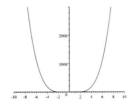

When you do the analysis, you realize that a crucial feature is missing from this picture, and you can find a wiser viewing window:

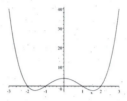

- The Intermediate Value Theorem for Polynomials is worth understanding. You will know you understand it when you think, "That is a completely trivial statement. Why on Earth are they making such a big fuss over something so obvious?" It is important, because it provides a theoretical basis for sketching polynomials based on what is happening between the zeros. It is also important because a more general version of it will come up in Calculus, and if you understand the polynomial version, the general version will not faze you.

Review

(A) Graphs of Polynomials.

The graph of a polynomial is always a smooth curve that has no breaks or sharp corners. The graphs of polynomials of degree 0 and degree 1 are **lines** and graphs of polynomials of degree 2 are **parabolas**.

- **Determining the end behavior of a polynomial.**

The **end behavior** of a function is what happens to the values of the function as $|x|$ becomes large. We use the following notation:

$$x \to \infty \quad \text{means} \quad \text{"x becomes large in the positive direction"}$$

$$x \to -\infty \quad \text{means} \quad \text{"x becomes large in the negative direction"}$$

The end behavior of the polynomial $P(x) = a_n x^n + a_{n-1} x^{n-1} + \cdots + a_1 x^1 + a_0$ is completely determined by the leading term, $a_n x^n$. Look at this term and think, "What happens if x is large and positive? What happens if x is large and negative?

Exercise:

(a) Describe the end behavior of
$$P(x) = -2x^5 + 4x^4 - 2x^3 - 3x.$$

Answer:

Since we are only interested in the end behavior, the term $-2x^5$ is the only important one. For large values of x, the other terms will fade into insignificance. When x is large and positive, then x^5 is large and positive and $-2x^5$ is large and negative. When x is large and negative, then x^5 is large and negative, and $-2x^5$ is large and positive. So when $x \to -\infty$, $y \to \infty$, and when $x \to \infty$, $y \to -\infty$.

(b) Confirm your description using a graphing calculator.

A graph confirms the end behavior of P.

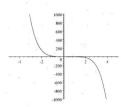

Exercise:

Answer:

(a) Describe the end behavior of
$Q(x) = 0.1x^8 + 5x^7 + x + 10$.

Since we are only interested in the end behavior, the term $0.1x^8$ is the only important one. For large values of x, the other terms will fade into insignificance. When x is large and positive, then x^8 is large and positive and $0.1x^8$ is large and positive. When x is large and negative, then x^8 is large and positive, and again $0.1x^8$ is large and positive. So when $x \to -\infty$, $y \to \infty$, and when $x \to \infty$, $y \to \infty$.

(b) Confirm your description using a graphing calculator.

We check the graph:

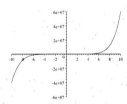

Something is wrong! The left side seems to be going the wrong way! The problem is rectified if we zoom out a bit.

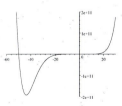

• **Finding the zeros of a polynomial.**

The graph of the polynomial $y = P(x)$ touches or crosses the x-axis at the point c if and only if $P(c) = 0$. The number c is a **zero** of $P(x)$. The statements in the following table are equivalent. (That is, if one of them is true, they are all true; and if one of them is false, they are all false.)

1.	c is a zero of P.
2.	$x = c$ is a solution to the equation $P(x) = 0$.
3.	$x - c$ is a factor of $P(x)$.
4.	c is an x-intercept of the graph of P.

Exercise: Find the zeros of $f(x) = x^4 + 3x^3 + 2x^2$.

Answer: We factor $f(x) = x^2(x + 1)(x + 2)$. Then we set each factor equal to zero: $x^2 = 0 \iff x = 0, x + 1 = 0 \iff x = -1$, and $x + 2 = 0 \iff x = -2$. The zeros of this polynomial are $x = 0$, $x = -1$, and $x = -2$.

• **Graphing a polynomial.**

To graph a polynomial, we find the zeros and then test points between the zeros to see where the polynomial is positive or negative. We then check end behavior, and sketch the polynomial.

Exercise: Sketch the graph of the polynomial $P(x) = 4x^2 - x^4$.

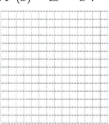

Answer: Start by finding the zeros of $P(x)$ by factoring. This polynomial is easy to factor:
$$P(x) = 4x^2 - x^4 = x^2(4 - x^2) = x^2(2 - x)(2 + x) \iff$$
$x = 0, 2, \text{ or } -2$. Thus, the x-intercepts are $x = 0$, $x = 2$, and $x = -2$. We test points in between the x intercepts: $P(-3)$ is negative, $P(-1)$ is positive, $P(1)$ is positive, and $P(3)$ is negative. We determine end behavior: when $x \to -\infty, y \to \infty$ and when $x \to \infty, y \to \infty$. We now use this information to draw a sketch. For a more accurate sketch, we plot a few additional points first.

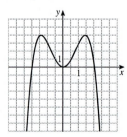

Exercise: Sketch the graph of the polynomial $P(x) = (x+2)(x-1)^2$.

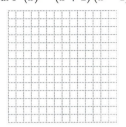

Answer: Set each factor equal to zero to find the x-intercepts:
$$P(x) = (x+2)(x-1)^2 = 0 \quad \Leftrightarrow \quad x = -2 \text{ or } x = 1.$$
Thus, the x-intercepts are $x = -2$ and $x = 1$. Since $P(0) = (2)(1)^2 = 2$, the y-intercept is $y = 2$. We test points in between the x-intercepts: $P(-3)$ is negative, $P(0)$ is positive, and $P(2)$ is positive. We determine end behavior: when $x \to -\infty$, $y \to -\infty$, and when $x \to \infty$, $y \to \infty$. We now use this information to draw a sketch. For a more accurate sketch, we plot a few additional points first.

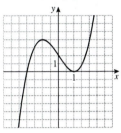

Exercise: Sketch the graph of the polynomial
$$P(x) = (2x-3)(x+1)(x+4).$$

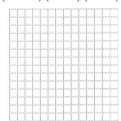

Answer: Set each factor equal to zero to find the x-intercepts:
$$P(x) = (2x-3)(x+1)(x+4) = 0 \quad \Leftrightarrow \quad x = \tfrac{3}{2}, -1, \text{ or}$$
-4. The x-intercepts are $x = \tfrac{3}{2}$, $x = -1$, and $x = -4$. The y-intercept is $P(0) = (-3)(1)(4) = -12$. We test points in between the x-intercepts: $P(-5)$ is negative, $P(-2)$ is positive, $P(0)$ is negative, and $P(2)$ is positive. We determine end behavior: when $x \to -\infty$, $y \to -\infty$, and when $x \to \infty$, $y \to \infty$. We now use this information to draw a sketch. For a more accurate sketch, we plot a few additional points first.

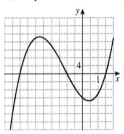

B **The Intermediate Value Theorem for Polynomials.**

The **Intermediate Value Theorem for Polynomials** states that if P is a polynomial function and $P(a)$ and $P(b)$ have opposite signs, then there exists at least one value c between a and b for which $P(c) = 0$. Between successive zeros, the values of the polynomial are either all positive or all negative. This theorem justifies our use of test points to characterize whole sections of a graph in the exercises above.

Notice in the above exercises that if c is a zero of $P(x)$, and the corresponding factor $(x-c)^m$ occurs *exactly* m times in the factorization of $P(x)$, then the graph crosses the x-axis at c when m is odd and does not cross the x-axis when m is even.

(C) Local Extrema of Polynomials.

The points at which the graph of a function changes from increasing to decreasing or from decreasing to increasing are called **local extrema**. Calculus tells us that a polynomial of degree n can have at most $n - 1$ local extrema. A local extremum is a **local maximum** when it is the highest point *within* some viewing rectangle and a **local minimum** when it is the lowest point *within* some viewing rectangle.

- **Finding the number of local extrema of a given polynomial from its graph.**

Exercise: Graph the following polynomials using a graphing device. Determine the y-intercept and approximate the x-intercepts. Also approximate the coordinates of all local extrema to 2 decimal places.

Answer: Pick a large viewing rectangle. Then either use the `Trace` function, zoom in, or reduce the size of the viewing rectangle to find the values to the nearest hundredth.

(a) $Q(x) = \frac{1}{4}x^4 - 3x^2 - 3x + 3$

The y-intercept is $Q(0) = 3$, the x-intercepts are approximately 0.62 and 3.79, there is a local maximum at approximately $(-0.52, 3.77)$, and there are local minima at approximately $(-2.15, 0.92)$ and $(2.67, -13.69)$.

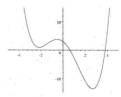

(b) $R(x) = -0.1x^5 + 2x^3 - 12x - 3$

The y-intercept is $R(0) = -3$, the x-intercept is approximately -0.25, there are local maxima at approximately $(-1.59, 9.06)$ and $(3.08, -9.24)$, and there are local minima at approximately $(-3.08, 3.24)$ and $(1.59, -15.06)$.

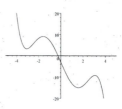

◢ Core Exercises

11, 37, 45, 81, 83

4.2   Dividing Polynomials

Concepts

(A) Long division of polynomials.

- Dividing one polynomial by another, and finding the remainder.
- Dividing a polynomial by $x - c$ using synthetic division.

(B) The Factor Theorem

- Factoring a polynomial using the Factor Theorem.
- Finding a polynomial with specified zeros.

Definitions

- **Dividend, divisor, quotient, remainder:** Long division for polynomials is a process similar to division for numbers. The **divisor** is divided into the quantity called the **dividend**. The result is called the **quotient** and what is left over is the **remainder**.
- **Synthetic division:** A shortcut that can be used to divide a polynomial by $x - c$, where c is a constant.

Questions to Ask Your Teacher

- If it is possible to use synthetic division on a problem, is it preferred?

Review

(A) Long Division of Polynomials.

- **Dividing one polynomial by another, and finding the remainder.**

  If $P(x)$ and $D(x)$ are polynomials with $D(x) \neq 0$, then there exist unique polynomials $Q(x)$ and $R(x)$ such that $P(x) = D(x) \cdot Q(x) + R(x)$, where $R(x) = 0$ or the degree of $R(x)$ is less than the degree of $D(x)$. $P(x)$ is the **dividend**, $D(x)$ is the **divisor**, $Q(x)$ is the **quotient**, and $R(x)$ is the **remainder**. The important thing to remember when dividing is to insert $0x^k$ for missing terms.

Exercise: Let

$P(x) = x^4 + 4x^3 + 6x^2 - x + 6$ and
$D(x) = x^2 + 2x$. Find the polynomials
$Q(x)$ and $R(x)$ such that
$P(x) = D(x) \cdot Q(x) + R(x)$.

Answer:

$$
\begin{array}{r}
x^2 \;+\; 2x \;+\; 2 \\
x^2 + 2x \overline{\smash{\big)}\; x^4 + 4x^3 + 6x^2 - x + 6} \\
\underline{x^4 + 2x^3} \\
2x^3 + 6x^2 \\
\underline{2x^3 + 4x^2} \\
2x^2 - x \\
\underline{2x^2 + 4x} \\
- 5x + 6
\end{array}
$$

So
$x^4 + 4x^3 + 6x^2 - x + 6$
$\qquad = \left(x^2 + 2x\right)\left(x^2 + 2x + 2\right) + (-5x + 6)$

Exercise: Let $P(x) = x^4 + x^3 + 4$ and $D(x) = x^2 - 3$. Find the polynomials $Q(x)$ and $R(x)$ such that
$$P(x) = D(x) \cdot Q(x) + R(x).$$

Answer: Don't forget to insert the terms $0x^2$ and $0x$ in the long division tableau.

$$
\begin{array}{r}
x^2 + x + 3 \\
x^2 - 3 \enclose{longdiv}{x^4 + x^3 + 0x^2 + 0x + 4} \\
\underline{x^4 \quad\quad - 3x^2} \\
x^3 + 3x^2 \\
\underline{x^3 \quad\quad - 3x} \\
3x^2 + 3x + 4 \\
\underline{3x^2 \quad\quad - 9} \\
3x + 13
\end{array}
$$

So $x^4 + x^3 + 4 = (x^2 - 3)(x^2 + x + 3) + (3x + 13)$.

- **Dividing a polynomial by $x - c$ using synthetic division.**

Synthetic division is a very important tool used in factoring when the divisor is of the form $x - c$. It is also useful when finding solutions to $P(x) = 0$ and in evaluating polynomials, that is, finding the value of $P(c)$. This is a skill that is developed by practice. Remember to insert 0s in the appropriate places if there are terms with coefficient 0. When $D(x) = x - c$, the division algorithm becomes the **Remainder Theorem**: $P(x) = (x - c) \cdot Q(x) + r$, where $r = P(c)$. This means that the remainder of $P(x)$ when divided by $x - c$ is the same as the function value at $x = c$.

Exercise: Verify the Remainder Theorem by dividing $P(x) = x^3 + 2x^2 - 10x$ by $x - 3$ and calculating $P(3)$ directly.

Answer: We use synthetic division:

$$
\begin{array}{r|rrrr}
3 & 1 & 2 & -10 & 0 \\
  &   & 3 & 15 & 15 \\
\hline
  & 1 & 5 & 5   & 15
\end{array}
$$

So $x^3 + 2x^2 - 10x = (x - 3)(x^2 + 5x + 5) + 15$. We verify the Remainder Theorem by finding $P(3)$:

$P(3) = (3)^3 + 2(3)^2 - 10(3) = 27 + 18 - 30 = 15$.

Exercise: Verify the Remainder Theorem by dividing
$$P(x) = x^5 - 3x^3 - 7x^2 - 5x + 2 \text{ by } x - 2$$
and calculating $P(2)$ directly.

Answer: Don't forget to insert 0 as the coefficient of x^4.

$$
\begin{array}{r|rrrrrr}
2 & 1 & 0 & -3 & -7 & -5 & 2 \\
  &   & 2 & 4 & 2 & -10 & -30 \\
\hline
  & 1 & 2 & 1 & -5 & -15 & -28
\end{array}
$$

Thus, the remainder is -28. We verify the Remainder Theorem by finding $P(2)$:

$P(2) = (2)^5 - 3(2)^3 - 7(2)^2 - 5(2) + 2$

$= 32 - 24 - 28 - 10 + 2 = -28$

(B) **The Factor Theorem.**

When the remainder is 0, the Remainder Theorem becomes the **Factor Theorem:** $P(c) = 0$ if and only if $x - c$ is a factor of $P(x)$.

- **Factoring a polynomial using the Factor Theorem.**

Exercise: Let
$P(x) = 6x^3 - x^2 - 31x - 24$. Show that
$P(-1) = 0$ and use this fact to factor $P(x)$
completely.

Answer:
$$
\begin{array}{r|rrrr}
-1 & 6 & -1 & -31 & -24 \\
  &   & -6 & 7 & 24 \\
\hline
  & 6 & -7 & -24 & 0
\end{array}
$$

So $P(-1) = 0$, $x = -1$ is a zero, and $x + 1$ is a factor. Then

$$P(x) = 6x^3 - x^2 - 31x - 24 = (x+1)\left(6x^2 - 7x - 24\right)$$
$$= (x+1)(3x-8)(2x+3)$$

Exercise: Let
$P(x) = 2x^3 - 7x^2 - 7x + 30$. Show that
$P(-2) = 0$ and use this fact to factor $P(x)$
completely.

Answer:
$$
\begin{array}{r|rrrr}
-2 & 2 & -7 & -7 & 30 \\
  &   & -4 & 22 & -30 \\
\hline
  & 2 & -11 & 15 & 0
\end{array}
$$

So $P(-2) = 0$, $x = -2$ is a zero, and $x + 2$ is a factor. From the synthetic division table, we have the factorization

$$P(x) = 2x^3 - 7x^2 - 7x + 30$$
$$= (x+2)\left(2x^2 - 11x + 15\right)$$
$$= (x+2)(2x-5)(x-3)$$

- **Finding a polynomial with specified zeros.**

Exercise: Find a polynomial of degree 5
that has zeros at $-5, -2, 1, 2,$ and 3.

Answer: Since $P(c) = 0$ if and only if $x - c$ is a factor,

$$P(x) = [x - (-5)]\,[x - (-2)]\,[x - 1]\,[x - 2]\,[x - 3]$$
$$= (x+5)(x+2)(x-1)(x-2)(x-3)$$
$$= x^5 + x^4 - 21x^3 + 11x^2 + 68x - 60$$

Core Exercises

1, 53, 67

4.3   Real Zeros of Polynomials

Concepts

(A) **Factoring polynomials.**

- Using the Rational Zeros Theorem to factor a polynomial.
- Using Descartes' Rule of Signs to help factor a polynomial.
- Using the Upper and Lower Bounds Theorem to factor a polynomial.

◤ Definitions

- **Rational zero of a polynomial:** A zero of a polynomial that is also a rational number.

◤ Hints and Tips

- If your class allows calculator use, it is worthwhile getting to know a wonderful friend called the `Poly` function. On some calculators, it is a key. On others, it is exiled to some submenu. It allows you to enter the coefficients of a polynomial, and it will list out the zeros for you. If it gives a zero as an ordered pair $(2, 5)$ then it is the complex root $2 + 5i$ (see Section 3.5). It is worth seeking out the `Poly` function!

◤ Review

(A) **Factoring polynomials.**

If $\dfrac{p}{q}$ is a rational zero of the polynomial equation $a_n x^n + a_{n-1} x^{n-1} + \cdots + a_1 x^1 + a_0 = 0$, where $a_n \neq 0$ (not

necessarily $a_n = 1$), $a_0 \neq 0$, and a_i is an integer, then $\dfrac{p}{q}$ is of the form $\pm \dfrac{\text{factor of } a_0}{\text{factor of } a_n}$.

- **Using the Rational Zeros Theorem to factor a polynomial.**

Exercise: Find all possible rational zeros of the equation $3x^3 - 5x^2 - 16x + 12 = 0$, and then solve it completely.

Answer: The possible zeros are numbers of the form $\pm \dfrac{\text{factor of } 12}{\text{factor of } 3}$, namely $\pm 1, \pm 2, \pm 3, \pm 4, \pm 6, \pm 12, \pm \frac{1}{3}, \pm \frac{2}{3}$, and $\pm \frac{4}{3}$. We try the positive integers first:

$$
\begin{array}{r|rrrr}
1 & 3 & -5 & -16 & 12 \\
  &   & 3 & -2 & -18 \\
\hline
  & 3 & -2 & -18 & -6
\end{array}
\qquad
\begin{array}{r|rrrr}
2 & 3 & -5 & -16 & 12 \\
  &   & 6 &   2 & -28 \\
\hline
  & 3 & 1 & -14 & -16
\end{array}
$$

$$
\begin{array}{r|rrrr}
3 & 3 & -5 & -16 & 12 \\
  &   & 9 & 12 & -12 \\
\hline
  & 3 & 4 & -4 & 0
\end{array}
\;\Rightarrow 3 \text{ is a zero.}
$$

Factoring the resulting polynomial yields the last two solutions:

$3x^2 + 4x - 4 = (3x - 2)(x + 2) = 0 \quad \Leftrightarrow \quad x = \frac{2}{3}$ or

$x = -2$. Thus, the zeros are $x = \frac{2}{3}$, 3, and -2.

Exercise: Find all possible rational zeros of the equation $x^4 - x^3 - 11x^2 + 9x + 18 = 0$, and then solve it completely.

Answer: The possible rational zeros are numbers of the form $\pm \dfrac{\text{factor of } 18}{\text{factor of } 1}$, namely $\pm 1, \pm 2, \pm 3, \pm 6, \pm 9, \pm 18$. There are 12 possible rational zeros. We try the positive integers first.

$$
\begin{array}{r|rrrrr}
1 & 1 & -1 & -11 & 9 & 18 \\
  &   & 1 & 0   & -11 & -2 \\
\hline
  & 1 & 0 & -11 & -2 & 16
\end{array}
\qquad
\begin{array}{r|rrrrr}
2 & 1 & -1 & -11 & 9 & 18 \\
  &   & 2 & 2   & -18 & -18 \\
\hline
  & 1 & 1 & -9 & -9 & 0
\end{array}
$$

So 2 is a zero. Since the resulting polynomial is $x^3 + x^2 - 9x - 9$, $\pm 2, \pm 6$, and ± 18 can be eliminated as possible roots. (Note that we have already eliminated 1 as a possible root.) So 3 and 9 are the only possible positive roots left.

$$
\begin{array}{r|rrrr}
3 & 1 & 1 & -9 & -9 \\
  &   & 3 & 12 & 9 \\
\hline
  & 1 & 4 & 3 & 0
\end{array}
\Rightarrow 3 \text{ is a zero.}
$$

Factoring the resulting polynomial yields the last two solutions:
$x^2 + 4x + 3 = (x + 1)(x + 3) = 0 \quad \Leftrightarrow \quad x = -1 \text{ or } -3$.
So the zeros are $x = -3, -1, 2$, and 3.

Exercise: Find all possible rational zeros of the equation
$10x^4 + 11x^3 - 51x^2 - 32x + 20 = 0$, and then solve it completely.

Answer: The possible rational zeros are numbers of the form $\pm \dfrac{\text{factor of } 20}{\text{factor of } 10}$, namely $\pm 1, \pm 2, \pm 4, \pm 5, \pm 10, \pm 20, \pm \frac{1}{2}, \pm \frac{1}{5}, \pm \frac{1}{10}, \pm \frac{2}{5}, \pm \frac{4}{5}$, and $\pm \frac{5}{2}$. There are 24 possible rational roots. We try the integers first.

$$
\begin{array}{r|rrrrr}
1 & 10 & 11 & -51 & -32 & 20 \\
  &   & 10 & 21 & -30 & -62 \\
\hline
  & 10 & 21 & -30 & -62 & -42
\end{array}
\qquad
\begin{array}{r|rrrrr}
2 & 10 & 11 & -51 & -32 & 20 \\
  &   & 20 & 62 & 22 & -20 \\
\hline
  & 10 & 31 & 11 & -10 & 0
\end{array}
$$

So 2 is a zero. We try $x = 2$ again in the resulting polynomial $10x^3 + 31x^2 + 11x - 10$:

$$
\begin{array}{r|rrrr}
2 & 10 & 31 & 11 & -10 \\
  &   & 20 & 102 & 226 \\
\hline
  & 10 & 51 & 113 & 216
\end{array}
$$

By the Upper and Lower Bounds Theorem, 2 is an upper bound for zeros. So we start trying negative numbers:

$$
\begin{array}{r|rrrr}
-1 & 10 & 31 & 11 & -10 \\
   &   & -10 & -21 & 10 \\
\hline
   & 10 & 21 & -10 & 0
\end{array}
$$

So -1 is a zero. Factoring the resulting polynomial gives
$10x^2 + 21x - 10 = (2x + 5)(5x - 2) = 0 \quad \Leftrightarrow \quad x = -\frac{5}{2}$
or $x = \frac{2}{5}$. Thus, the roots are $x = -\frac{5}{2}, -1, \frac{2}{5}$, and 2.

• **Using Descartes' Rule of Signs to help factor a polynomial.**

Descartes' Rule of Signs states that if $P(x)$ is a polynomial with real coefficients, then

---

1. The number of positive real solutions of $P(x)$ is at most equal to the number of variations in sign in $P(x)$ or is less than that by an even number.
2. The number of negative real solutions of $P(x)$ is at most equal to the number of variations in sign in $P(-x)$ or is less than that by an even number.

This is very useful when there is an odd number of variations in sign. In this case you are guaranteed at least one real solution, either negative or positive.

Exercise: Use Descartes' Rule of Signs to help find all rational zeros of the polynomial $P(x) = x^4 - 2x^3 - x^2 - 4x - 6$. Then find all of the rational zeros of P.

Answer: $P(x)$ has one variation in sign, so there is one positive real zero. $P(-x) = x^4 + 2x^3 - x^2 + 4x - 6$ has three variations in sign, so there are either three or one negative real zeros. The possible zeros are ± 1, ± 2, ± 3, and ± 6.

$$\begin{array}{r|rrrrr} -1 & 1 & -2 & -1 & -4 & -6 \\ & & -1 & 3 & -2 & 6 \\ \hline & 1 & -3 & 2 & -6 & 0 \end{array}$$

So -1 is a zero. Looking at the resulting degree-three polynomial, $Q(x) = x^3 - 3x^2 + 2x - 6$ and $Q(-x) = -x^3 - 3x^2 - 2x - 6$, we can see that there is still one positive zero (we cannot increase the number of positive zeros), but no more negative zero. We try the possible positive zeros:

$$\begin{array}{r|rrrr} 1 & 1 & -3 & 2 & -6 \\ & & 1 & -2 & 0 \\ \hline & 1 & -2 & 0 & -6 \end{array} \qquad \begin{array}{r|rrrr} 2 & 1 & -3 & 2 & -6 \\ & & 2 & -2 & 0 \\ \hline & 1 & -1 & 0 & -6 \end{array}$$

$$\begin{array}{r|rrrr} 3 & 1 & -3 & 2 & -6 \\ & & 3 & 0 & 6 \\ \hline & 1 & 0 & 2 & 0 \end{array}$$

So 3 is a zero. Since the result $x^2 + 2$ does not factor over the real numbers, we are finished:

$$P(x) = x^4 - 2x^3 - x^2 - 4x - 6 = (x+1)(x-3)(x^2+2)$$

has rational zeros -1 and 3.

● **Using the Upper and Lower Bounds Theorem to factor a polynomial.**

The number a is a **lower bound** and the number b is an **upper bound** for the roots of a polynomial equation if every real root c of the equation satisfies $a \leq c \leq b$. Bounds a and b are found in the following way:

Let $P(x)$ be a polynomial with real coefficients.

> 1. If we divide $P(x)$ by $x - b$ (where $b > 0$) using synthetic division, and if the row that contains the quotient and remainder has no negative entries, then b is an upper bound for the real roots for $P(x) = 0$.
>
> 2. If we divide $P(x)$ by $x - a$ (where $a < 0$) using synthetic division, and if the row that contains the quotient and remainder has entries that are alternately nonpositive and nonnegative, then a is a lower bound for the real roots for $P(x) = 0$.

The roots of the equation $P(x) = 0$ are the same as the zeros of the polynomial $P(x)$. We summarize the steps in finding the roots of a polynomial $P(x)$:

> 1. List all possible rational roots.
>
> 2. Use Descartes' Rule of Signs to determine the possible number of positive and negative roots.
>
> 3. Use synthetic division to test for possible roots (in order). Stop when you have reached an upper or lower bound or when all predicted positive or negative roots have been found.
>
> 4. If you find a root, repeat the process with the quotient. Remember that you do not have to check roots that did not work before. If you reach a quotient that you can factor or on which you can use the quadratic formula, use those techniques to find solutions.

Exercise: Find all rational roots of the equation $x^3 - 3x^2 - 10x + 24 = 0$.

Answer:

1. List all possible rational roots: $\pm 1, \pm 2, \pm 3, \pm 4, \pm 6, \pm 8, \pm 12, \pm 24$.

2. There are two variations in sign, so there are two or no positive root(s). $P(-x) = -x^3 - 3x^2 + 10x + 24$, so there is only one possible negative root.

3.

```
1 | 1  -3  -10   24        2 | 1  -3  -10   24
  |      1   -2  -12         |      2   -2  -24
  ---------------------      ---------------------
    1  -2  -12   12            1  -1  -12    0
```

So $x = 2$ is a root.

4. We factor the quotient: $x^2 - x - 12 = (x - 4)(x + 3) = 0$
    $\Leftrightarrow \quad x = 4$ or $x = -3$. The roots are $x = 2$, $x = 4$, and $x = -3$.

Exercise: Find all rational zeros of the equation $2x^4 + 11x^3 + 4x^2 - 44x - 48 = 0$.

Answer:

1. List all possible rational roots: $\pm\frac{1}{2}$, ±1, $\pm\frac{3}{2}$, ±2, ±3, ±4, ±6, ±8, ±12, ±16, ±24, ±48.

2. There is one variation in sign, so there is one positive root. $P(-x) = 2x^4 - 11x^3 + 4x^2 + 44x - 48$, so there are either three or one negative root(s).

3.

$$\begin{array}{r|rrrrr} \frac{1}{2} & 2 & 11 & 4 & -44 & -48 \\ & & 1 & 6 & 5 & \frac{39}{2} \\ \hline & 2 & 12 & 10 & -39 & - \end{array} \qquad \begin{array}{r|rrrrr} 1 & 2 & 11 & 4 & -44 & -48 \\ & & 2 & 13 & 17 & -27 \\ \hline & 2 & 13 & 17 & -27 & -75 \end{array}$$

$$\begin{array}{r|rrrrr} \frac{3}{2} & 2 & 11 & 4 & -44 & -48 \\ & & 3 & 21 & \frac{75}{2} & - \\ \hline & 2 & 14 & 25 & - & - \end{array} \qquad \begin{array}{r|rrrrr} 2 & 2 & 11 & 4 & -44 & -48 \\ & & 4 & 30 & 68 & 48 \\ \hline & 2 & 15 & 34 & 24 & 0 \end{array}$$

Thus, $x = 2$ is a root. We know that this is the only positive root, so we look for negative roots:

$$\begin{array}{r|rrrr} -\frac{1}{2} & 2 & 15 & 34 & 24 \\ & & -1 & -7 & -\frac{27}{2} \\ \hline & 2 & 14 & 27 & - \end{array} \qquad \begin{array}{r|rrrr} -1 & 2 & 15 & 34 & 24 \\ & & -2 & -13 & -21 \\ \hline & 2 & 13 & 21 & 3 \end{array}$$

$$\begin{array}{r|rrrr} -\frac{3}{2} & 2 & 15 & 34 & 24 \\ & & -3 & -18 & -24 \\ \hline & 2 & 12 & 16 & 0 \end{array}$$

Thus, $x = -\frac{3}{2}$ is a root.

4. We factor the quotient:
$2x^2 + 12x + 16 = 2(x+4)(x+2) = 0 \quad \Leftrightarrow \quad x = -4$ or $x = 2$. Thus, the zeros are $x = -2, -4, -\frac{3}{2}$, and 2.

Core Exercises

3, 55, 69, 77, 97

4.4   Complex Zeros and the Fundamental Theorem of Algebra

Concepts

(A) Four key theorems.

- Understanding the Complete Factorization Theorem.
- Understanding the Zeros Theorem.
- Understanding the Linear and Quadratic Factors Theorem.
- Understanding the Conjugate Zeros Theorem.

(B) Complete factorization.

- Factoring a polynomial completely.
- Finding polynomials with specified zeros.

Hints and Tips

- You will notice that Item A above is more theoretical than usual.   It is important to understand the theory of how factoring works, and the four theorems mentioned tell the story.   Take any polynomial of degree n, and it has n zeros.   Said another way, you can factor any nth-degree polynomial into n linear factors: $P(x) = a(x - c_1)(x - c_2)\cdots(x - c_n)$, where the c_is are real or complex numbers. If the polynomial has real coefficients, we can say more: we know that the complex roots will come in conjugate pairs, and if we want to factor it into pieces that have only real coefficients, we can break it down into linear and quadratic factors. Study the theorems until you can get a coherent picture in your mind of how polynomials factor, and what their zeros are like.

Review

(A) Four key theorems.

This section contains many important theorems and ideas. Notice that these theorems say only that these zeros *exist* — the theorems do not explain how to find them.

- **Understanding the Complete Factorization Theorem.**

  This is our most general factor theorem. Take any polynomial of degree n, with real or complex coefficients, and factor it. The Complete Factorization Theorem says you can always factor it down to n linear terms, although the constants may be complex. The following are not problems for you to solve, but examples for you to look at to help you understand this theorem.

  - If $P(x) = 3x^3 + 18x^2 + 33x + 18$, we can write $P(x) = 3(x+1)(x+2)(x+3)$. Notice that there are three (non-constant) factors.

  - If $P(x) = x^5 + (-7+i)x^4 + (18-7i)x^3 + (-26+16i)x^2 + (32-12i)x - 24$ we can write $P(x) = (x-2)(x-2)(x-3)(x-i)(x+2i)$. Here there are five factors.

  - If $P(x) = x^4 - 1$, we can write $P(x) = (x-1)(x+1)(x-i)(x+i)$. There are four factors.

- **Understanding the Zeros Theorem.**

  In Section 3.1 we learned that if c is a zero of $P(x)$, then $x - c$ is a factor of $P(x)$, and vice versa. So the Zeros Theorem can be thought of as equivalent to the Complete Factorization Theorem. Take any polynomial of degree n, real or complex coefficients. The zeros theorem says that there will be exactly n zeros, counting multiplicity. [That is, if the factorization contains $(x - 1)^2$, we count 1 as two zeros.]

  The following are not problems for you to solve, but examples for you to look at to help you understand this theorem. Notice that the polynomials are the same ones we used in the previous set of examples.

  - $P(x) = 3x^3 + 18x^2 + 33x + 18$ has zeros -1, -2, and -3.

  - $P(x) = x^5 + (-7 + i)x^4 + (18 - 7i)x^3 + (-26 + 16i)x^2 + (32 - 12i)x - 24$ has zeros 2, 3, i, and $-2i$, where 2 has multiplicity two. So we can say that the five zeros are 2, 2, 3, i, and $-2i$.

  - $P(x) = x^4 - 1$ has zeros i, $-i$, 1, and -1.

- **Understanding the Linear and Quadratic Factors Theorem.**

  The Linear and Quadratic Factors Theorem deals only with polynomials with real coefficients. Take any polynomial of degree n, with real coefficients. We have seen that we can factor it down to linear factors. What if we want the factors to also have real coefficients? In other words, we want to forbid writing $x^2 + 1$ as $(x + i)(x - i)$. The Linear and Quadratic Factors Theorem says that we can always factor our polynomial down to linear and quadratic terms with real coefficients. Again, the following are not problems for you to solve, but examples for you to look at to help you understand this theorem.

  - $P(x) = x^4 + x^3 - 2x^2 - 6x - 4 = (x + 1)(x - 2)(x^2 + 2x + 2)$. Notice that the quadratic cannot be factored into real linear factors.

  - $P(x) = x^6 - x^5 + x^4 - 2x^3 + x^2 - x + 1 = (x^2 + 1)(x^2 + x + 1)(x - 1)(x - 1)$. Notice that neither quadratic can be factored into real linear factors.

  - $P(x) = x^5 - 12x^4 + 59x^3 - 158x^2 + 250x - 200 = (x - 4)(x^2 - 6x + 10)(x^2 - 2x + 5)$. Once again, the quadratics cannot be factored further over the real numbers.

- **Understanding the Conjugate Zeros Theorem.**

  The Conjugate Zeros Theorem says that a polynomial of degree n has exactly n zeros (counting multiplicities). These zeros, as we have seen, can be either real or complex. If the polynomial has real coefficients, we know a bit more: that the complex zeros come in conjugate pairs. To illustrate this theorem, we'll use the same polynomials from the last set of examples.

  - $P(x) = x^4 + x^3 - 2x^2 - 6x - 4$ has zeros -1, 2, $-1 + i$, and $-1 - i$. Notice that $-1 + i$ and $-1 - i$ are a conjugate pair.

  - $P(x) = x^6 - x^5 + x^4 - 2x^3 + x^2 - x + 1$ has zeros 1, 1, i, $-i$, $-\frac{1}{2} + \frac{\sqrt{3}}{2}i$, and $-\frac{1}{2} - \frac{\sqrt{3}}{2}i$. There are two sets of conjugate pairs: $\pm i$ and $-\frac{1}{2} \pm \frac{\sqrt{3}}{2}i$.

  - $P(x) = x^5 - 12x^4 + 59x^3 - 158x^2 + 250x - 200$ has zeros 4, $3 + i$, $3 - i$, $1 + 2i$, $1 - 2i$. Again, notice that the complex roots occur in conjugate pairs, as guaranteed by the Conjugate Zeros Theorem.

(B) Complete factorization.

• **Factoring a polynomial completely.**

Exercise: Let $P(x) = 5x^6 - 80x^2$.

Answer:

(a) Factor P into linear and irreducible quadratic factors. Identify these factors.

$$P(x) = 5x^6 - 80x^2 = 5x^2 (x^4 - 16)$$
$$= 5x^2 (x^2 - 4) (x^2 + 4)$$
$$= 5x^2 (x - 2) (x + 2) (x^2 + 4)$$

The linear factors are x, $x - 2$, and $x + 2$. The factor x has multiplicity 2. $x^2 + 4$ is an irreducible quadratic factor. Note that x^2 is *not* an irreducible quadratic factor, but rather a product of two linear factors.

(b) Factor P completely.

$$P(x) = 5x^6 - 80x^2$$
$$= 5x^2 (x - 2) (x + 2) (x^2 + 4)$$
$$= 5x^2 (x - 2) (x + 2) (x - 2i) (x + 2i)$$

Exercise: $-1 - 3i$ is a zero of the polynomial
$$P(x) = 4x^5 + 4x^4 + 25x^3 - 56x^2 - 74x - 20.$$
Find the other zeros.

Answer: Since $-1 - 3i$ is a zero of $P(x)$, $-1 + 3i$ is also a zero of $P(x)$. So one irreducible quadratic factor of $P(x)$ is

$$Q(x) = [x - (-1 - 3i)] [x - (-1 + 3i)]$$
$$= [(x + 1) + 3i] [(x + 1) - 3i] = (x + 1)^2 - (3i)^2$$
$$= x^2 + 2x + 1 - 9i^2 = x^2 + 2x + 10$$

We use long division:

$$
\begin{array}{r}
4x^3 - 4x^2 - 7x - 2 \\
x^2 + 2x + 10 \overline{\smash{\big)}\ 4x^5 + 4x^4 + 25x^3 - 56x^2 - 74x - 20} \\
\underline{4x^5 + 8x^4 + 40x^3} \\
-4x^4 - 15x^3 - 56x^2 \\
\underline{-4x^4 - 8x^3 - 40x^2} \\
-7x^3 - 16x^2 - 74x \\
\underline{-7x^3 - 14x^2 - 70x} \\
-2x^2 - 4x - 20 \\
\underline{-2x^2 - 4x - 20} \\
0
\end{array}
$$

We concentrate on the quotient: $R(x) = 4x^3 - 4x^2 - 7x - 2$ has one variation in sign and $R(-x) = -4x^3 - 4x^2 + 7x - 2$ has two variations in sign. So $R(x)$ has exactly one positive root and two or zero negative roots. Possible rational roots are $\pm\frac{1}{4}, \pm\frac{1}{2}, \pm1$, and ±2. We check for the positive root first:

$$
\begin{array}{r|rrrr}
2 & 4 & -4 & -7 & -2 \\
  &   & 8 & 8 & 2 \\
\hline
  & 4 & 4 & 1 & 0
\end{array}
$$

Thus 2 is a zero. Factoring the resulting polynomial gives us $4x^2 + 4x + 1 = (2x+1)^2 = 0 \quad \Leftrightarrow \quad x = -\frac{1}{2}$. So the zeros of $P(x)$ are $-1 - 3i$, $-1 + 3i$, 2, and $-\frac{1}{2}$, where $-\frac{1}{2}$ has multiplicity 2.

● **Finding polynomials with specified zeros.**

Exercise: Find a polynomial that satisfies the given description.

Answer:

(a) A polynomial $P(x)$ of degree 4 with zeros -2, -1, 3, and 5, and constant coefficient -60.

$$
\begin{aligned}
P(x) &= a\,[x - (-2)]\,[x - (-1)]\,[x - 3]\,[x - 5] \\
&= a(x+2)(x+1)(x-3)(x-5) \\
&= a(x^2 + 3x + 2)(x^2 - 8x + 15) \\
&= a(x^4 - 5x^3 - 7x^2 + 29x + 30)
\end{aligned}
$$

Since $-60 = a(30)$, $a = -2$; hence

$$
\begin{aligned}
P(x) &= -2(x^4 - 5x^3 - 7x^2 + 29x + 30) \\
&= -2x^4 + 10x^3 + 14x^2 - 58x - 60
\end{aligned}
$$

(b) A polynomial $Q(x)$ of degree 6 with zeros 0, 1, and -3, where 1 has multiplicity 2 and -3 has multiplicity 3.

$$
\begin{aligned}
Q(x) &= a(x-0)(x-1)^2\,[x - (-3)]^3 \\
&= ax(x-1)^2(x+3)^3 \\
&= ax(x^2 - 2x + 1)(x^3 + 9x^2 + 27x + 27) \\
&= ax(x^5 + 7x^4 + 10x^3 - 18x^2 - 27x + 27) \\
&= a(x^6 + 7x^5 + 10x^4 - 18x^3 - 27x^2 + 27x)
\end{aligned}
$$

Since no information is given about Q other than its zeros, we choose $a = 1$:
$Q(x) = x^6 + 7x^5 + 10x^4 - 18x^3 - 27x^2 + 27x$.

(c) A polynomial $R(x)$ of degree 5 with zeros $2 - i$, $2 + i$ and 2, where 2 is zero of multiplicity 3. The leading coefficient is -1.

$$R(x) = a\left[x - (2 - i)\right]\left[x - (2 + i)\right](x - 2)^3$$
$$= a\left[(x - 2) + i\right]\left[(x - 2) - i\right](x - 2)^3$$
$$= a\left[(x^2 - 4x + 4) - i^2\right](x^3 - 6x^2 + 12x - 8)$$
$$= a\left(x^2 - 4x + 5\right)(x^3 - 6x^2 + 12x - 8)$$
$$= a\left(x^5 - 10x^4 + 41x^3 - 86x^2 + 92x - 40\right)$$

Since the leading term is ax^5 and we are given that the leading coefficient is -1, $a = -1$. So
$$R(x) = -x^5 + 10x^4 - 41x^3 + 86x^2 - 92x + 40.$$

Exercise: Find a polynomial $P(x)$ of degree 5 with real coefficients that has zeros $3 - 2i$, 0, 2, and 4.

Answer: Since $3 - 2i$ is a zero, so is its conjugate, $3 + 2i$. Again we will assume $a = 1$.

$$P(x) = x\left[x - (3 - 2i)\right]\left[x - (3 + 2i)\right](x - 2)(x - 4)$$
$$= x\left[(x - 3) + 2i\right]\left[(x - 3) - 2i\right](x - 2)(x - 4)$$
$$= x\left[(x - 3)^2 - (2i)^2\right](x^2 - 6x + 8)$$
$$= x\left(x^2 - 6x + 9 - 4i^2\right)(x^2 - 6x + 8)$$
$$= x\left(x^2 - 6x + 13\right)(x^2 - 6x + 8)$$
$$= x\left(x^4 - 12x^3 + 57x^2 - 126x + 104\right)$$
$$= x^5 - 12x^4 + 57x^3 - 126x^2 + 104x$$

Core Exercises

33, 41, 57

4.5   Rational Functions

Concepts

(A) Rational functions and asymptotes.

- Graphing rational functions.

(B) Slant asymptotes.

- Graphing functions with slant asymptotes.

(C) Vertical asymptotes and end behavior on graphing devices.

- Using graphing calculators to graph rational functions.

(D) Applications.

- Solving applied problems involving asymptotes.

Definitions

- **Rational function:** A function of the form $r(x) = \dfrac{P(x)}{Q(x)}$, where P and Q are polynomials.

- **Horizontal asymptote:** A horizontal line that is approached by a graph, as x gets arbitrarily large and positive, or arbitrarily large and negative.

- **Vertical asymptote:** A vertical line that is approached by a graph as x approaches some finite number a.

- **Slant or oblique asymptote:** A line, neither horizontal or vertical, that is approached by a graph in a manner similar to a horizontal asymptote.

- **Arrow notation:**

Symbol	Meaning		
$x \to a^-$	x approaches a from the left	or	x approaches a from below
$x \to a^+$	x approaches a from the right	or	x approaches a from above
$x \to -\infty$	x goes to negative infinity	or	x decreases without bound
$x \to \infty$	x goes to infinity	or	x increases without bound

Hints and Tips

- If a function $f(x)$ models some quantity that changes over time, then the horizontal asymptote tells you what will happen to the quantity in the long run ($x \to \infty$) or long before the experiment began ($x \to -\infty$).

Review

(A) Rational functions and asymptotes.

A rational function is a function of the form $r(x) = \dfrac{P(x)}{Q(x)}$, where P and Q are polynomials and $Q(x)$ is not the zero polynomial. The domain of $r(x)$ is the set of real numbers where $Q(x) \neq 0$. The x-intercepts are the real numbers where $P(x) = 0$. The y-intercept is $r(0)$, provided 0 is in the domain of $r(x)$.

- When determining vertical asymptotes, look for places where the denominator of the rational function is zero and the numerator is not. When determining horizontal asymptotes, it is only necessary to look at the ratio of the largest term in the numerator divided by the largest term in the denominator. For example, as $x \to \infty$, $\dfrac{3x^2 - x + 1}{x^2 + 2x - 7}$ will look like $\dfrac{3x^2}{x^2} = 3$, so $y = 3$ is a horizontal asymptote. As $x \to \infty$, $\dfrac{3x + 1}{x^2 + 2x - 7}$ will look like $\dfrac{3x}{x^2} = \dfrac{3}{x}$. Now as x gets larger and larger, that fraction gets smaller and smaller, so $y = 0$ is a horizontal asymptote. As $x \to \infty$, $\dfrac{3x^5 - x + 1}{x^2 + 2x - 7}$ will look like $\dfrac{3x^5}{x^2} = 3x^3$, and there is no horizontal asymptote.

The text summarizes these cases in the following table:

Case 1: $n < m$	In this case, the denominator goes to infinity faster than the numerator does, so $y \to 0$ as $x \to -\infty$ and as $x \to \infty$. Thus, $y = 0$ is a horizontal asymptote.	
Case 2: $n = m$	In this case, $y \to \dfrac{a_n}{b_m}$ as $x \to -\infty$ and as $x \to \infty$. So $y = \dfrac{a_n}{b_m}$ is a horizontal asymptote.	
Case 3: $n > m$	Here the end behavior is the same as the end behavior of $y = \dfrac{a_n}{b_m} x^{n-m}$. (See slant asymptotes in Item B for more information about this case.)	

- **Graphing rational functions.**

Exercise: Sketch the graph of the function

$$y = \frac{x^2 - 4}{x^2 - 2x - 3}.$$

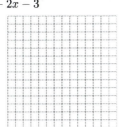

Answer: First factor the numerator and denominator to find the domain, x-intercepts, y-intercepts, and asymptotes:

$$y = \frac{x^2 - 4}{x^2 - 2x - 3} = \frac{(x - 2)(x + 2)}{(x - 3)(x + 1)},$$ so the domain is

$\{x \mid x \neq 3 \text{ and } x \neq -1\}$. There are vertical asymptotes at $x = 3$ and at $x = -1$, the x-intercepts are 2 and -2, and the y-intercept is $y(0) = \frac{4}{3}$. Since $n = m = 2$, numerator and denominator have the same degree, so the horizontal asymptote is the ratio of the leading coefficients, namely $y = \frac{1}{1} = 1$.

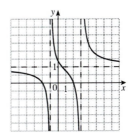

Exercise: Sketch the graph of the function

$$y = \frac{x^3 - 3x^2 + 2x}{x^4 + 1}.$$

Answer: First factor the numerator and denominator to find the domain, x-intercepts, y-intercepts, and asymptotes:

$$y = \frac{x^3 - 3x^2 + 2x}{x^4 + 1} = \frac{x(x-2)(x-1)}{x^4 + 1},$$ so the domain is all

real numbers. There is no vertical asymptote, the x-intercepts are 0, 1, and 2, and the y-intercept is $y(0) = 0$. Since $n < m$, the horizontal asymptote is $y = 0$.

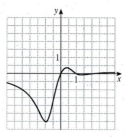

(B) Slant asymptotes.

When the degree of the numerator of a rational function is one more than the degree of the denominator, we have a

slant asymptote. So if $r(x) = \dfrac{P(x)}{Q(x)}$ and the degree of $P(x)$ is one more than the degree of $Q(x)$, we use long

division to write $r(x) = D(x) + \dfrac{R(x)}{Q(x)}$. The line $y = D(x)$ is an equation of our slant asymptote.

• **Graphing functions with slant asymptotes.**

Exercise: Sketch the graph of the function

$$y = \frac{x^3 - 4x}{x^2 - 1}.$$

Answer: $y = \dfrac{x^3 - 4x}{x^2 - 1} = \dfrac{x(x-2)(x+2)}{(x-1)(x+1)}$ has domain

$\{x \mid x \neq -1 \text{ and } x \neq 1\}$. The vertical asymptotes are $x = -1$ and $x = 1$, the x-intercepts are 0, 2, and -2, and the y-intercept is 0. The slant asymptote is $y = x$ since

$$y = \frac{x^3 - 4x}{x^2 - 1} = x - \frac{3x}{x^2 - 1}.$$

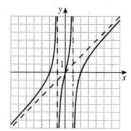

ⓒ Vertical asymptotes and end behavior on graphing devices.

Some graphing calculators and computer algebra systems do not graph function with vertical asymptotes properly. Sometimes vertical asymptotes will appear as vertical lines. Other problems are caused by viewing rectangles that are too wide. In this case, vertical asymptotes might disappear altogether. However, wide viewing rectangles are useful in showing horizontal asymptotes as well as other end behavior. The key is to use a variety of sizes of viewing rectangle to determine vertical asymptotes as well as end behavior.

- **Using graphing calculators to graph rational functions.**

Exercise: Graph each rational function in each of the given viewing rectangles. Determine all vertical and horizontal asymptotes from the graphs.

Answer:

(a) $y = \dfrac{5x + 7}{x - 3}$, $[-5, 5]$ by $[-20, 20]$ and $[-50, 50]$ by $[-1, 10]$

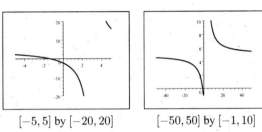

$[-5, 5]$ by $[-20, 20]$ $[-50, 50]$ by $[-1, 10]$

This function has vertical asymptote $x = 3$ and horizontal asymptote $y = 5$.

(b) $y = \dfrac{x^2 - 4}{x^3 + 3x}$, $[-5, 5]$ by $[-5, 5]$ and $[-20, 20]$ by $[-5, 5]$

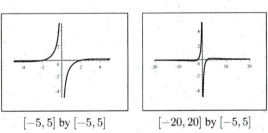

$[-5, 5]$ by $[-5, 5]$ $[-20, 20]$ by $[-5, 5]$

This function has vertical asymptote $x = 0$ and horizontal asymptote $y = 0$.

(c) $y = \dfrac{8 + 3x + 2x^2}{9 - x^2}$, $[-5, 5]$ by $[-5, 5]$ and $[-50, 50]$ by $[-5, 5]$

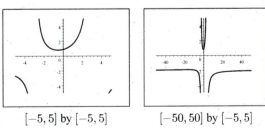

$[-5, 5]$ by $[-5, 5]$ $[-50, 50]$ by $[-5, 5]$

This function has vertical asymptotes $x = -3$ and $x = 3$, and horizontal asymptote $y = -2$.

Exercise: Graph the rational function

$$y = \frac{3x^2 + x - 4}{x^2 - 5x + 6}$$ in appropriate viewing

rectangles. Determine all vertical and horizontal asymptotes, x- and y-intercepts, and all local extrema correct to two decimal places.

Answer:

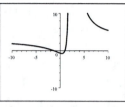 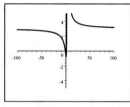

$[-10, 10]$ by $[-10, 10]$ $[-100, 100]$ by $[-5, 5]$

This function has vertical asymptotes $x = 2$ and $x = 3$, horizontal asymptote $y = 3$, x-intercepts of approximately $x = -1.33$ and $x = 1.00$, and y-intercept of approximately $y = -0.67$.

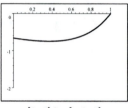

$[0, 1]$ by $[-2, 0]$

Zooming in, we see that there is a local minimum at approximately $(0.37, -0.75)$.

D **Applications**

Exercise: A greebly is a microscopic organism whose population dynamics are very unusual. (Actually, it isn't; I just made it up for the purposes of this exercise. But it doesn't matter.) I put some greeblies in a petri dish, and after x days the population was $f(x) = \dfrac{500x^2 + 200}{x^2 + 1}$ million greeblies.

Answer:

(a) How many greeblies did I start with?

$f(0) = 200$ million greeblies

(b) What will happen if I leave the dish unattended for a very long time?

There is a horizontal asymptote at $y = 500$. So as time goes on, the greebly population will get closer and closer to 500 million.

◤ **Core Exercises**

5, 33, 49, 51, 71, 83

5 Exponential and Logarithmic Functions

5.1 Exponential Functions

Concepts

(A) **Graphs of exponential functions.**
- Identifying graphs of exponential functions.
- Transforming graphs of exponential functions.

(B) **The natural exponential function.**
- Evaluating values using a calculator.
- Graphing using a calculator.

(C) **Applications.**
- Solving applied problems using exponential functions.
- Compound interest.

Definitions

- e: The natural exponential growth constant, or Euler's constant. Like π, it is an irrational number. $e \approx 2.7183$.
- **Principal:** The amount of money invested in an investment plan.
- **Frequency of compounding:** Number of times per year that interest is computed. The more times per year it is compounded, the more interest the investor earns on earlier interest.
- **Continuous compounding:** Compounding occurring frequently enough to be indistinguishable from constant compounding.

Questions to Ask Your Teacher

- To how many decimal places should we present our answers when calculating answers numerically? You may have asked this question before, but decimal places of accuracy are very important when studying exponential functions, because their growth rate can be very large.

Hints and Tips

- To really understand how exponential functions differ from any function you've seen so far, go through the Discovery Project after this section.
- You might be asking yourself the question, "Why e?" The text talks of how e is derived from continuous compounding, but that's not the only reason we mathematicians love it so. It turns out that there are a lot of formulas that come up when studying exponential growth that have awful constants associated with them. When we use e as a base, those constants go away. Like magic. I forgive you if, right now, you still prefer 2^x and 10^x over e^x, but give e a chance. After a bit of calculus, it will grow on you.

◤ Review

(A) **Graphs of exponential functions.**

For $a > 0$, the **exponential function with base** a is defined as $f(x) = a^x$ for every real number x. The value of the base determines the general shape of the graph of $f(x) = a^x$. For $0 < a < 1$, the graph of $f(x) = a^x$ decreases rapidly. When $a = 1$, $f(x) = 1^x = 1$ is a constant function. And for $a > 1$, the function $f(x) = a^x$ increases rapidly. Remember that $a^{-1} = \dfrac{1}{a}$, so $a^{-x} = \dfrac{1}{a^x}$, $a^0 = 1$, and a positive number to any power is always positive. As a result, the domain of $f(x) = a^x$ is the real numbers and the range is the positive real numbers.

• **Identifying graphs of exponential functions.**

Exercise: Find the exponential function $f(x) = a^x$ whose graph is given.

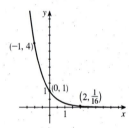

Answer: We know that $a^{-1} = 4$, $a^0 = 1$, and $a^2 = \frac{1}{16}$. So $a = \frac{1}{4}$ and $f(x) = \left(\frac{1}{4}\right)^x$.

• **Transforming graphs of exponential functions.**

Exercise: Sketch the graph of $y = 3^x$ and use it to sketch the graphs of $y = -2 + 3^x$ and $y = 4 - 3^x$ on on the same set of axes.

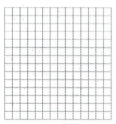

Answer: First graph $y = 3^x$. To graph $y = -2 + 3^x$, shift the graph of $y = 3^x$ down 2 units. To graph $y = 4 - 3^x$, reflect the graph of $y = 3^x$ about the x-axis, and then shift the resulting graph up 4 units.

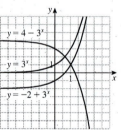

(B) **The natural exponential function.**

As n becomes large, the values of $\left(1 + \dfrac{1}{n}\right)^n$ approaches an irrational number called $e \approx 2.71828$. The function $f(x) = e^x$ is called the **natural exponential function**. Its domain is the real numbers and its range is the positive real numbers.

- **Evaluating values using a calculator.**

Exercise: Complete the table and graph the function $f(x) = -2e^x$.

x	$f(x)$
-3	
-2	
-1	
0	
1	
2	
3	

Answer:

x	$f(x)$
-3	-0.10
-2	-0.27
-1	-0.74
0	-2
1	-5.44
2	-14.78
3	-40.17

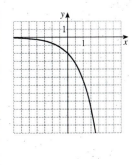

- **Graphing using a calculator.**

Exercise: Graph the function $y = -e^{3x} + 1$ using your calculator. Make sure the viewing window is appropriate.

Answer: You want to choose a window that doesn't result in a graph that looks like a right angle. One good viewing rectangle is $[-1, 1.5]$ by $[-40, 5]$:

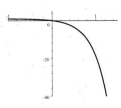

© **Applications.**

- **Solving applied problems using exponential functions.**

Exercise: The temperature (in °F) of a hot liquid x seconds after it is poured is given by the function $T(x) = 85 + 115e^{-0.01x}$.

(a) What is the temperature of the liquid as it is poured?

$T(0) = 200°$ F

(b) What is the temperature of the liquid one minute after it is poured?

$T(60) = 85 + 115e^{-0.6} \approx 148.11°$ F

(c) What is the temperature of the liquid one hour after it is poured?

$T(3600) = 85 + 115e^{-37} \approx 85.00°$ F.

(d) What is the temperature of the room the cup is in? How do you know?

The room's temperature is 85° F. Hot liquid will cool to the ambient temperature and no cooler. The graph of the liquid's temperature versus time is shown below.

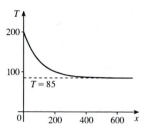

- **Compound interest.**

If an amount of money P, called the **principal**, is invested at a rate r, compounded n times per year, then the amount of money after t years is given by $A = P\left(1 + \dfrac{r}{n}\right)^{nt}$. The rate $\dfrac{r}{n}$ is the rate per compounding period and nt is the number of compounding periods. As the number of compounding periods per year become large, the value of the amount approaches $A = Pe^{rt}$. This is **continuous compounding of interest**.

Exercise: Compare the balance after 5 years when $1000 is invested in each type of account.

Answer:

(a) 10% compounded monthly

$P = 1000$, $r = 0.1$, $n = 12$, and $t = 5$, so

$$A = P\left(1 + \frac{r}{n}\right)^{nt} = 1000\left(1 + \frac{0.1}{12}\right)^{12 \cdot 5}$$

$$\approx 1000\,(1.008333333)^{60}$$

$$\approx 1000\,(1.645308935) \approx \$1645.31$$

(b) 9.9% compounded weekly, assuming 52 weeks per year

$P = 1000$, $r = 0.099$, $n = 52$, and $t = 5$, so

$$A = P\left(1 + \frac{r}{n}\right)^{nt} = 1000\left(1 + \frac{0.099}{52}\right)^{52 \cdot 5}$$

$$\approx 1000\,(1.001903846)^{260}$$

$$\approx 1000\,(1.639726394) \approx \$1639.73$$

(c) 9.9% compounded continuously

$P = 1000$, $r = 0.099$, and $t = 5$, so

$$A = Pe^{rt} = 1000e^{0.099 \cdot 5} = 1000e^{0.495}$$

$$\approx 1000\,(1.640498239) \approx \$1640.50$$

Exercise: How much money needs to be invested in an account paying 7.8% compounded monthly in order to have \$5000 in 5 years?

Answer: We solve the equation $A = P\left(1+\frac{r}{n}\right)^{rt}$ for P: $P = \frac{A}{(1+r/n)^{rt}}$. Substituting $A = 5000$, $r = 0.078$, $n = 12$, and $t = 5$, we find

$$P = \frac{5000}{(1+0.078/12)^{12\cdot 5}} = \frac{5000}{(1.0065)^{60}} \approx \$3389.56.$$

Core Exercises

3, 17, 23, 41, 43, 53, 67, 77, 83

5.2 Logarithmic Functions

Concepts

(A) Definition of logarithm.

- Evaluating logarithms.
- Transforming graphs of logarithms.

(B) Properties of logarithms.

- Using the properties of logarithms.

Definitions

- **Logarithm:** A function whose domain is all positive real numbers defined by the property $\log_a x = y$ if and only if $a^y = x$.

Hints and Tips

- Every logarithmic function has a base associated with it. That's the a in the definition above. In theory, we could use any base we want when playing with log functions. However, time has whittled it down to three bases that most people use. Many people in the applied sciences use the common logarithm $\log_{10} x$. Others (including mathematicians) often prefer the natural logarithm $\log_e x$. If you just see $\log x$, then that is assumed to be the common logarithm. If you see $\ln x$ that is the natural logarithm. (Computer scientists and numerical analysts also use the base 2 logarithm, $\log_2 x$.)

Review

(A) Definition of logarithm.

For $a \neq 1$, the inverse of the exponential function $f(x) = a^x$ is a function called the **logarithm**, written as $g(x) = \log_a x$. The function $\log_a x$ is read as "log base a of x". So $\log_a x$ is the *exponent* to which the base a must be raised to give x.

Logarithmic Form: $\log_a x = y$	Exponential Form: $a^y = x$

In both forms, a is called the **base** and y is called the **exponent**. It is often useful to switch between the different forms of the definition. Notice that $\log_a 1 = 0$ and $\log_a a = 1$.

• Evaluating logarithms.

Exercise: Express the equation in exponential form.

(a) $\log_4 64 = 3$

(b) $\log_5 0.04 = -2$

(c) $\log_9 243 = \frac{5}{2}$

Answer:

$\log_4 64 = 3 \quad \Leftrightarrow \quad 4^3 = 64$

$\log_5 0.04 = -2 \quad \Leftrightarrow \quad 5^{-2} = 0.04$

$\log_9 243 = \frac{5}{2} \quad \Leftrightarrow \quad 9^{5/2} = 243 \quad \Leftrightarrow \quad \left(9^{1/2}\right)^5 = 243$
$\Leftrightarrow \quad 3^5 = 243$

Exercise: Express the equation in logarithmic form.

(a) $6^3 = 216$

(b) $8^{-5/3} = \frac{1}{32}$

Answer:

$6^3 = 216 \quad \Leftrightarrow \quad \log_6 216 = 3$

$8^{-5/3} = \frac{1}{32} \quad \Leftrightarrow \quad \log_8\left(\frac{1}{32}\right) = -\frac{5}{3}$

Exercise: Evaluate.

(a) $\log_5 25$

(b) $\log_2\left(\frac{1}{8}\right)$

(c) $\log_{(1/3)} 27$

Answer:

$\log_5 25 = 2$ since $5^2 = 25$.

$\log_2\left(\frac{1}{8}\right) = -3$ since $2^{-3} = \frac{1}{8}$.

$\log_{(1/3)} 27 = -3$ since $\left(\frac{1}{3}\right)^{-3} = 3^3 = 27$.

Exercise: Use the definition of the logarithmic function to find x.

(a) $\log_3 x = 4$

(b) $\log_x 121 = 2$

(c) $\log_{(1/5)} 25 = x$

Answer:

$\log_3 x = 4 \quad \Leftrightarrow \quad x = 3^4 = 81$

$\log_x 121 = 2 \quad \Leftrightarrow \quad x^2 = 121$. Since $x > 0$, we have $x = \sqrt{121} = 11$.

$\log_{(1/5)} 25 = x \quad \Leftrightarrow \quad \left(\frac{1}{5}\right)^x = 25 \quad \Leftrightarrow \quad 5^{-x} = 25 \quad \Leftrightarrow$
$5^{-x} = 5^2 \quad \Leftrightarrow \quad -x = 2 \quad \Leftrightarrow \quad x = -2$

- **Transforming graphs of logarithms.**

For $a > 0$ and $a \neq 1$, the domain for $f(x) = \log_a x$ is the positive real numbers and the range is the real numbers. As with exponential functions, the graph of $y = \log_a x$ is directly dependent on the value of a. The graphs of $f(x) = \log_a x$ and $f^{-1}(x) = a^x$ for various values of a are shown below.

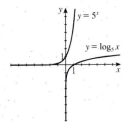
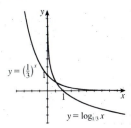

Exercise: Determine the domain and range of $f(x) = \log_3(x - 2)$ and sketch its graph.

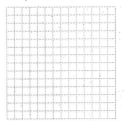

Answer: f has domain $\{x \mid x - 2 > 0\} = (2, \infty)$ and range $(-\infty, \infty)$.

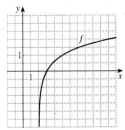

Exercise: Determine the domain and range of $f(x) = \log_5(x^2 + 1)$ and sketch its graph.

Answer: Because $x^2 + 1 > 0$ for all x, f has domain $(-\infty, \infty)$. Because $x^2 + 1 \geq 1$, $\log_5(x^2 + 1) \geq \log_5 1 = 0$, so the range is $[0, \infty)$, the nonnegative real numbers.

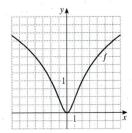

B **Properties of logarithms.**

The logarithm with base e is called the **natural logarithm** and is given a special name and notation, $\log_e x = \ln x$. The logarithm with base 10 is called the **common logarithm** and given an abbreviated notation, $\log_{10} x = \log x$.

$$\ln x = y \quad \Leftrightarrow \quad e^y = x \quad \text{and} \quad \log x = y \quad \Leftrightarrow \quad 10^y = x$$

Logarithms have the following properties:

Property	Reason
$\log_a 1 = 0$	We must raise a to the power 0 to get 1.
$\log_a a = 1$	We must raise a to the power 1 to get a.
$\log_a a^x = x$	We must raise a to the power x to get a^x.
$a^{\log_a x} = x$	$\log_a x$ is the power to which a must be raised to get x.

- **Using the properties of logarithms.**

Exercise: Evaluate. $\|$ **Answer:**

(a) $\log_8 1$ $\|$ 0

(b) $\log 1$ $\|$ 0

(c) $\log_2 2$ $\|$ 1

(d) $\ln e$ $\|$ 1

(e) $\log_3 \left(3^8\right)$ $\|$ 8

(f) $e^{\ln\left(x^2 + 2x + 2\right)}$ $\|$ $x^2 + 2x + 2$

(g) $\log \left(10^{37.32}\right)$ $\|$ 37.32

Core Exercises

1, 5, 9, 27, 39, 47, 67, 83, 87

5.3   Laws of Logarithms

Concepts

A **Laws of Logarithms.**

- Using the Laws of Logarithms.
- Using the Change of Base Formula.

Hints and Tips

- The table containing the Laws of Logarithms is another one of those things that you really need to memorize — if you do so, your future mathematical life will be *much* easier. One of the most annoying things about teaching calculus and more advanced courses is taking off points for students who have forgotten their log rules. I always have my students memorize two additional rules, which I call the *non-rules*:

$$\log(a+b) = \log(a+b) \qquad \frac{\log a}{\log b} = \frac{\log a}{\log b}$$

These rules look silly, but memorizing them may help you to avoid the mistakes that the text warns you about after Example 4.

Review

(A) **Laws of Logarithms.**

Suppose that $x > 0$, $y > 0$, and r is any real number. Then the Laws of Logarithms are:

Rule	Meaning
1. $\log_a (xy) = \log_a x + \log_a y$	The log of a product is the sum of the logs.
2. $\log_a \left(\dfrac{x}{y}\right) = \log_a x - \log_a y$	The log of a quotient is the difference of the logs.
3. $\log_a (x^r) = r \log_a x$	The log of a power of a number is the exponent times the logarithm of the number.

- **Using the Laws of Logarithms.**

Exercise: Use the Laws of Logarithms to rewrite the expression in a form that contains no logarithm of a product, a quotient, or a power.

Answer:

(a) $\log_2 9x$

The log of a product is the sum of the logs, so
$\log_2 9x = \log_2 9 + \log_2 x$.

(b) $\log_5 \left(\frac{2x}{y}\right)$

The log of a quotient is the difference of the logs, so
$\log_5 \left(\dfrac{2x}{y}\right) = \log_5 (2x) - \log_5 y$. Now the log of a product is the sum of the logs, so this simplifies to
$\log_5 2 + \log_5 x - \log_5 y$.

(c) $\log_3 (x^4 y^3)$

The log of a product is the sum of the logs, so
$\log_3 (x^4 y^3) = \log_3 x^4 + \log_3 y^3$. Now the log of a power of a number is the exponent times the logarithm of the number, so this simplifies to $4 \log_3 x + 3 \log_3 y$.

(d) $\log_4\left(\dfrac{16w^5}{\sqrt[3]{x^2y}}\right)$

$\log_4\left(\dfrac{16w^5}{\sqrt[3]{x^2y}}\right) = \log_4\left(16w^5\right) - \log_4\left(\sqrt[3]{x^2y}\right)$

$= \log_4 16 + \log_4 w^5 - \frac{1}{3}\log_4\left(x^2y\right)$

$= \log_4 4^2 + \log_4 w^5 - \frac{1}{3}\left(\log_4 x^2 + \log_4 y\right)$

$= \log_4 4^2 + \log_4 w^5 - \frac{1}{3}\left(2\log_4 x + \log_4 y\right)$

$= 2 + 5\log_4 w - \frac{2}{3}\log_4 x - \frac{1}{3}\log_4 y$

Exercise: Use the Laws of Logarithms to rewrite each of the following as a single logarithm.

Answer:

(a) $4\log_3 x - \frac{3}{7}\log_3 y$

$4\log_3 x - \frac{3}{7}\log_3 y = \log_3 x^4 - \log_3\sqrt[7]{y^3} = \log_3\left(\dfrac{x^4}{\sqrt[7]{y^3}}\right)$

(b) $2\log_{10} x + 3\log_{10} y - \frac{1}{2}\log_{10} 8$

$2\log_{10} x + 3\log_{10} y - \frac{1}{2}\log_{10} 8$

$= \log_{10} x^2 + \log_{10} y^3 - \log_{10}\sqrt{8}$

$= \log_{10}\left(\dfrac{x^2y^3}{\sqrt{8}}\right)$

(c) $4\log_5 x + 2 - \log_5 y - 2\log_5 w$

To convert 2 to the form $\log_5 a$, we note that $25 = 5^2$, so $2 = \log_5 25$. Thus,

$4\log_5 x + 2 - \log_5 y - 2\log_5 w$

$= \log_5 x^4 + \log_5 25 - \log_5 y - \log_5 w^2$

$= \log_5\left(25x^4\right) - \log_5\left(yw^2\right)$

$= \log_5\left(\dfrac{25x^4}{yw^2}\right)$

- **Using the Change of Base Formula.**

Since most calculators have only the common log and natural log functions, it is important to be able to change the base of a logarithm. The **Change of Base Formula** is

$$\log_b x = \frac{\log_a x}{\log_a b}$$

This allows us to evaluate a logarithm of any base (provided the base is greater than 0 and not equal to 1). The Change of Base Formula can also be expressed as $\log_a b \cdot \log_b x = \log_a x$. Of particular importance are the formulas $\log_b x = \dfrac{\log x}{\log b}$ and $\log_b x = \dfrac{\ln x}{\ln b}$, which allows us to use our calculators to find logarithms with uncommon bases.

Exercise: Evaluate $\log_6 7$ to six decimal places.

Answer: Using common logarithms, we calculate

$$\log_6 7 = \frac{\log 7}{\log 6} \approx \frac{0.84509804}{0.778151250} = 1.086033.$$

Or we could use natural logarithms:

$$\log_6 7 = \frac{\ln 7}{\ln 6} \approx \frac{1.945910149}{1.791759469} = 1.086033.$$

Exercise: Simplify $(\log_3 8)(\log_8 7)$.

Answer: Use the Change of Base Formula to express all logarithms to the base 3:

$$(\log_3 8)(\log_8 7) = \log_3 8 \cdot \frac{\log_3 7}{\log_3 8} = \log_3 7.$$

Core Exercises

7, 19

5.4 Exponential and Logarithmic Equations

Concepts

(A) Equations involving exponents and logarithms.

- Solving exponential equations.
- Solving logarithmic equations.

(B) Applications.

- Solving compound interest problems.

Hints and Tips

- The main rule for solving equations never changes. You are always going from one equation to an equivalent equation — hopefully a simpler one. So if you are going to use logarithms to get something out of an exponent, you must remember to take the logarithm of both sides of the equation, so your new equation is equivalent to the old one.

Review

(A) **Equations involving exponents and logarithms.**

- **Solving exponential equations.**

  Guidelines for solving exponential equations:

  1. Isolate the exponential expression on one side of the equation.
  2. Take the logarithm of each side, then use the Laws of Logarithms to bring down the exponent.
  3. Solve for the variable.

Exercise: Solve $2^{3x+1} = 32$ for x.

Answer: We take logarithms of both sides and use Law 3:

$$2^{3x+1} = 32 \quad \Leftrightarrow \quad \log_2\left(2^{3x+1}\right) = \log_2 32 \quad \Leftrightarrow$$

$$\log_2\left(2^{3x+1}\right) = \log_2 2^5 \quad \Leftrightarrow \quad (3x+1)\log_2 2 = 5\log_2 2.$$

Solving, we get $3x + 1 = 5 \quad \Leftrightarrow \quad 3x = 4 \quad \Leftrightarrow \quad x = \frac{4}{3}$.

Exercise: Solve $x^5 e^x + 3x^4 e^x = 4x^3 e^x$.

Answer: $x^5 e^x + 3x^4 e^x = 4x^3 e^x$ $\Leftrightarrow$ $x^5 e^x + 3x^4 e^x - 4x^3 e^x = 0$ $\Leftrightarrow$ $\left(x^5 + 3x^4 - 4x^3\right) e^x = 0$ $\Leftrightarrow$ $x^3 \left(x^2 + 3x - 4\right) e^x = 0$ $\Leftrightarrow$ $x^3 \left(x + 4\right) \left(x - 1\right) = 0$ (since $e^x \neq 0$). Thus, the solutions are $x = 0, -4$, or 1.

Exercise: Use a graphing calculator to solve $e^{2x} = 5e^x$ to two decimal places.

Answer: $e^{2x} = 5e^x$ $\Leftrightarrow$ $e^{2x} - 5e^x = 0$, so we graph $f(x) = e^{2x} - 5e^x$ and find where it intersects the x-axis:

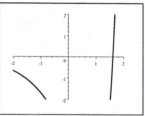

We see that the only solution is $x \approx 1.61$. A larger viewing rectangle shows that $f(x)$ approaches the x-axis as $x \to -\infty$, but there is no solution for $x < 0$.

Exercise: Use a graphing calculator to solve $xe^x - 3x^3 = 0$ to two decimal places.

Answer: We graph $f(x) = xe^x - 3x^3$ and find where it intersects the x-axis.

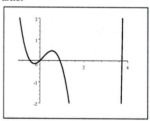

We see that the solutions are $x \approx -0.46$, $x = 0$, $x \approx 0.91$, and $x \approx 3.73$.

Exercise: Solve the equation $4e^{2x-1} = 7$.

Answer: $4e^{2x-1} = 7$ $\Leftrightarrow$ $e^{2x-1} = \frac{7}{4}$ $\Leftrightarrow$ $\ln\left(e^{2x-1}\right) = 2x - 1 = \ln\left(\frac{7}{4}\right)$ $\Leftrightarrow$ $2x = 1 + \ln\left(\frac{7}{4}\right)$ $\Leftrightarrow$ $x = \frac{1}{2} + \frac{1}{2}\ln\left(\frac{7}{4}\right) \approx 0.78$

Exercise: Solve the equation $e^{2x} - 4e^x + 3 = 0$.

Answer: We factor: $e^{2x} - 4e^x + 3 = 0$ $\Leftrightarrow$ $\left(e^x - 1\right)\left(e^x - 3\right) = 0$ $\Leftrightarrow$ $e^x - 1 = 0$ or $e^x - 3 = 0$ $\Leftrightarrow$ $x = \ln 1 = 0$ or $x = \ln 3$. Thus, the solutions are $x = 0$ and $x = \ln 3$.

● **Solving logarithmic equations.**

Guidelines for solving logarithmic equations:

> 1. Isolate the logarithmic term on one side of the equation; you may need to combine the logarithmic terms first.
> 2. Write the equation in exponential form (or raise the base to each side of the equation).
> 3. Solve for the variable.

Exercise: Solve $\log_5 (62 - x) = 3$ for x.

Answer: We rewrite the equation in exponential form:

$\log_5 (62 - x) = 3 \quad \Leftrightarrow \quad 62 - x = 5^3 \quad \Leftrightarrow$
$62 - x = 125 \quad \Leftrightarrow \quad -x = 63 \quad \Leftrightarrow \quad x = -63.$
Check: $\log_5 (62 - (-63)) = \log_5 125 = \log_5 5^3 = 3.$

Exercise: Solve $\log_3 x + \log_9 x = 6$ for x.

Answer: We need to convert to a common base. Since $9 = 3^2$, we convert $\log_3 x$ to base 9. Since $(\log_9 3)(\log_3 x) = \log_9 x$ and $\log_9 3 = \frac{1}{2}$, we have

$\log_3 x = \dfrac{\log_9 x}{\log_9 3} = \dfrac{\log_9 x}{1/2} = 2\log_9 x.$ Thus,

$\log_3 x + \log_9 x = 6 \quad \Leftrightarrow \quad 2\log_9 x + \log_9 x = 6 \quad \Leftrightarrow$
$3\log_9 x = 6 \quad \Leftrightarrow \quad \log_9 x = 2 \quad \Leftrightarrow \quad x = 9^2 = 81.$ Check:
$\log_3 81 = \log_3 3^4 = 4$ and $4\log_9 81 = \log_9 9^2 = 2$, so
$\log_3 81 + \log_9 81 = 4 + 2 = 6.$

Exercise: Solve
$\log_8 (x + 2) + \log_8 (x + 5) = \frac{2}{3}.$

Answer: $\log_8 (x + 2) + \log_8 (x + 5) = \frac{2}{3} \quad \Leftrightarrow$
$\log_8 ((x + 2)(x + 5)) = \frac{2}{3} \quad \Leftrightarrow \quad \log_8 (x^2 + 7x + 10) = \frac{2}{3}$
$\Leftrightarrow \quad 8^{\log_8 (x^2 + 7x + 10)} = 8^{2/3} \quad \Leftrightarrow$
$x^2 + 7x + 10 = 8^{2/3} = \left(8^{1/3}\right)^2 \quad \Leftrightarrow$
$x^2 + 7x + 10 = 2^2 = 4 \quad \Leftrightarrow \quad x^2 + 7x + 6 = 0 \quad \Leftrightarrow$
$(x + 6)(x + 1) = 0 \quad \Leftrightarrow \quad x = -6 \text{ or } -1.$
Check $x = -6$: $\log_8 ((-6) + 2) + \log_8 ((-6) + 5) \overset{?}{=} \frac{2}{3}$, but
$\log_8 (-4)$ and $\log_8 (-1)$ are not defined, so $x = -6$ is *not* a
solution.
Check $x = -1$: $\log_8 ((-1) + 2) + \log_8 ((-1) + 5) =$
$\log_8 1 + \log_8 4 = 0 + \log_8 \left(\left(\sqrt[3]{8} \right)^2 \right) = \frac{2}{3}.$ Thus, the only
solution is $x = -1.$

(B) Applications.

The three main interest equations are exponential equations. So if you know all the variables but one, you can solve for the missing quantity using the techniques from this section.

> If a principal P is invested at an interest rate r for a period of t years, then the amount A of the investment is given by
>
> $A = P(1 + r)$ Simple interest, one year
>
> $A = P\left(1 + \dfrac{r}{n}\right)^{nt}$ Interest compounded n times per year
>
> $A = Pe^{rt}$ Interest compounded continuously

• **Solving compound interest problems.**

Exercise: Suppose I invest \$20,000 in a bank account at 5% interest compounded continuously.

Answer:

(a) How long will it take to double my money?

We write $40,000 = 20,000e^{0.05t}$ and solve for t: $2 = e^{0.05t}$

$\Leftrightarrow \quad \ln 2 = \ln\left(e^{0.05t}\right) \quad \Leftrightarrow \quad \ln 2 = 0.05t \quad \Leftrightarrow$

$t = \dfrac{\ln 2}{0.05} \approx 13.863$ years.

(b) If I instead invest \$5000, how long will it take to double?

We write $10,000 = 5000e^{0.05t}$ and solve for t. $2 = e^{0.05t} \quad \Leftrightarrow$

$t = \frac{\ln 2}{0.05} \approx 13.863$ years.

(c) If I invest N dollars, how long will it take to double?

We write $2N = Ne^{0.05t}$ and solve for t: $2 = e^{0.05t} \quad \Leftrightarrow$

$t = \dfrac{\ln 2}{0.05} \approx 13.863$ years.

▶ Core Exercises

11, 19, 27, 33, 65, 67, 71, 75, 85

5.5   Modeling with Exponential and Logarithmic Functions

Concepts

(A) **Applications of exponential and logarithmic functions.**

- Working with exponential growth models.
- Working with exponential decay models.
- Working with Newton's Law of Cooling.
- Working with logarithmic models.

Hints and Tips

There are many models presented in this section. Try to see beyond the specifics, and see how they all are similar.

Review

(A) **Applications of exponential and logarithmic functions.**

- **Working with exponential growth models.**

  A population that experiences **exponential growth** increases according to the formula $n(t) = n_0 e^{rt}$ where n_0 is the initial size of the population, r is the relative rate of growth (expressed as a proportion of the population), t is time, and $n(t)$ is the population at time t.

Exercise: Under ideal conditions a certain bacteria population is known to increase at a relative growth rate of 7.8% per hour. Suppose that there are initially 2000 bacteria.	**Answer:**
(a) What will be the size of the population in 18 hours?	We substitute $n_0 = 2000$, $r = 0.078$, and $t = 18$ into the equation $n(t) = n_0 e^{rt}$: $n(18) = 2000 e^{(0.078)(18)} = 2000 e^{1.404} \approx 8143$, so there will be about 8100 bacteria after 18 hours.
(b) How many bacteria will there be after 15 days? Do you think this model is appropriate for this problem? Why or why not?	We substitute $n_0 = 2000$, $r = 0.078$, and $t = 15(24) = 360$: $n(360) = 2000 e^{(0.078)(360)} = 2000 e^{28.08} \approx 3.1334 \times 10^{15}$. The model predicts that there will be about 3.1334×10^{15} bacteria in 15 days. This seems like an awful lot: remember that this model holds only *under ideal conditions* — most likely the culture will run out of nutrients before reaching this population.

- **Working with exponential decay models.**

Logarithms are used to extract the power in exponential functions.   This is an especially useful tool when the exponent contains the unknown variable.

Exercise: At a certain time, a bacterial culture contains 115,200 bacteria. Four hours later, the culture has 1,440,000 bacteria. How many bacteria will there be after another 11 hours?

Answer: $n\left(t\right) = n_0 e^{rt}$. We use $n_0 = 115{,}200$ and $t = 4$ to find r: $n\left(4\right) = 1{,}440{,}000 = 115{,}200\, e^{4r}$ $\Leftrightarrow$

$$e^{4r} = \frac{1{,}440{,}000}{115{,}200} = 12.5 \quad \Leftrightarrow$$

$\ln\left(e^{4r}\right) = 4r = \ln 12.5 \approx 2.526 \quad \Leftrightarrow \quad r \approx 0.631$. Thus,

$n\left(15\right) = 115{,}200 e^{(0.631)(15)} = 115{,}200 \cdot 12984.12$
$$= 1{,}495{,}770{,}721 \approx 1.5 \times 10^9$$

- **Working with Newton's Law of Cooling.**

Newton's Law of Cooling states that the rate of cooling of an object is proportional to the temperature difference between the object and its surroundings, provided that the temperature difference is not too large. It can be shown that $T\left(t\right) = T_s + D_0 e^{-kt}$, where T_s is the temperature of the surrounding environment, D_0 is the initial temperature difference, and k is a positive constant that is associated with the cooling object.

Exercise: A bowl of soup at $115°$ F is placed in an air-conditioned room at $70°$ F. After 5 minutes, the temperature of the soup is $100°$ F.

Answer:

(a) What is the temperature after 10 minutes?

First find k: $T\left(5\right) = 100°$ F, $T_s = 70°$ F, and $D_0 = 115 - 70 = 45°$ F, so substituting gives $T\left(t\right) = T_s + D_0 e^{-kt} \quad \Leftrightarrow \quad 100 = 70 + 45 e^{-5k} \quad \Leftrightarrow$

$30 = 45 e^{-5k} \quad \Leftrightarrow \quad \frac{2}{3} = e^{-5k} \quad \Leftrightarrow$

$\ln\left(\frac{2}{3}\right) = \ln\left(e^{-5k}\right) = -5k \quad \Leftrightarrow \quad k = -\frac{1}{5} \cdot \ln\left(\frac{2}{3}\right) \approx 0.081$.

Thus, $T\left(t\right) = 70 + 45 e^{-0.081t}$, so

$T\left(10\right) = 70 + 45 e^{-0.081(10)} \approx 90°$ F.

(b) How long will it take the soup cool to $80°$ F?

We find t such that $T\left(t\right) = 80$: $T\left(t\right) = 70 + 45 e^{-0.081t} \quad \Leftrightarrow$

$80 = 70 + 45 e^{-0.081t} \quad \Leftrightarrow \quad 10 = 45 e^{-0.081t} \quad \Leftrightarrow$

$\frac{10}{45} = \frac{2}{9} = e^{-0.081t} \quad \Leftrightarrow$

$\ln\left(\frac{2}{9}\right) = \ln\left(e^{-0.081t}\right) = -0.081t \quad \Leftrightarrow$

$t = \dfrac{\ln\left(\frac{2}{9}\right)}{-0.081} \approx 19$ minutes.

• **Working with logarithmic models.**

Logarithms make quantities which vary over very large ranges more manageable. Three commonly used logarithmic scales are outlined below.

1. **pH scale**	Used to measure acids and bases. $\text{pH} = -\log \left[H^+ \right]$, where $\left[H^+ \right]$ is the concentration of hydrogen ions, measured in moles per liter.
2. **Richter scale**	Used to measure earthquakes. $M = \log \dfrac{I}{S} = \log I - \log S$, where I is the intensity of the earthquake and S is the intensity of a "standard" earthquake.
2. **The Decibel scale**	Used to measure sounds. $B = 10 \log \dfrac{I}{I_0}$, where I_0 is a reference intensity of $10^{-12} \dfrac{\text{watts}}{\text{m}^2}$ at a frequency of 1000 Hz.

Exercise: The hydrogen ion concentration of a certain shampoo is 8.2×10^{-8}. Find its pH.

Answer:
$$\begin{aligned}
\text{pH} &= -\log \left(8.2 \times 10^{-8} \right) \\
&= -\left[\log (8.2) + \log \left(10^{-8} \right) \right] \\
&= -(0.91 - 8) = 7.09
\end{aligned}$$

Exercise: On October 17, 1989, the Loma Prieta Earthquake rocked the Santa Cruz mountains. Researchers initially measured the earthquake as 6.9 on the Richter scale. Later, the earthquake was reevaluated as 7.1 on the Richter scale. How do the two measures compare in intensity?

Answer: We want to compare the intensities of the two Richter measurements, so we solve $M = \log I - \log S$ for I:
$$M = \log I - \log S \quad \Leftrightarrow \quad M + \log S = \log I \quad \Leftrightarrow$$
$$10^{(M + \log S)} = 10^{\log I} \quad \Leftrightarrow \quad I = S \cdot 10^M. \text{ Comparing the}$$
two intensities, we have $\dfrac{S \cdot 10^{7.1}}{S \cdot 10^{6.9}} = 10^{0.2} \approx 1.58$. So the earthquake was about 1.58 as intense as originally thought.

Exercise: Find the intensity level of a computer speaker system if the system is advertised to deliver 30 watts per square meter at a distance of one meter from the speaker.

Answer: Using the formula for decibel intensity level, we find that the intensity level is
$$\begin{aligned}
B &= 10 \log \frac{I}{I_0} = 10 \log \left(\frac{3 \times 10}{10^{-12}} \right) \\
&= 10 \log \left(3 \times 10^{13} \right) = 10 \cdot 13 \cdot \log 3 \\
&\approx 62 \text{ dB}
\end{aligned}$$

▐ Core Exercises

1, 13, 19, 27, 41

6 Trigonometric Functions of Angles

6.1 Angle Measure

Concepts

(A) **Angle measure.**
- Estimating the measure of a given angle.
- Converting between radians and degrees.

(B) **Area.**
- Finding the area of a circular sector.

(C) **Circular motion.**
- Finding linear and angular speed.

Definitions

- **Initial side / Terminal side:** We consider an angle as two rays with a common vertex, going from the first ray, the initial side, to the second ray, the terminal side.
- **Measure of an angle:** The amount of rotation about the common vertex to get from the initial side to the terminal side. If the rotation is counterclockwise, this is a positive number. If it is clockwise, it is a negative number.
- **Radian measure:** There are 2π radians in an angle that makes a complete revolution about a circle.
- **Degree measure:** There are 360 degrees ($°$) in an angle that makes a complete revolution about a circle.
- **Standard position:** An angle is in standard position if it is drawn in the xy-plane with its vertex at the origin and its initial side on the positive x-axis.
- **Coterminal angles:** Two angles in standard position are coterminal if their terminal sides coincide.
- **Angular speed / linear speed:** If an object moves along a circle, the angular speed is the rate at which the angle is changing, and the linear speed is the rate at which the distance traveled is changing.

Questions to Ask Your Teacher

- In general, should answers be given in degrees or radians?

Hints and Tips

- Radian measure isn't that hard to get used to. First, realize that there are 2π radians in a $360°$ angle. Then look at the right and straight angles:

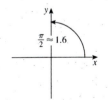
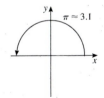
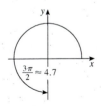
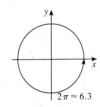

Once you get a feel for those four angles, divide the circle up into 8 pieces and get used to the other four divisions:

 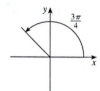 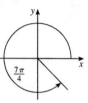

• Note that because radian measure is defined as a ratio of lengths, it technically has no units attached to it. So sometimes you will see (perfectly correct) equations like $\pi = 180°$ or $3 = \left(\frac{540}{\pi}\right)°$.

▚ Review

(A) **Angle measure.**

The **measure** of an angle is the amount of rotation about the vertex required to rotate the **initial side** to the **terminal side**. One unit of measure for angles is the **degree**. One degree is $\frac{1}{360}$ of a complete rotation. Another unit of measure is the **radian**, defined as the length of the arc subtended on a circle of radius 1 whose center is at the vertex of the angle. The two measures are related by the following equations:

$$180° = \pi \text{ rad} \qquad 1 \text{ rad} = \left(\frac{180}{\pi}\right)° \qquad 1° = \frac{\pi}{180} \text{ rad}$$

Convert **degrees to radians** by multiplying by $\frac{\pi}{180}$.

Convert **radians to degrees** by multiplying by $\frac{180}{\pi}$.

• **Estimating the measure of a given angle.**

Exercise: Estimate the measures of the following angles, in both degrees and radians.

Answer:

(a)

$\frac{\pi}{2}$ radians; 90°

(b)

$\frac{\pi}{3}$ radians; 60°

(c)

$\frac{9\pi}{4}$ radians; 405°

(d)

1 radian; about 57.3°

- **Converting between radians and degrees.**

Exercise: Find the radian measure of the angle with the given degree measure.

Answer:

(a) $75°$

Multiplying by $\frac{\pi}{180}$, we have $75° = 75\left(\frac{\pi}{180}\right) = \frac{5\pi}{12}$ (radians)

(b) $300°$

Multiplying by $\frac{\pi}{180}$, we have $300° = 300\left(\frac{\pi}{180}\right) = \frac{5\pi}{3}$. No units are needed; see Hint 2 above.

Exercise: Find the degree measure of the angle with the given radian measure.

Answer:

(a) $\frac{5\pi}{9}$

Multiplying by $\frac{180}{\pi}$, we have $\frac{5\pi}{9} = \frac{5\pi}{9}\left(\frac{180}{\pi}\right)° = 100°$.

(b) 5

Multiplying by $\frac{180}{\pi}$, we have
$5 = 5\left(\frac{180}{\pi}\right)° = \left(\frac{900}{\pi}\right)° \approx 286.5°$.

B **Area.**

In a circle of radius r the area A of the sector shown is given by $A = \frac{1}{2}r^2\theta$.

- **Finding the area of a circular sector.**

Exercise: Find the area A of the sector in each figure.

Answer:

(a)

First express $204°$ in radians: $\theta = 204° = 204 \cdot \frac{\pi}{180} = \frac{17\pi}{15}$.
From the figure, $r = 8$, so
$A = \frac{1}{2}r^2\theta = \frac{1}{2}\left(8^2\right)\frac{17\pi}{15} = \frac{544\pi}{15} \approx 113.94$.

(b)

From the figure, $\theta = 2\pi - \frac{\pi}{6} = \frac{11\pi}{6}$ and $r = 5$, so
$A = \frac{1}{2}r^2\theta = \frac{1}{2}\left(5^2\right)\frac{11\pi}{6} = \frac{275\pi}{12} \approx 71.99$.

Ⓒ Circular motion.

Suppose a particle moves along a circle. We can describe the motion of the particle either in terms of its **linear speed** or its **angular speed**. Linear speed is the rate at which the distance traveled is changing, and angular speed is the rate at which the central angle changes.

Angular speed: $\omega = \dfrac{\theta}{t}$; units are $\dfrac{\text{radians}}{\text{time}}$	Linear speed: $v = \dfrac{s}{t}$; units are $\dfrac{\text{distance}}{\text{time}}$

• Finding linear and angular speed.

Exercise: A carousel at a seaside amusement park revolves five times per minute.

Answer:

(a) What is the angular speed of this carousel?

Using $\omega = \dfrac{\theta}{t}$, we get $\omega = \dfrac{5 \cdot 2\pi}{1} = 10\pi$ rad/min.

(b) How fast (in mi/h) is a person traveling if they are riding a horse 12 ft from the center of the carousel?

We first find the linear speed in ft/min, and then convert to mi/h. Using $v = r\omega$ with $r = 12$ ft and $\omega = 10\pi$ rad/min [from part (a)], we get $v = 12 \cdot 10\pi = 120\pi \approx 377$ ft/min. Converting, the person's speed is

$$v \approx 377 \ \frac{\text{ft}}{\text{min}} \cdot \frac{60 \ \text{min}}{\text{h}} \cdot \frac{1 \ \text{mi}}{5280 \ \text{ft}} \approx 4.3 \ \frac{\text{mi}}{\text{h}}.$$

(c) How fast (in mi/h) is a person traveling if they are riding a horse 25 ft from the center of the carousel?

In this case, the speed is $v = 25 \cdot 10\pi \cdot 60 \cdot \frac{1}{5280} \approx 8.9 \ \dfrac{\text{mi}}{\text{h}}$.

Exercise: A hillside restaurant runs a cable car from the parking lot to the restaurant. The cable for the cable car is driven by a one foot diameter wheel attached to a motor at the base of the hill. How fast must the wheel rotate in order for the cable car to make the 300 foot trip in 2 minutes?

Answer: We solve $v = r\omega$ for ω, so $\omega = v/r$. The radius of the wheel is $\frac{1}{2}$ ft and the desired speed is $\dfrac{300 \ \text{ft}}{2 \ \text{min}} = 150 \ \dfrac{\text{ft}}{\text{min}}$, so

$\omega = \dfrac{150}{1/2} = 300 \ \dfrac{\text{rad}}{\text{min}}$. Since there are 2π radians in each revolution, the wheel must turn at $\dfrac{300}{2\pi} \approx 47.75$ rpm.

▐ Core Exercises

39, 45, 71, 75

6.2   Trigonometry of Right Triangles

Concepts

(A) Trigonometric functions as ratios.

- Computing functions from given right triangles.
- Computing functions of special angles.
- Computing trigonometric functions using a calculator.

(B) Applications of trigonometric functions.

- Solving right triangles.
- Using right triangle models.

Definitions

- **sine, cosine, etc.:** From the perspective of this chapter, we define the trigonometric functions as ratios of sides of a right triangle with a specified angle θ as follows:

$$\sin \theta = \frac{\text{opposite}}{\text{hypotenuse}} \qquad \csc \theta = \frac{\text{hypotenuse}}{\text{opposite}}$$

$$\cos \theta = \frac{\text{adjacent}}{\text{hypotenuse}} \qquad \sec \theta = \frac{\text{hypotenuse}}{\text{adjacent}}$$

$$\tan \theta = \frac{\text{opposite}}{\text{adjacent}} \qquad \cot \theta = \frac{\text{adjacent}}{\text{opposite}}$$

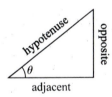

Since any two right triangles with angle θ are similar, these ratios are the same regardless of the size of the triangle.

- **Line of sight:** The line between an observer and an object.
- **Angle of elevation:** The angle between the line of sight and the horizontal. If negative, its absolute value is called the **angle of depression**.
- **Angle of inclination:** The angle between an inclined plane and an object.

Questions to Ask Your Teacher

- Chapters 5 and 6 present different perspectives on the trigonometric functions. When you see "$\sin \frac{\pi}{3}$", what are you picturing in your mind?

Hints and Tips

- Your teacher has probably told you that when you are looking at an application problem involving distances and angles, you should draw a picture. Your teacher has never been more right. Do it!

Review

(A) **Trigonometric functions as ratios.**

- • **Computing functions from given right triangles.**

Exercise: Find the values of the six trigonometric ratios of the angle θ in the triangle shown.

(a)

Answer:

$\sin\theta = \frac{7}{\sqrt{113}}$, $\cos\theta = \frac{8}{\sqrt{113}}$, $\tan\theta = \frac{7}{8}$, $\csc\theta = \frac{\sqrt{113}}{7}$,

$\sec\theta = \frac{\sqrt{113}}{8}$, $\cot\theta = \frac{8}{7}$

(b)

$\sin\theta = \frac{5}{13}$, $\cos\theta = \frac{12}{13}$, $\tan\theta = \frac{5}{12}$, $\csc\theta = \frac{13}{5}$, $\sec\theta = \frac{13}{12}$,

$\cot\theta = \frac{12}{5}$

(c)

$\sin\theta = \frac{2\sqrt{10}}{7}$, $\cos\theta = \frac{3}{7}$, $\tan\theta = \frac{2\sqrt{10}}{3}$, $\csc\theta = \frac{7}{2\sqrt{10}}$,

$\sec\theta = \frac{7}{3}$, $\cot\theta = \frac{3}{2\sqrt{10}}$

Exercise: If $\tan\theta = \frac{12}{5}$, find the other five trigonometric ratios.

Answer: Draw a right triangle and label one angle θ. Since $\tan\theta = \frac{12}{5} = \frac{\text{opposite}}{\text{adjacent}}$, the side with length 12 is opposite θ and the side with length 5 is adjacent to θ.

The hypotenuse is found with the Pythagorean theorem: $\sqrt{12^2 + 5^2} = \sqrt{144 + 25} = \sqrt{169} = 13$, so $\sin\theta = \frac{12}{13}$, $\cos\theta = \frac{5}{13}$, $\csc\theta = \frac{13}{12}$, $\sec\theta = \frac{13}{5}$, and $\cot\theta = \frac{5}{12}$.

Exercise: If $\csc\theta = 3$, find the other five trigonometric ratios.

Answer: Draw a right triangle and label one angle θ. Since $\csc\theta = 3 = \frac{3}{1} = \frac{\text{hypotenuse}}{\text{opposite}}$, the side with length 1 is opposite θ and the side with length 3 is the hypotenuse.

By the Pythagorean theorem, $1^2 + x^2 = 3^2 \Rightarrow x^2 = 8$ $\Rightarrow x = 2\sqrt{2}$, so $\sin\theta = \frac{1}{3}$, $\cos\theta = \frac{2\sqrt{2}}{3}$,

$\tan\theta = \frac{1}{2\sqrt{2}} = \frac{\sqrt{2}}{4}$, $\sec\theta = \frac{3}{2\sqrt{2}} = \frac{3\sqrt{2}}{4}$, and

$\cot\theta = \frac{2\sqrt{2}}{1} = 2\sqrt{2}$.

- **Computing functions of special angles.**

$30°$-$60°$-$90°$ right triangles and $45°$ right triangles have ratios that can be calculated directly from the Pythagorean theorem. These ratios appear frequently, so it is worth memorizing at least the sine and cosine values.

θ (°)	θ (rad)	$\sin\theta$	$\cos\theta$	$\tan\theta$	$\csc\theta$	$\sec\theta$	$\cot\theta$
30	$\frac{\pi}{6}$	$\frac{1}{2}$	$\frac{\sqrt{3}}{2}$	$\frac{\sqrt{3}}{3}$	2	$\frac{2\sqrt{3}}{3}$	$\sqrt{3}$
45	$\frac{\pi}{4}$	$\frac{\sqrt{2}}{2}$	$\frac{\sqrt{2}}{2}$	1	$\sqrt{2}$	$\sqrt{2}$	1
60	$\frac{\pi}{3}$	$\frac{\sqrt{3}}{2}$	$\frac{1}{2}$	$\sqrt{3}$	$\frac{2\sqrt{3}}{3}$	2	$\frac{\sqrt{3}}{3}$

Exercise: Find the length of the side labeled x, correct to six decimal places.

Answer:

(a)

The sides with lengths x and 12 are related by the tangent function: $\tan 30° = 12/x$ $\Leftrightarrow$

$x = 12/(\tan 30°) = 12\cot 30° = 12\sqrt{3} \approx 20.784610$.

(b)

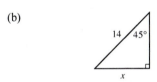

The sides with lengths x and 14 are related by the sine function:

$\sin 45° = x/14$ $\Leftrightarrow$

$x = 14\sin 45° = 14 \cdot \frac{\sqrt{2}}{2} = 7\sqrt{2} \approx 9.899495$.

- **Computing trigonometric functions using a calculator.**

Exercise: Find the given value, accurate to five decimal places.

Answer:

(a) $\sin\frac{\pi}{8}$

$\sin\frac{\pi}{8} \approx 0.38268$

(b) $\tan 32$

$\tan 32 \approx 0.66101$

(c) $\sec 3$

$\sec 3 = \dfrac{1}{\cos 3} = -1.01011$

(B) **Applications of trigonometric functions.**

- **Solving right triangles.**

A triangle consists of three angles and three sides. To **solve a triangle** means to determine all six quantities. Applications that involve solving triangles frequently use the terms "line of sight" and "angle of elevation/depression" defined above.

Exercise: Express x and y in terms of θ.	**Answer:**

(a)

x and θ are related by the equation $\cos\theta = x/4$, so $x = 4\cos\theta$.
y and θ are related by $\sin\theta = y/4$, so $y = 4\sin\theta$.

(b)

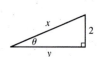

x and θ are related b $\sin\theta = 2/x$, so $x = 2/\left(\sin\theta\right) = 2\csc\theta$.
y and θ are related by $\cot\theta = y/2$, so $y = 2\cot\theta$.

• **Using right triangle models.**

Exercise: A builder needs to determine the height of a house from a nearby hill. Using surveying equipment and a tape measure, she determines the horizontal, the angles of elevation and depression shown in the diagram, and the distance from the hill to the base of the house.

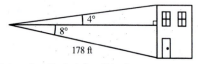

Determine the height of the house.

Answer: To find the height of the house, we need to find the lengths labeled x and y in the diagram below.

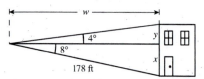

We find x using the sine function: $x/178 = \sin 8°$ $\Leftrightarrow$ $x = 178\sin 8° \approx 24.8$ ft. To find y, we must first find w:
$w/178 = \cos 8°$ $\Leftrightarrow$ $w = 178\cos 8° \approx 176.3$ ft, so
$y/w = \tan 4°$ $\Leftrightarrow$
$y = w\tan 4° \approx 176.3\tan 4° \approx 12.3$ ft.
Adding x and y, the height of the house is about $24.8 + 12.3 = 37.1$ ft.

Core Exercises

7, 25, 33, 43, 49

6.3 Trigonometric Functions of Angles

Concepts

(A) Trigonometric functions of arbitrary angles.
  • Finding reference angles.
  • Finding trigonometric functions of arbitrary angles.

(B) Trigonometric identities and formulas.
  • Using trigonometric identities to evaluate trigonometric functions.
  • Finding areas of triangles.

◤ Definition

- **Reference angle:** Given an angle θ, the reference angle $\overline{\theta}$ is the acute angle formed by the terminal side of θ and the x-axis.

◤ Hints and Tips

- As I suggested in Section 5.2, you can save yourself a lot of memorization if you master a couple of fundamental concepts. For example, I have only the first of the three Pythagorean identities memorized: $\sin^2 x + \cos^2 x = 1$. If I need another one, which I often do, I just divide both sides of the equation by $\sin^2 x$ or $\cos^2 x$, as done in the text. For example, if we divide both sides by $\sin^2 x$ we get $1 + \cot^2 x = \csc^2 x$.

- It is not really necessary to memorize phrases like All Students Take Calculus, Aunt Sally Tips Cows, etc. You are going to have to learn the definitions of sine, cosine and tangent anyway. If you have an angle in, say, quadrant II, its x-coordinate is negative and its y-coordinate is positive. So if you understand the definitions of sine, cosine and tangent, you will be able to use your coconut to tell you that sine is positive, cosine is negative, and tangent (sine over cosine) is negative.

◤ Review

(A) **Trigonometric functions of arbitrary angles.**

Let θ be an angle in standard position and $P(x,y)$ a point on the terminal side. If $r = \sqrt{x^2 + y^2}$ is the distance from the origin to P, then

$$\sin\theta = \frac{y}{r} \qquad\qquad \cos\theta = \frac{x}{r} \qquad\qquad \tan\theta = \frac{y}{x}\ (x \neq 0)$$

$$\csc\theta = \frac{r}{y}\ (y \neq 0) \qquad\qquad \sec\theta = \frac{r}{x}\ (x \neq 0) \qquad\qquad \cot\theta = \frac{x}{y}\ (y \neq 0)$$

- **Finding reference angles.**

  The trick to finding reference angles is to subtract (or add) 2π to the given angle until you have an angle between 0 and 2π, and then (with the aid of a sketch if necessary) find the smallest acute angle between the terminal side and the x-axis.

Exercise: Find the reference angle.   ‖ **Answer:**

(a) $\theta = \frac{16\pi}{3}$

$\frac{16\pi}{3} - 4\pi = \frac{4\pi}{3}$.

The reference angle $\overline{\theta}$ is $\frac{\pi}{3}$.

(b) $\theta = -\frac{\pi}{6}$

$-\frac{\pi}{6} + 2\pi = \frac{11}{6}\pi.$

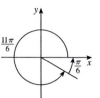

The reference angle is $\overline{\theta} = \frac{\pi}{6}$.

• **Finding trigonometric functions of arbitrary angles.**

Finding the value of a trigonometric function often involves determining the absolute value of the function using the acute **reference angle**, then determining the sign of the function from the quadrant in which the angle lies.

Exercise: Find the exact value of each trigonometric function.

Answer:

(a) $\sin \frac{7\pi}{4}$

Since $\frac{7\pi}{4}$ is in the fourth quadrant, the sine function is negative. The reference angle is $2\pi - \frac{7\pi}{4} = \frac{\pi}{4}$, so since $\sin \frac{\pi}{4} = \frac{\sqrt{2}}{2}$, we have $\sin \frac{7\pi}{4} = -\frac{\sqrt{2}}{2}$.

(b) $\tan \frac{7\pi}{6}$

Since $\frac{7\pi}{6}$ is in the third quadrant, the tangent function is positive. The reference angle is $\frac{7\pi}{6} - \pi = \frac{\pi}{6}$, and so $\tan \frac{7\pi}{6} = \frac{\sqrt{3}}{3}$.

(c) $\cos \left(-\frac{3\pi}{4}\right)$

Since $-\frac{3\pi}{4}$ is in the third quadrant, the cosine function is negative. The reference angle is $-\frac{3\pi}{4} + \pi = \frac{\pi}{4}$, so $\cos \left(-\frac{3\pi}{4}\right) = -\frac{\sqrt{2}}{2}$.

(d) $\sec \frac{79\pi}{6}$

We first strip out multiples of 2π. Since $\frac{79\pi}{6} = \frac{72\pi}{6} + \frac{7\pi}{6} = 12\pi + \frac{7\pi}{6}$, we have $\sec \frac{79\pi}{6} = \sec \frac{7\pi}{6}$. $\frac{7\pi}{6}$ is in the third quadrant, so cosine (and thus secant) are negative. The reference angle is $\frac{\pi}{6}$, so $\sec \frac{79\pi}{6} = -\sec \frac{\pi}{6} = -\frac{2\sqrt{3}}{3}$.

(B) **Trigonometric identities and formulas.**

The trigonometric functions are related to one another through equations called **trigonometric identities**. An important class of identities is the **reciprocal identities**:

$$\csc t = \frac{1}{\sin t} \qquad \sec t = \frac{1}{\cos t} \qquad \tan t = \frac{\sin t}{\cos t} = \frac{1}{\cot t} \qquad \cot t = \frac{\cos t}{\sin t} = \frac{1}{\tan t}$$

Another important class is the **Pythagorean identities**:

$$\sin^2 t + \cos^2 t = 1 \qquad \tan^2 t + 1 = \sec^2 t \qquad 1 + \cot^2 t = \csc^2 t$$

Note the usual convention of writing $\sin^n t$ for $(\sin t)^n$ for all integers n *except* $n = -1$. $\sin^{-1} t$ is a different function entirely.

• **Using trigonometric identities to evaluate trigonometric functions.**

Exercise: Find the values of the trigonometric functions of t.

Answer:

(a) $\cos t = \frac{7}{8}$ and the terminal point of t is in quadrant IV.

We find the easy one first: $\sec t = \dfrac{1}{\cos t} = \frac{8}{7}$. Next we find sine: t is in quadrant IV, so $\sin t < 0$ and $\sin^2 t + \cos^2 t = 1$

$$\Leftrightarrow \quad \sin^2 t + \left(\tfrac{7}{8}\right)^2 = 1 \quad \Leftrightarrow \quad \sin^2 t = 1 - \tfrac{49}{64} = \tfrac{15}{64} \quad \Leftrightarrow$$

$\sin t = -\dfrac{\sqrt{15}}{8}$. Then $\csc t = \dfrac{1}{\sin t} = -\dfrac{8}{\sqrt{15}} = -\dfrac{8\sqrt{15}}{15}$,

$\tan t = \dfrac{\sin t}{\cos t} = \dfrac{-\sqrt{15}/8}{7/8} = -\dfrac{\sqrt{15}}{7}$, and

$\cot t = \dfrac{1}{\tan t} = -\dfrac{7}{\sqrt{15}} = -\dfrac{7\sqrt{15}}{15}$.

(b) $\csc t = \frac{25}{7}$ and the terminal point of t is in quadrant II.

We find the easy one first: $\sin t = \dfrac{1}{\csc t} = \frac{7}{25}$. Next we find cosine: t is in quadrant II, so $\cos t < 0$ and $\sin^2 t + \cos^2 t = 1$

$$\Leftrightarrow \quad \left(\tfrac{7}{25}\right)^2 + \cos^2 t = 1 \quad \Leftrightarrow \quad \cos^2 t = 1 - \tfrac{49}{625} = \tfrac{576}{625}$$

$$\Leftrightarrow \quad \cos t = -\tfrac{24}{25}, \sec t = \dfrac{1}{\cos t} = -\tfrac{25}{24},$$

$\tan t = \dfrac{\sin t}{\cos t} = \dfrac{7/25}{-24/25} = -\tfrac{7}{24}$, and $\cot t = \dfrac{1}{\tan t} = -\tfrac{24}{7}$.

(c) $\cot t = 3$ and $\cos t < 0$.

Since $\cos t < 0$ and $\cot t > 0$, t is in quadrant III. We find the easy one first: $\tan t = \dfrac{1}{\cot t} = \frac{1}{3}$. Now we use the Pythagorean identity to find $\csc t$: $1 + \cot^2 t = \csc^2 t \quad \Leftrightarrow$ $\csc^2 t = 1 + 3^2 = 10 \quad \Leftrightarrow \quad \csc t = -\sqrt{10}$ (cosecant is negative in quadrant III). Then $\sin t = \dfrac{1}{\csc t} = \dfrac{1}{-\sqrt{10}} = -\dfrac{\sqrt{10}}{10}$, $\cos t = -\sqrt{1 - \sin^2 t} = -\sqrt{\frac{9}{10}} = -\dfrac{3\sqrt{10}}{10}$ (since cosine is negative in quadrant III), and $\sec t = \dfrac{1}{\cos t} = -\dfrac{\sqrt{10}}{3}$.

- **Finding areas of triangles.**

The area A of a triangle whose sides have lengths a and b with included angle θ is

$$A = \tfrac{1}{2} ab \sin \theta$$

Exercise: Find the area of a triangle whose sides have lengths 17 and 16, with included angle $157°$.

Answer: Apply the formula:
$A = \frac{1}{2} ab \sin \theta = \frac{1}{2} (17)(16) \sin 157° \approx 53.14$.

Exercise: Find the area of an isosceles triangle whose equal sides have length 235 and whose equal angles measure $22°$.

Answer: Before we can apply the formula, we need to find the value of the included angle using the fact that a triangle's angles sum to $180°$: $\theta = 180° - 2(22°) = 136°$, so
$A = \frac{1}{2}(235)^2 \sin 136° \approx 19{,}181$.

Exercise: The area of a triangle is 75 ft^2 and two of its sides have lengths 16 ft and 20 ft. Find the angle included by these two sides.

Answer: We solve the formula for θ: $A = \frac{1}{2} ab \sin \theta \quad \Leftrightarrow$
$75 = \frac{1}{2}(16)(20) \sin \theta \quad \Leftrightarrow \quad 75 = 160 \sin \theta \quad \Leftrightarrow$
$\sin \theta \approx 0.46875$. By trial and error or by using the $\sin^{-1}$ key on a calculator, we find that $\theta \approx 28°$. Note that there is another possibility: $\theta = 180° - 28° = 152°$ has the same reference angle, so $\sin 152° \approx 0.46875$ as well.

◢ Core Exercises

3, 21, 41, 49, 57, 63, 65

6.4   The Law of Sines

Concepts

(A) **The Law of Sines.**

- Solving triangles given one side and two angles (ASA or SAA)
- Solving triangles given two sides and one angle (SSA)

Hints and Tips

- It should not be necessary to memorize all of the different cases. When solving a triangle, I always use the following steps:
  - The sum of the angles in a triangle is $180°$. That is, if I have two angles, I find the third.
  - Try the Pythagorean theorem. If one of the angles is a right angle, then I can use the Pythagorean theorem and definition of sine and cosine to solve it.
  - If necessary, use the Law of Sines. Remember to be careful in the SSA case.
  - Finally, if necessary, use the Law of Cosines, which is covered in the next section.

Review

(A) **The Law of Sines.**

The **Law of Sines** says that in any triangle, the lengths of the sides are proportional to the sines of the corresponding opposite angles.

- **Solving triangles given one side and two angles (ASA or SAA).**

Exercise: Solve the triangle.

Answer:

(a)

Here we are given that $\angle A = 39°$, $\angle B = 105°$, and $b = 60$. Thus, we must find $\angle C$, a, and c. We calculate $\angle C = 180° - 39° - 105° = 36°$. Since we have b and $\angle B$, we use the Law of Sines to find a and c: $\dfrac{\sin 39°}{a} = \dfrac{\sin 105°}{60}$

$\Leftrightarrow \quad a = \dfrac{60 \sin 39°}{\sin 105°} \approx 39$ and $\dfrac{\sin 36°}{c} = \dfrac{\sin 105°}{60} \quad \Leftrightarrow$

$c = \dfrac{60 \sin 36°}{\sin 105°} \approx 36.5.$

(b)

Here we are given that $\angle A = 38°$, $\angle B = 75°$, and $c = 25$. Thus, we must find $\angle C$, a, and b. We calculate $\angle C = 180° - 38° - 75° = 67°$. Now that we have c and $\angle C$, we use the Law of Sines to find a and b: $\dfrac{\sin 38°}{a} = \dfrac{\sin 67°}{25}$

$\Leftrightarrow \quad a = \dfrac{25 \sin 38°}{\sin 67°} \approx 16.7$ and $\dfrac{\sin 75°}{b} = \dfrac{\sin 67°}{25} \quad \Leftrightarrow$

$b = \dfrac{25 \sin 75°}{\sin 67°} \approx 26.2.$

Exercise: Sketch the triangle, then solve it using the Law of Sines.

Answer:

(a) $\angle A = 64°$, $\angle B = 64°$, $c = 15$

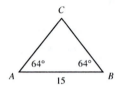

We find $\angle C$ and then use the Law of Sines to find a and b.

$\angle C = 180° - 64° - 64° = 52°$, so $\dfrac{\sin 64°}{a} = \dfrac{\sin 52°}{15}$ $\Leftrightarrow$

$a = \dfrac{15 \sin 64°}{\sin 52°} \approx 17.1$. By symmetry, $b = a \approx 17.1$.

(b) $\angle B = 73°$, $\angle C = 28°$, $a = 258$

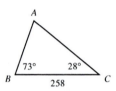

$\angle A = 180° - 73° - 28° = 79°$, so using the Law of Sines, we

have $\dfrac{\sin 73°}{b} = \dfrac{\sin 79°}{258}$ $\Leftrightarrow$ $b = \dfrac{258 \sin 73°}{\sin 79°} \approx 251.3$

and $\dfrac{\sin 28°}{c} = \dfrac{\sin 79°}{258}$ $\Leftrightarrow$ $c = \dfrac{258 \sin 28°}{\sin 79°} \approx 123.4$.

• **Solving triangles given two sides and one angle (SSA).**

This case is ambiguous in the sense that there can be two solutions, one solution, or no solution.

Exercise: Use the Law of Sines to solve for all possible triangles.

Answer:

(a) $a = 73$, $b = 58$, $\angle A = 128°$

Since $\angle A > 90°$, at most one triangle is possible. We find $\angle B$:

$\dfrac{\sin B}{58} = \dfrac{\sin 128°}{73}$ $\Leftrightarrow$ $\sin B = \dfrac{58 \sin 128°}{73} \approx 0.6261$

$\Leftrightarrow$ $\angle B \approx 38.8°$, so $\angle C \approx 180° - 128° - 38.8° = 13.2°$.

Now we can calculate c: $\dfrac{\sin 13.2°}{c} = \dfrac{\sin 128°}{73}$ $\Leftrightarrow$

$c = \dfrac{73 \sin 13.2°}{\sin 128°} \approx 21.2$.

(b) $a = 56$, $c = 71$, $\angle A = 45°$

Since $\angle A < 90°$, two triangles are possible. $\dfrac{\sin C}{71} = \dfrac{\sin 45°}{56}$

$\Leftrightarrow \quad \sin C = \dfrac{71 \sin 45°}{56} \approx 0.8965$, so $\angle C \approx 63.7°$ or

$180° - 63.7° \approx 116.3$.

Case 1: $\angle C \approx 63.7°$. In this case,

$\angle B \approx 180° - 45° - 63.7° = 71.3°$. Thus,

$\dfrac{\sin 71.3°}{b} = \dfrac{\sin 45°}{56} \quad \Leftrightarrow \quad b = \dfrac{56 \sin 71.3°}{\sin 45°} \approx 75.0$.

Case 2: $\angle C \approx 116.3°$. Here

$\angle B \approx 180° - 45° - 116.3° = 18.7°$, so $\dfrac{\sin 18.7°}{b} = \dfrac{\sin 45°}{56}$

$\Leftrightarrow \quad b = \dfrac{56 \sin 18.7°}{\sin 45°} \approx 25.4$.

Core Exercises

7, 13, 19, 27, 35

6.5 The Law of Cosines

Concepts

(A) The Law of Cosines.

- Solving triangles given two sides and one angle (SAS).
- Solving triangles given three sides (SSS).

(B) Finding areas with Heron's Formula.

Review

(A) The Law of Cosines.

The Law of Sines cannot be used directly to solve a triangle when we know two sides and the angle between them, or when we know all three sides. In these cases the **Law of Cosines** is used:

In any triangle ABC,

$$a^2 = b^2 + c^2 - 2bc \cos A$$
$$b^2 = a^2 + c^2 - 2ac \cos B$$
$$c^2 = a^2 + b^2 - 2ab \cos C$$

- **Solving triangles given two sides and one angle (SAS).**

Exercise: Solve triangle ABC if $a = 19$, $b = 34$, and $\angle C = 75°$.

Answer: We need to find $\angle A$, $\angle B$, and c. We use the Law of Cosines to find c: $c^2 = a^2 + b^2 - 2ab\cos C = 19^2 + 34^2 - 2(19)(34)\cos 75° \approx 1182.61$, so $c \approx \sqrt{1182.61} \approx 34.4$. Now we can use the Law of Sines to find $\angle A$: $\dfrac{\sin C}{c} = \dfrac{\sin A}{a} \quad \Leftrightarrow$

$\sin A = a\dfrac{\sin C}{c} \approx 19\dfrac{\sin 75°}{34.4} \approx 0.5335$, so $\angle A \approx 32.2°$.

Then $\angle B \approx 180° - 75° - 32.2°$.

Exercise: A homeowner needs to replace the light bulb above a staircase. The dimensions are shown in the figure. How long is the ladder?

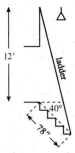

Answer: First we label the vertices of the triangle, as shown.

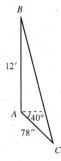

Before using the Law of Cosines, we convert side b to ft: 78 in. $= 6.5$ ft. Next we find $\angle A = 90° + 40° = 130°$, so

$$a^2 = b^2 + c^2 - 2bc\cos A$$
$$= 6.5^2 + 12^2 - 2 \cdot 6.5 \cdot 12 \cos 130°$$
$$\approx 286.52$$

and thus $a \approx 16.9$ ft.

- **Solving triangles given three sides (SSS).**

Exercise: The sides of a triangle have lengths $a = 9$, $b = 10$, and $c = 11$. Find the angles of the triangle.

Answer: We solve $a^2 = b^2 + c^2 - 2bc\cos A$ for $\cos A$:

$2bc\cos A = b^2 + c^2 - a^2 \quad \Leftrightarrow \quad \cos A = \dfrac{b^2 + c^2 - a^2}{2bc}$. In

this case, $\cos A = \dfrac{10^2 + 11^2 - 9^2}{2 \cdot 10 \cdot 11} \approx 0.6364$ and

$\angle A \approx 50.5°$. Similarly, $\cos B = \dfrac{9^2 + 11^2 - 10^2}{2 \cdot 9 \cdot 11} \approx 0.5152$,

so $\angle B \approx 59°$. Then $C \approx 180 - 50.5° - 59° = 70.5°$.

(B) **Finding areas with Heron's Formula.**

The semiperimeter s of a triangle is half its perimeter.

> The area A of a triangle with side lengths a, b, and c is given by **Heron's Formula:**
> $A = \sqrt{s(s-a)(s-b)(s-c)}$, where $s = \frac{1}{2}(a+b+c)$

Exercise: Find the area of the triangle with the given side lengths.

(a) $a = 45$, $b = 28$, $c = 57$

(b) $a = 23$, $b = 37$, $c = 21$

Answer:

The semiperimeter is $s = \frac{1}{2}(45 + 28 + 57) = 65$, so Heron's Formula gives $A = \sqrt{65(65-45)(65-28)(65-57)} = \sqrt{384{,}800} \approx 620.32$

The semiperimeter is $s = \frac{1}{2}(23 + 37 + 21) = 40.5$, so Heron's Formula gives
$A = \sqrt{40.5(40.5-23)(40.5-37)(40.5-21)} \approx 219.94$

Exercise: An irregular lot has the dimensions (in feet) shown in the figure below. Find the area of the lot.

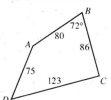

Answer: We draw the diagonal AC and use the Law of Cosines to find its length. Then we find add the areas of the triangles ABC and ADC to find the total area of the lot.

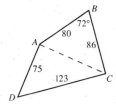

$$|AC|^2 = |AB|^2 + |BC|^2 - 2|AB||BC|\cos B$$
$$= 80^2 + 86^2 - 2 \cdot 80 \cdot 86 \cdot \cos 72°$$
$$\approx 9543.93$$

so $|AC| \approx 97.7$. We can find the area of triangle ABC using the formula from Section 6.3:
$A_1 = \frac{1}{2}ac\sin B = \frac{1}{2}(80)(86)\sin 72° \approx 3271.63$. Now we use Heron's Formula to find the area of triangle ADC. The semiperimeter is $s \approx \frac{1}{2}(75 + 123 + 97.7) = 147.85$, so
$A_2 = \sqrt{147.85(72.85)(24.85)(50.15)} \approx 3663.74$. The total area is approximately
$A_1 + A_2 \approx 3271.63 + 3663.74 = 6935.37$ ft^2.

Core Exercises

1, 25, 27, 41, 51

7 Trigonometric Functions of Real Numbers

7.1 The Unit Circle

Concepts

(A) **The unit circle.**

- Finding terminal points on the unit circle.
- Finding reference numbers.
- Using reference numbers to find terminal points.

Definitions

- **The unit circle:** The circle with radius 1 and center through the origin. Its equation is $x^2 + y^2 = 1$.
- **Terminal point:** If one starts at the point $(1, 0)$ and moves t units around the unit circle, one will end at a point on the circle. This point is called the **terminal point** determined by the real number t.
- **Reference number:** If t is a real number, then the **reference number** $\bar{t}$ is the shortest distance along the unit circle between the terminal point determined by the t and the x-axis. So $0 \leq \bar{t} \leq \frac{\pi}{2}$.

Hints and Tips

- If you are like me, you are looking at this section and thinking, "Why on earth do we care about reference numbers and terminal points?" They are certainly not interesting things on their own merits. But later in this section, we will be discussing trigonometric functions. So far, the functions we've been thinking about have been functions of real numbers — you put a real number in, you get a real number out. Trigonometric functions are often thought of as functions of angles — you put an angle in, you get a real number out. Terminal points and reference numbers help us to make the link between real numbers and points on the unit circle, and later we will be linking points on the unit circle to angles.

Review

(A) **The unit circle.**

- **Finding terminal points on the unit circle.**

The **terminal point** determined by the real number t is the point which lies a distance t along the unit circle, starting at the point $(1, 0)$ and moving in a counterclockwise direction for $t > 0$ (or a clockwise direction for $t < 0$).

Examples:

t	0	$\frac{\pi}{6}$	$\frac{\pi}{4}$	$\frac{\pi}{3}$	$\frac{\pi}{2}$
Terminal point determined by t	$(1, 0)$	$\left(\frac{\sqrt{3}}{2}, \frac{1}{2}\right)$	$\left(\frac{\sqrt{2}}{2}, \frac{\sqrt{2}}{2}\right)$	$\left(\frac{1}{2}, \frac{\sqrt{3}}{2}\right)$	$(0, 1)$

Exercise: Find the terminal point on the unit circle determined by the real number t.

Answer:

(a) $t = \frac{3\pi}{2}$

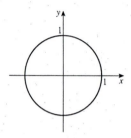

We draw the terminal point determined by $\frac{3\pi}{2}$ on the unit circle. We find that the terminal point is $(0, -1)$.

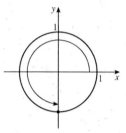

(b) $t = 5\pi$

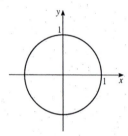

We draw the terminal point determined by 5π on the unit circle. We find that the terminal point is $(-1, 0)$.

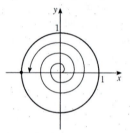

Exercise: Suppose that the terminal point determined by t is $\left(-\frac{5}{13}, \frac{12}{13}\right)$. Find the terminal point determined by

Answer: As in the previous problem, it is helpful to draw t on the unit circle.

(a) $\pi - t$

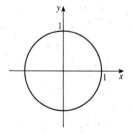

By drawing the point determined by t and the point determined by $\pi - t$ on the same circle, we find that t and $\pi - t$ have the same y-coordinate. The terminal point determined by $\pi - t$ lies in quadrant I, so its coordinates are $\left(\frac{5}{13}, \frac{12}{13}\right)$.

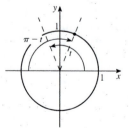

(b) $\pi + t$

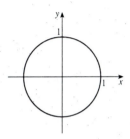

By drawing the point determined by t and the point determined by $\pi + t$ on the same circle, we find that the terminal point determined by $\pi + t$ lies in quadrant IV, so its coordinates are $\left(\frac{5}{13}, -\frac{12}{13}\right)$.

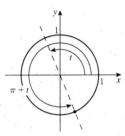

● Finding reference numbers.

With terminal points, we started with a real number and found a corresponding point on the unit circle. With reference numbers, we start with a point on the unit circle, and wind up with a real number.

Exercise: Find the reference number determined by the real number t.

Answer: First plot the relative position of t on the unit circle. Then find the reference number.

(a) $t = -\frac{3\pi}{4}$

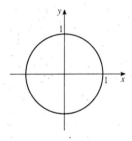

The terminal point determined by t is in quadrant III, and $\bar{t} = \frac{\pi}{4}$.

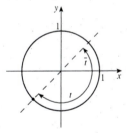

(b) $t = \frac{11\pi}{6}$

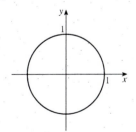

$\bar{t} = \frac{\pi}{6}$

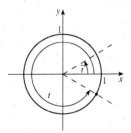

(c) $t = 4$

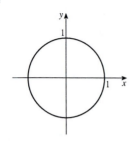

Since $\pi \le t \le \frac{3\pi}{2}, \overline{t} = 4 - \pi \approx 0.858$.

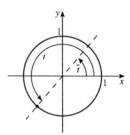

- **Using reference numbers to find terminal points.**

Exercise: Find the terminal point determined by each real number t.

(a) $t = \frac{101\pi}{2}$.

(b) $t = \frac{15\pi}{4}$.

Answer:

The reference number is $\overline{t} = \frac{\pi}{2}$, so the terminal point is $(0, 1)$.

The reference number is $\overline{t} = \frac{\pi}{4}$. The terminal point is in the fourth quadrant; it is $\left(\frac{\sqrt{2}}{2}, -\frac{\sqrt{2}}{2} \right)$.

Core Exercises

15, 19, 31, 41, 55

7.2   Trigonometric Functions of Real Numbers

Concepts

(A) **The six trigonometric functions**

- Evaluating trigonometric functions for special angles.
- Evaluating trigonometric functions using a calculator.
- Using identities to find values of all trigonometric functions given the value of one function.

Definitions

- **sine, cosine, tangent, secant, cotangent, cosecant:** Let t be a real number, and let (x, y) be its corresponding terminal point. Then $\sin t = y$, $\cos t = x$, $\tan t = \dfrac{y}{x}$, $\sec t = \dfrac{1}{x}$, $\cot t = \dfrac{x}{y}$, and $\csc t = \dfrac{1}{y}$.

Hints and Tips

- You can save yourself a lot of memorization if you master a couple of fundamental concepts. For example, I have only the first of the three Pythagorean identities memorized: $\sin^2 x + \cos^2 x = 1$. If I need another one, which I often do, I just divide both sides of the equation by $\sin^2 x$ or $\cos^2 x$, as done in the text. For example, if we divide both sides by $\sin^2 x$ we get $1 + \cot^2 x = \csc^2 x$.

- It is not really necessary to memorize phrases like All Students Take Calculus, Aunt Sally Tips Cows, etc. You are going to have to learn the definitions of sine, cosine and tangent anyway. If you have an angle in, say, quadrant II, its x-coordinate is negative and its y-coordinate is positive. So if you understand the definitions of sine, cosine and tangent, you will be able to use your coconut to tell you that sine is positive, cosine is negative, and tangent (sine over cosine) is negative.

Review

(A) **The six trigonometric functions**

- **Evaluating trigonometric functions for special angles.**

Exercise: The terminal point $P(x, y)$ determined by t is $\left(\frac{\sqrt{3}}{3}, -\frac{\sqrt{6}}{3}\right)$. Find all six trigonometric functions.

Answer: $\sin t = -\frac{\sqrt{6}}{3}, \cos t = \frac{\sqrt{3}}{3},$

$$\tan t = \frac{-\sqrt{6}/3}{\sqrt{3}/3} = -\frac{\sqrt{6}}{\sqrt{3}} = \sqrt{2}, \sec t = \frac{1}{\sqrt{3}/3} = \frac{3}{\sqrt{3}} = \sqrt{3},$$

$$\csc t = \frac{1}{-\sqrt{6}/3} = -\frac{3}{\sqrt{6}} = -\frac{\sqrt{6}}{2},$$

$$\cot t = \frac{\sqrt{3}/3}{-\sqrt{6}/3} = -\frac{\sqrt{3}}{\sqrt{6}} = -\frac{\sqrt{2}}{2}$$

Exercise: The terminal point $P(x, y)$ determined by t is $\left(\frac{2}{5}, \frac{\sqrt{21}}{5}\right)$. Find all six trigonometric functions.

Answer: $\sin t = \frac{\sqrt{21}}{5}, \cos t = \frac{2}{5}, \tan t = \frac{\sqrt{21}/5}{2/5} = \frac{\sqrt{21}}{2},$

$$\sec t = \frac{1}{2/5} = \frac{5}{2}, \csc t = \frac{1}{\sqrt{21}/5} = \frac{5}{\sqrt{21}} = \frac{5\sqrt{21}}{21},$$

$$\cot t = \frac{2/5}{\sqrt{21}/5} = \frac{2}{\sqrt{21}} = \frac{2\sqrt{21}}{21}$$

Exercise: Find all six trigonometric functions of $t = \frac{11\pi}{3}$.

Answer: The reference number is $\bar{t} = \frac{\pi}{3}$ and the terminal point is in quadrant IV, so $\sin t = -\frac{\sqrt{3}}{2}, \cos t = \frac{1}{2},$

$$\tan t = -\frac{\sqrt{3}/2}{1/2} = -\sqrt{3}, \csc t = \frac{1}{\sqrt{3}/2} = \frac{2\sqrt{3}}{3},$$

$$\sec t = \frac{1}{1/2} = 2, \cot t = -\frac{1}{\sqrt{3}} = -\frac{\sqrt{3}}{3}.$$

- **Evaluating trigonometric functions using a calculator.**
  You will notice your calculator has buttons for only the sine, cosine, and tangent functions. So approximating $\sin x$, $\cos x$, and $\tan x$ is easy with a calculator. The only tricky part is that *you must make sure that your calculator is in radian mode.* Test it out by finding sin 90. Your calculator should display $0.8939966\ldots$. If it displays 1, you are in degree mode, and need to change modes.

  Note also that the "$\sin^{-1}$" key does not give you $\dfrac{1}{\sin x}$. It is a different function called arcsine that will be covered later. To evaluate the other functions, use their definitions.

Exercise: Use a calculator to approximate the values of all six trigonometric functions.

Answer:

(a) $t = 1.5$

$\sin 1.5 \approx 0.99749$, $\cos 1.5 \approx 0.07074$, and $\tan 1.5 \approx 14.101$. Now we use the reciprocal identities:

$$\csc 1.5 = \frac{1}{\sin 1.5} \approx 1.00251, \ \sec 1.5 = \frac{1}{\cos 1.5} \approx 14.13683,$$

and $\cot 1.5 = \dfrac{1}{\tan 1.5} \approx 0.07091$.

(b) $t = 200$

$\sin 200 \approx -0.87330$, $\cos 200 \approx 0.48719$, and $\tan 200 \approx -1.79253$. Now we use the reciprocal identities:

$$\csc 200 = \frac{1}{\sin 1.5} \approx -1.14509,$$

$$\sec 200 = \frac{1}{\cos 1.5} \approx 2.05260, \text{ and}$$

$$\cot 200 = \frac{1}{\tan 1.5} \approx -0.55787.$$

- **Using identities to find values of all trigonometric functions given the value of one function.**

  The trigonometric functions are related to one another through equations called **trigonometric identities**. An important class of identities is the **reciprocal identities:**

$$\csc t = \frac{1}{\sin t} \qquad \sec t = \frac{1}{\cos t} \qquad \tan t = \frac{\sin t}{\cos t} = \frac{1}{\cot t} \qquad \cot t = \frac{\cos t}{\sin t} = \frac{1}{\tan t}$$

Another important class is the **Pythagorean identities:**

$$\sin^2 t + \cos^2 t = 1 \qquad \tan^2 t + 1 = \sec^2 t \qquad 1 + \cot^2 t = \csc^2 t$$

Note the usual convention of writing $\sin^n t$ for $(\sin t)^n$ for all integers n *except* $n = -1$ (see above; $\sin^{-1} t$ is a different function entirely).

Exercise: Find the values of the trigonometric functions of t.

Answer:

(a) $\cos t = \frac{7}{8}$ and the terminal point of t is in quadrant IV.

We find the easy one first: $\sec t = \dfrac{1}{\cos t} = \dfrac{8}{7}$. Next we find

sine: t is in quadrant IV, so $\sin t < 0$ and $\sin^2 t + \cos^2 t = 1$

$\Leftrightarrow \quad \sin^2 t + \left(\frac{7}{8}\right)^2 = 1 \quad \Leftrightarrow \quad \sin^2 t = 1 - \frac{49}{64} = \frac{15}{64} \quad \Leftrightarrow$

$\sin t = -\dfrac{\sqrt{15}}{8}$. Then $\csc t = \dfrac{1}{\sin t} = -\dfrac{8}{\sqrt{15}} = -\dfrac{8\sqrt{15}}{15}$,

$\tan t = \dfrac{\sin t}{\cos t} = \dfrac{-\sqrt{15}/8}{7/8} = -\dfrac{\sqrt{15}}{7}$, and

$\cot t = \dfrac{1}{\tan t} = -\dfrac{7}{\sqrt{15}} = -\dfrac{7\sqrt{15}}{15}$.

(b) $\csc t = \frac{25}{7}$ and the terminal point of t is in quadrant II.

We find the easy one first: $\sin t = \dfrac{1}{\csc t} = \frac{7}{25}$. Next we find cosine: t is in quadrant II, so $\cos t < 0$ and $\sin^2 t + \cos^2 t = 1$

$\Leftrightarrow \quad \left(\frac{7}{25}\right)^2 + \cos^2 t = 1 \quad \Leftrightarrow \quad \cos^2 t = 1 - \frac{49}{625} = \frac{576}{625}$

$\Leftrightarrow \quad \cos t = -\frac{24}{25},\ \sec t = \dfrac{1}{\cos t} = -\frac{25}{24},$

$\tan t = \dfrac{\sin t}{\cos t} = \dfrac{7/25}{-24/25} = -\frac{7}{24},$ and $\cot t = \dfrac{1}{\tan t} = -\frac{24}{7}.$

Core Exercises

1, 17, 47, 73, 79, 81

7.3 Trigonometric Graphs

Concepts

(A) **Graphs of the sine and cosine functions.**

- Sketching transformed sine and cosine functions.
- Determining the amplitude, period, and phase shift of a sine or cosine curve.
- Using a calculator to graph sine and cosine functions.

Definitions

- **Periodic function:** A function is periodic if its values repeat in a cycle. For example, the function graphed below is periodic.

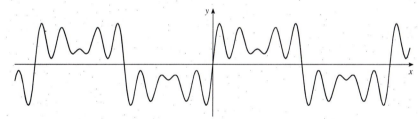

- **Period:** If a function is periodic, then its period is the smallest positive number p such that $f(x+p) = f(x)$. For example, the period of the function graphed below is 2.

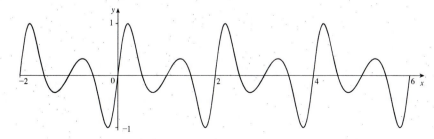

- **Amplitude of a sine or cosine curve:** The maximum displacement of a sine or cosine curve from its average value.

Questions to Ask Your Teacher

- Is it acceptable to write $y = \sin(3x + 6)$, or is it always preferred to write $y = \sin(3(x + 2))$?

Hints and Tips

- Notice that the ideas of transforming sine and cosine graphs are not all that different from the rules discussed in Section 2.5. We are applying the same rules to a new set of functions.

Review

(A) **Graphs of the sine and cosine functions.**

Sine and cosine each have a period of 2π. The graph of the sine function is symmetric with respect to the origin (sine is odd), and the graph of the cosine function is symmetric with respect to the y-axis (cosine is even).

- **Determining the amplitude, period, and phase shift of a sine or cosine curve.**

The graphs of functions of the form $y = a \sin k(x - b)$ and $y = a \cos k(x - b)$ are simply sine and cosine curves that have been transformed. The number $|a|$ is called the **amplitude** and is the largest value attained by the function. The number $k > 0$ causes the **period** to be $2\pi/k$, and b is the **phase shift**: for $b > 0$, the graph is shifted to the right, and for $b < 0$ the graph is shifted to the left.

Exercise: Find the amplitude, period, and phase shift of $y = 3 \sin \pi x$.

|| **Answer:** Amplitude 3, period 2, phase shift 0.

- **Sketching transformed sine and cosine functions.**

Exercise: Sketch the graph of the given function.

|| **Answer:**

(a) $f(x) = 2 \sin x$

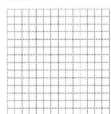

|| The graph of f is obtained by stretching the graph of $\sin x$ vertically by a factor of 2.

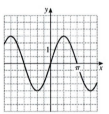

(b) $g(x) = 3 - \cos x$

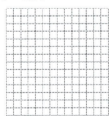

|| The graph of g is obtained by reflecting the graph of $\cos x$ about the x-axis, then shifting the resulting graph up 3 units.

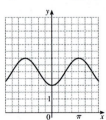

Exercise: Find the amplitude, period, and phase shift of $y = 3\sin(3x - \pi)$, and sketch its graph.

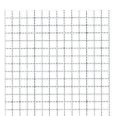

Answer: $y = 3\sin(3x - \pi) = 3\sin 3\left(x - \frac{\pi}{3}\right)$ has amplitude 3, period $\frac{2\pi}{3}$, and phase shift $\frac{\pi}{3}$.

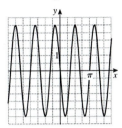

Exercise: Find the amplitude, period, and phase shift of $y = 3\cos\left(x + \frac{\pi}{2}\right)$, and sketch its graph.

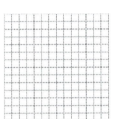

Answer: $y = 3\cos\left(x + \frac{\pi}{2}\right)$ has amplitude 3, period 2π, and phase shift $-\frac{\pi}{2}$.

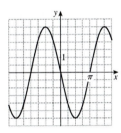

- **Using a calculator to graph sine and cosine functions.**

As stated before, choosing an appropriate viewing rectangle is one of the most important aspects of using a graphing calculator. In graphing periodic functions, a viewing rectangle should generally be only a few periods wide.

Exercise: Use a graphing calculator to draw the graph of $f(x) = \sin 30x$ in the given viewing rectangle.

Answer:

(a) $[-5, 5]$ by $[-2, 2]$

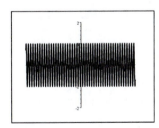

(b) $[-0.2, 0.2]$ by $[-2, 2]$

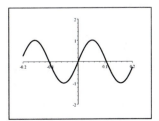

Exercise: Graph the functions
$y = \sqrt{x^2 + 1}$, $y = -\sqrt{x^2 + 1}$, and
$y = \sqrt{x^2 + 1}\sin x$ on a common screen.
How are the graphs related?

Answer:

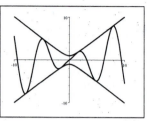

The graph of $y = \sqrt{x^2 + 1}\sin x$ oscillates between the values of the other two functions.

Exercise: Let $f(x) = \dfrac{1}{\sin^2 x + 1}$.

(a) Use a graphing device to graph f.

Answer:

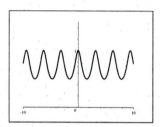

(b) Determine from the graph whether the function is periodic and, if so, determine the period.

Yes, this function is periodic. Its period is π.

⬤ Core Exercises

3, 31, 49, 69

7.4 More Trigonometric Graphs

⬤ Concepts

(A) **Graphs involving tangents and cotangents.**
  • Graphing transformed tangent and cotangent curves.

(B) **Graphs involving secants and cosecants.**
  • Graphing transformed secant and cosecant curves.

⬤ Hints and Tips

• The trigonometric curves you will see most often in calculus will be sine or cosine curves, and they have period 2π. The cosecant and secant functions are reciprocals of the sine and cosine functions, so they also have period 2π. Not too hard to remember so far, right? The only other two trigonometric functions are tangent and cotangent, and they have period π.

◤ Review

(A) **Graphs involving tangents and cotangents.**

Recall that the functions $\tan t$ and $\sec t$ are not defined at $t = \frac{\pi}{2} + n\pi$, n an integer; and the functions $\cot t$ and $\csc t$ are not defined at $t = n\pi$, n an integer.

- **Graphing transformed tangent and cotangent curves.**

Functions of the form $y = a \tan k\,(x - b)$ and $y = a \cot k\,(x - b)$ are transformations of the tangent and cotangent functions. As in Section 5.3, b is the phase shift; however, now the period is π/k. To sketch a complete period, it is convenient to select an interval between vertical asymptotes.

Exercise: Find the phase shift and period of the function, and sketch at least 3 periods of its graph

Answer:

(a) $y = \frac{1}{4} \cot x$

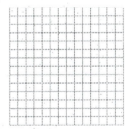

Phase shift 0, period π.

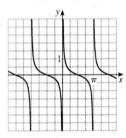

(b) $y = \tan 3x$

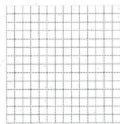

Phase shift 0, period $\frac{\pi}{3}$.

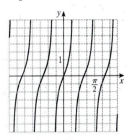

Ⓑ **Graphs involving secants and cosecants.**

- **Graphing transformed secant and cosecant curves.**

Functions of the form $y = a \sec k(x - b)$ and $y = a \csc k(x - b)$ are transformations of the secant and cosecant functions. Here b is the phase shift and the period is $2\pi/k$. As with other trigonometric functions, it is a good idea to determine the phase shift and period before sketching the graph.

Exercise: Find the phase shift and period of the function, and sketch at least one period of its graph.

Answer:

(a) $y = \frac{1}{4} \csc\left(\frac{1}{2}x\right)$

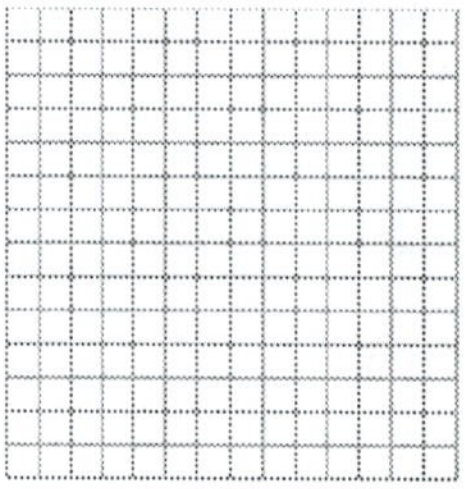

Phase shift 0, period $\dfrac{2\pi}{1/2} = 4\pi$.

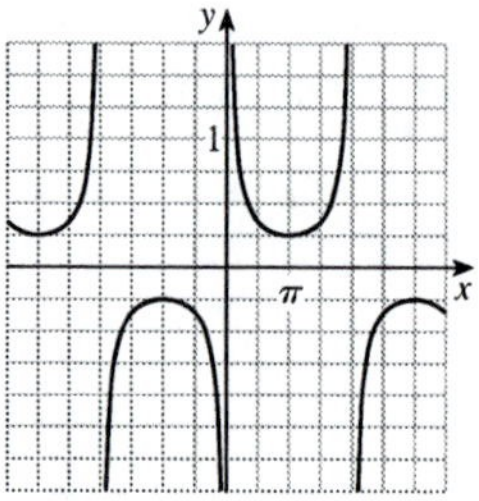

(b) $y = 3 \sec\left(x + \frac{\pi}{3}\right)$

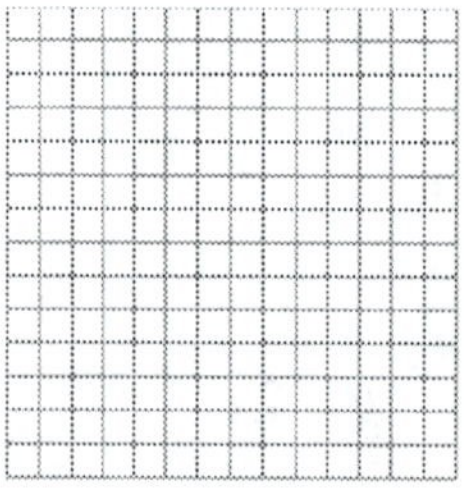

Phase shift $-\frac{\pi}{3}$, period 2π.

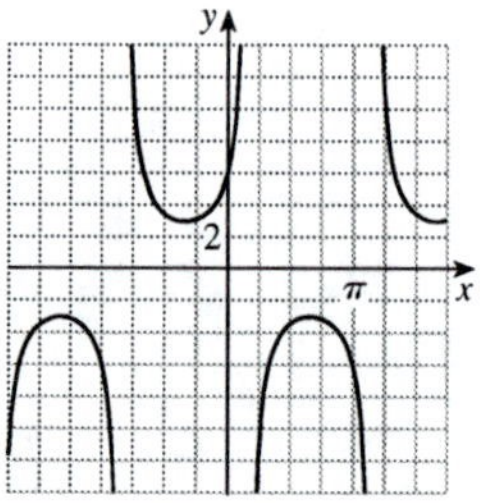

Core Exercises

11, 39, 57

7.5 Modeling Harmonic Motion

Concepts

Ⓐ **Models involving harmonic motion.**

- Working with simple harmonic motion.
- Working with damped harmonic motion.

Definitions

- **Simple harmonic motion:** Motion that can be described by $y = a \sin \omega t$ or $a \cos \omega t$. The curves discussed in Section 5.3 can be interpreted as curves illustrating simple harmonic motion.
- **Damped harmonic motion:** Motion that oscillates in the same way as simple harmonic motion, except that the motion dies out exponentially.
- **Period:** In the context of simple harmonic motion, the period of the motion is the time to complete one cycle. Notice that this is compatible with our earlier definition of the word.
- **Frequency:** The number of cycles per unit time. The frequency is the reciprocal of the period.

Hints and Tips

- Once you understand simple and damped harmonic motion, you will be surprised by how many examples you see in your daily life. A swinging door that swings back and forth a few times before coming to rest, your head bobbing up and down as you walk, your cat running to its food dish, pausing, running to you to touch your knee with its head, running back to the food dish — all of these phenomena can be modeled by simple or damped harmonic motion. Enjoy noticing these things; it is fun when your course work helps you to see reality in a new way!

Review

(A) **Models involving harmonic motion.**

- **Working with simple harmonic motion.**

  When we model with simple harmonic motion we use $y = a \sin \omega t$ or $a \cos \omega t$. The **amplitude** $|a|$ of the motion is the maximum displacement of the object. The **period** $\dfrac{2\pi}{\omega}$ is the time required to complete one cycle. The **frequency** $\dfrac{\omega}{2\pi}$ is the number of cycles per unit of time. If the maximum variation occurs at time 0, we use $y = a \cos \omega t$; if the moment of least variation occurs at time 0, we use $y = a \sin \omega t$; otherwise we use a horizontal shift, such as $y = a \sin (\omega (t - c))$.

Exercise: Basho the cat runs back and forth between you and his food dish, and his distance from the halfway point is given by $y = 5 \sin \left(\frac{\pi}{4} t \right)$, where t is measured in seconds and y in feet.	**Answer:**
(a) What is the distance between you and Basho's food bowl?	The amplitude is 5 feet, so the distance between you and the halfway point is 5 feet and the distance between you and the food bowl is 10 feet.
(b) How long does it take for Basho to get from you to his food bowl and back?	This time is equal to one period, namely $\frac{2\pi}{\pi/4} = 8$ seconds.
(c) At time $t = 0$, where is Basho? After one minute, is he closer to you or to the food bowl?	At $t = 0$, we have $y = 0$, so Basho is at the halfway point. At $t = 60$, we have $y = 0$ and Basho is once again at the halfway point.

- **Working with damped harmonic motion.**

The height of a swinging pendulum can be modelled by simple harmonic motion, if we ignore the effects of friction. If we do not ignore these effects, the pendulum will eventually stop, and we have to add a bit of complexity to our model. A model of the form $y = ke^{-ct} \sin \omega t$ or $y = ke^{-ct} \cos \omega t$ is called **damped harmonic motion**. A typical graph of such a model looks like this:

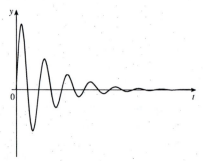

Exercise: An object rocks with damped harmonic motion. The following is a graph of its motion versus time. Find the damping constant c.

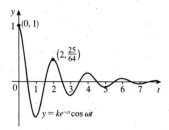

Answer: From the labeled points, we know that $1 = ke^{-c(0)}$ and $\frac{25}{64} = ke^{-c(2)}$. From the first equation, we have $k = 1$, so we solve the second equation $\frac{25}{64} = e^{-2c}$ for c:

$$\ln \frac{25}{64} = \ln \left(e^{-2c}\right) \quad \Leftrightarrow \quad \ln \frac{25}{64} = -2c \quad \Leftrightarrow$$

$$c = -\frac{1}{2} \ln \frac{25}{64} \approx 0.47.$$

▰ Core Exercises

11, 15, 17, 27

8   Analytic Trigonometry

8.1   Trigonometric Identities

⬛ Concepts

(A) Using trigonometric identities.
- Simplifying a trigonometric expression.

(B) Proving trigonometric identities.
- Proving a given trigonometric identity.

⬛ Definitions

- **Identity:** An identity is an equation which is true for all values of variables for which both sides are defined. A simple identity is $x + x = 2x$.

⬛ Questions to Ask Your Teacher

- Which identities should be memorized?

⬛ Review

(A) Using trigonometric identities.

Identities enable us to write the same expression in different ways. **Trigonometric identities** are equations which are true for all values of variables for which both sides are defined. Some of the important classes of trigonometric identities are shown below.

Reciprocal Identities	$\csc t = \dfrac{1}{\sin t} \qquad \sec t = \dfrac{1}{\cos t} \qquad \cot t = \dfrac{1}{\tan t}$ $\tan t = \dfrac{\sin t}{\cos t} \qquad \cot t = \dfrac{\cos t}{\sin t}$
Pythagorean Identities	$\sin^2 t + \cos^2 t = 1 \qquad \tan^2 t + 1 = \sec^2 t$ $1 + \cot^2 t = \csc^2 t$
Even-Odd Identities	$\sin(-t) = -\sin t \qquad \cos(-t) = \cos t$ $\tan(-t) = -\tan t \qquad \csc(-t) = -\csc t$ $\sec(-t) = \sec t \qquad \cot(-t) = -\cot t$
Cofunction Identities	$\sin\left(\frac{\pi}{2} - t\right) = \cos t \qquad \cos\left(\frac{\pi}{2} - t\right) = \sin t$ $\tan\left(\frac{\pi}{2} - t\right) = \cot t \qquad \cot\left(\frac{\pi}{2} - t\right) = \tan t$ $\sec\left(\frac{\pi}{2} - t\right) = \csc t \qquad \csc\left(\frac{\pi}{2} - t\right) = \sec t$

• **Simplifying a trigonometric expression.**

Exercise: Write the expression
$\sin \theta \sec \theta + \cos \theta \csc \theta$ in terms of sine and cosine, and then simplify.

Answer:

$$\sin \theta \sec \theta + \cos \theta \csc \theta = \sin \theta \frac{1}{\cos \theta} + \cos \theta \frac{1}{\sin \theta}$$
$$= \frac{\sin \theta}{\cos \theta} \cdot \frac{\sin \theta}{\sin \theta} + \frac{\cos \theta}{\sin \theta} \cdot \frac{\cos \theta}{\cos \theta}$$
$$= \frac{\sin^2 \theta}{\cos \theta \sin \theta} + \frac{\cos^2 \theta}{\sin \theta \cos \theta}$$
$$= \frac{\sin^2 \theta + \cos^2 \theta}{\cos \theta \sin \theta} = \frac{1}{\cos \theta \sin \theta}$$
$$= \sec \theta \csc \theta$$

Exercise: Write the expression
$$\frac{1}{\cot x - \csc x}$$
in terms of sine and cosine, and then simplify.

Answer:
$$\frac{1}{\cot x - \csc x} = \frac{1}{\frac{\cos x}{\sin x} - \frac{1}{\sin x}}$$
$$= \frac{1}{\frac{\cos x - 1}{\sin x}} = \frac{\sin x}{\cos x - 1}$$

Exercise: Simplify
$\left(\tan^2 \alpha - 1 \right) \left(\sec^2 \alpha + 1 \right)$.

Answer: $\left(\tan^2 \alpha - 1 \right) \left(\sec^2 \alpha + 1 \right)$
$$= \tan^2 \alpha \sec^2 \alpha + \tan^2 \alpha - \sec^2 \alpha - 1$$
$$= \tan^2 \alpha \sec^2 \alpha + \tan^2 \alpha - \left(\tan^2 \alpha + 1 \right) - 1$$
$$= \tan^2 \alpha \sec^2 \alpha - 2$$

B **Proving trigonometric identities.**

To show that an equation is not an identity, all you need to do is show that the equation is false for some choice of the variable(s). This choice is called a **counterexample**. To prove that an equation *is* an identity, however, we must show that it is true for *every* choice of the variable(s). To do this, we start with one side of the equation and transform it into the other side. At each step, a trigonometric or algebraic identity is used. If you get stuck, start by rewriting all functions in terms of sine and cosine.

In general, it is easier to transform the more complicated side of the equation into the simpler side.

It is very important to understand that when proving a trigonometric identity, we are not solving an equation, so we do not perform the same operation on both sides of the equation.

• **Proving a given trigonometric identity.**

Exercise: Prove the identity
$\cos \left(\frac{\pi}{4} + x \right) = \sin \left(\frac{\pi}{4} - x \right)$.

Answer: We use the cofunction identity for cosine:
$\cos \left(\frac{\pi}{4} + x \right) = \sin \left(\frac{\pi}{2} - \left(\frac{\pi}{4} + x \right) \right) = \sin \left(\frac{\pi}{4} - x \right)$.

Exercise: Prove the identity.

(a) $\sin w + \cos w = \dfrac{1 + \cot w}{\csc w}$

Answer:

Since the right-hand side is more complicated, we transform it into the left-hand side. First, we express it in terms of sine and cosine:

$$\frac{1 + \cot w}{\csc w} = \frac{1 + \dfrac{\cos w}{\sin w}}{\dfrac{1}{\sin w}} = \frac{1 + \dfrac{\cos w}{\sin w}}{\dfrac{1}{\sin w}} \cdot \frac{\sin w}{\sin w}$$

$$= \frac{\sin w + \cos w}{1} = \sin w + \cos w$$

(b) $\csc^4 \gamma - \cot^4 \gamma = \csc^2 \gamma + \cot^2 \gamma$

We transform the left-hand side into the right-hand side. At the second line, we use the fact that $1 + \cot^2 \gamma = \csc^2 \gamma$ $\Leftrightarrow$ $\csc^2 \gamma - \cot^2 \gamma = 1$:

$$\csc^4 \gamma - \cot^4 \gamma = \left(\csc^2 \gamma - \cot^2 \gamma\right)\left(\csc^2 \gamma + \cot^2 \gamma\right)$$

$$= 1\left(\csc^2 \gamma + \cot^2 \gamma\right) = \csc^2 \gamma + \cot^2 \gamma$$

(c) $\sec y - \cos y = \sin y \tan y$

We transform the left-hand side into the right-hand side:

$$\sec y - \cos y = \frac{1}{\cos y} - \cos y = \frac{1}{\cos y} - \cos y\left(\frac{\cos y}{\cos y}\right)$$

$$= \frac{1 - \cos^2 y}{\cos y} = \frac{\sin^2 y}{\cos y} = \sin y \cdot \frac{\sin y}{\cos y}$$

$$= \sin y \tan y$$

Exercise: Show that the equation is not an identity.

(a) $\dfrac{\cos \theta + 1}{\sin \theta} = \dfrac{\sin \theta}{\cos \theta - 1}$

Answer:

We show that $\theta = \frac{\pi}{3}$ gives a counterexample:

$$\frac{\cos \frac{\pi}{3} + 1}{\sin \frac{\pi}{3}} = \frac{\frac{1}{2} + 1}{\frac{\sqrt{3}}{2}} = \sqrt{3}, \text{ but } \frac{\sin \frac{\pi}{3}}{\cos \frac{\pi}{3} - 1} = \frac{\frac{\sqrt{3}}{2}}{\frac{1}{2} - 1} = -\sqrt{3}.$$

(b) $\dfrac{\sin \gamma \cos \gamma}{\sin \gamma + \cos \gamma} = \cot \gamma + \tan \gamma$

$\gamma = \frac{3\pi}{4}$ gives a counterexample:

$$\frac{\sin \frac{3\pi}{4} \cos \frac{3\pi}{4}}{\sin \frac{3\pi}{4} + \cos \frac{3\pi}{4}} = \frac{\frac{\sqrt{2}}{2} \cdot \left(-\frac{\sqrt{2}}{2}\right)}{\frac{\sqrt{2}}{2} - \frac{\sqrt{2}}{2}} \text{ which is not defined, but}$$

$$\cot \frac{3\pi}{4} + \tan \frac{3\pi}{4} = -1 - 1 = -2.$$

▰ Core Exercises

3, 67, 101

8.2   Addition and Subtraction Formulas

Concepts

(A) Addition and subtraction formulas.

- Using the addition and subtraction formulas to calculate sines and cosines.
- Using the addition and subtraction formulas to prove identities.
- Using the addition formula for sine to simplify $A \sin x + B \cos x$.

Hints and Tips

- If you already know that $\sin(-t) = -\sin t$, $\cos(-t) = \cos t$, and $\tan(-t) = -\tan t$, then you don't need to memorize the subtraction formulas — you can derive them from the addition formulas by (for example) writing $\sin(s - t)$ as $\sin(s + (-t))$ and then using an addition formula.

Review

(A) Addition and subtraction formulas.

The identities for sine, cosine, and tangent of sums and differences are shown in the table below. These identities are used to increase the number of angles for which we know the exact values of the trigonometric function. The identities are also very useful in calculus.

$$\sin(s + t) = \sin s \cos t + \cos s \sin t \qquad \sin(s - t) = \sin s \cos t - \cos s \sin t$$

$$\cos(s + t) = \cos s \cos t - \sin s \sin t \qquad \cos(s - t) = \cos s \cos t + \sin s \sin t$$

$$\tan(s + t) = \frac{\tan s + \tan t}{1 - \tan s \tan t} \qquad \tan(s - t) = \frac{\tan s - \tan t}{1 + \tan s \tan t}$$

- **Using the addition and subtraction formulas to calculate sines and cosines.**

Exercise: Find the exact value.

(a) $\sin 75°$

Answer:

We must first express $75°$ in terms of angles for which we know the exact values of sine and cosine. Since $75° = 45° + 30°$ and we know the exact values for these two angles, we calculate

$$\sin 75° = \sin(45° + 30°)$$
$$= \sin 45° \cos 30° + \cos 45° \sin 30°$$
$$= \frac{\sqrt{2}}{2} \cdot \frac{\sqrt{3}}{2} + \frac{\sqrt{2}}{2} \cdot \frac{1}{2} = \frac{\sqrt{6} + \sqrt{2}}{4}$$

(b) $\cos 15°$

$15° = 45° - 30°$ and we know the exact values of sine and cosine for these angles, so

$$\cos 15° = \cos(45° - 30°)$$
$$= \cos 45° \cos 30° + \sin 45° \sin 30°$$
$$= \frac{\sqrt{2}}{2} \cdot \frac{\sqrt{3}}{2} + \frac{\sqrt{2}}{2} \cdot \frac{1}{2} = \frac{\sqrt{6}+\sqrt{2}}{4}$$

We can verify that $\sin 75° = \cos 15°$ using one of the cofunction identities in Section 8.1.

Exercise: Express $\sin 3x$ in terms of $\sin x$ and $\cos x$.

Answer: We first apply the sine addition formula with $3x = 2x + x$. Then we apply both the sine and cosine addition formulas with $2x = x + x$:

$$\sin 3x = \sin(2x + x) = \sin 2x \cos x + \cos 2x \sin x$$
$$= (\sin x \cos x + \cos x \sin x)\cos x$$
$$+ (\cos x \cos x - \sin x \sin x)\sin x$$
$$= 2\sin x \cos^2 x + \sin x \cos^2 x - \sin^3 x$$
$$= 3\sin x \cos^2 x - \sin^3 x$$

Exercise: If α and β are acute angles such that $\cos \alpha = \frac{4}{5}$ and $\sin \beta = \frac{5}{13}$, find $\sin(\alpha + \beta)$, $\sin(\alpha - \beta)$, $\cos(\alpha + \beta)$, and $\cos(\alpha - \beta)$.

Answer: We first need to determine $\sin \alpha$ and $\cos \beta$. Since α and β are acute, we draw diagrams of right triangles and calculate the other quantities:

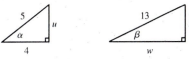

$$u^2 + 4^2 = 5^2 \quad \Leftrightarrow \quad u^2 = 9 \quad \Rightarrow \quad u = 3, \text{ so } \sin \alpha = \tfrac{3}{5}.$$
$$w^2 + 5^2 = 13^2 \quad \Leftrightarrow \quad w^2 = 144 \quad \Leftrightarrow \quad w = 12, \text{ so}$$
$\cos \beta = \frac{12}{13}$. Thus,

$$\sin(\alpha + \beta) = \sin \alpha \cos \beta + \cos \alpha \sin \beta$$
$$= \tfrac{3}{5} \cdot \tfrac{12}{13} + \tfrac{4}{5} \cdot \tfrac{5}{13} = \tfrac{36}{65} + \tfrac{20}{65} = \tfrac{56}{65}$$

$$\sin(\alpha - \beta) = \sin \alpha \cos \beta - \cos \alpha \sin \beta$$
$$= \tfrac{3}{5} \cdot \tfrac{12}{13} - \tfrac{4}{5} \cdot \tfrac{5}{13} = \tfrac{36}{65} - \tfrac{20}{65} = \tfrac{16}{65}$$

$$\cos(\alpha + \beta) = \cos \alpha \cos \beta - \sin \alpha \sin \beta$$
$$= \tfrac{4}{5} \cdot \tfrac{12}{13} - \tfrac{3}{5} \cdot \tfrac{5}{13} = \tfrac{48}{65} - \tfrac{15}{65} = \tfrac{33}{65}$$

$$\cos(\alpha - \beta) = \cos \alpha \cos \beta + \sin \alpha \sin \beta$$
$$= \tfrac{4}{5} \cdot \tfrac{12}{13} + \tfrac{3}{5} \cdot \tfrac{5}{13} = \tfrac{48}{65} + \tfrac{15}{65} = \tfrac{63}{65}$$

- **Using the addition and subtraction formulas to prove identities.**

Exercise: Prove the identity

$$\frac{\sin (x + y)}{\cos x \cos y} = \tan x + \tan y.$$

Answer: Expand the left-hand side using the addition formula for sine:

$$\frac{\sin (x + y)}{\cos x \cos y} = \frac{\sin x \cos y + \cos x \sin y}{\cos x \cos y}$$
$$= \frac{\sin x \cos y}{\cos x \cos y} + \frac{\cos x \sin y}{\cos x \cos y}$$
$$= \frac{\sin x}{\cos x} + \frac{\sin y}{\cos y} = \tan x + \tan y$$

Exercise: Verify that the period of tangent is π by applying an addition formula to $\tan (x + \pi)$.

Answer: Applying the addition formula for tangent,

$$\tan (x + \pi) = \frac{\tan x + \tan \pi}{1 - \tan x \tan \pi} = \frac{\tan x + 0}{1 - \tan x \cdot 0} = \tan x.$$

- **Using the addition formula for sine to simplify $A \sin x + B \cos x$.**

  We can use the addition formula for sine to derive a theorem that shows how expressions of the form $A \sin x + B \cos x$ can be expressed as a single trigonometric function:

  $$A \sin x + B \cos x = k \sin (x + \phi)$$

  where $k = \sqrt{A^2 + B^2}$ and ϕ satisfies $\cos \phi = \dfrac{A}{\sqrt{A^2 + B^2}}$ and $\sin \phi = \dfrac{B}{\sqrt{A^2 + B^2}}$

Exercise: Write $\frac{\sqrt{3}}{3} \sin x + \cos x$ in terms of sine only.

Answer: Here $A = \frac{\sqrt{3}}{3}$ and $B = 1$, so

$$k = \sqrt{\left(\frac{\sqrt{3}}{3}\right)^2 + 1^2} = \sqrt{\frac{4}{3}} = \frac{2\sqrt{3}}{3}. \text{ Thus, } \cos \phi = \frac{\frac{\sqrt{3}}{3}}{\frac{2\sqrt{3}}{3}} = \frac{1}{2}$$

and $\sin \phi = \dfrac{1}{\frac{2\sqrt{3}}{3}} = \frac{\sqrt{3}}{2}$. Therefore $\phi = \frac{\pi}{3}$ and

$$\frac{\sqrt{3}}{3} \sin x + \cos x = \frac{2\sqrt{3}}{3} \sin \left(x + \frac{\pi}{3}\right).$$

Exercise: Write $-5 \sin 2x + 5 \cos 2x$ in terms of sine only.

Answer: Here $A = -5$ and $B = 5$, so

$$k = \sqrt{(-5^2) + 5^2} = 5\sqrt{2}. \; \cos \phi = \frac{-5}{5\sqrt{2}} = -\frac{\sqrt{2}}{2} \text{ and}$$

$\sin \phi = \frac{5}{5\sqrt{2}} = \frac{\sqrt{2}}{2}$, so $\phi = \frac{3\pi}{4}$ and

$$-5 \sin 2x + 5 \cos 2x = 5\sqrt{2} \sin \left(2x + \frac{3\pi}{4}\right).$$

◢ Core Exercises

5, 13, 23, 43, 49, 57

8.3   Double-Angle, Half-Angle, and Product-Sum Formulas

▧ Concepts

(A) Double-angle formulas.

- Using the double-angle formulas.

(B) Half-angle formulas.

- Using half-angle formulas to lower powers in a trigonometric expression.

(C) Product-to-sum and sum-to-product formulas.

- Expressing a trigonometric product as a sum and vice versa.

▧ Hints and Tips

- I find myself using some of these formulas more than others. In second semester calculus, I use half-angle formulas for lowering powers of sine and cosine often, and the double-angle formula for sine tends to pop up in the most unlikely of places. If you are going to be taking calculus, these are the most important to retain, although all of the formulas can be useful.

▧ Review

(A) Double-angle formulas.

The double-angle formulas are special cases of the addition formulas. They allow us to express the values of trigonometric functions of $2x$ in terms of trigonometric functions of x.

$$\sin 2x = 2 \sin x \cos x$$

$$\cos 2x = \cos^2 x - \sin^2 x = 1 - \sin^2 x = 2 \cos^2 x - 1$$

$$\tan 2x = \frac{2 \tan x}{1 - \tan^2 x}$$

Notice that there are three variations of the formula for cosine.

- **Using the double-angle formulas.**

Exercise: Find $\sin 2x$, $\cos 2x$ and $\tan 2x$ from the given information.

Answer:

(a) $\cos x = -\frac{12}{13}$ and x is in quadrant III.

First find $\sin x$. Since x is in quadrant III, $\sin x < 0$, so applying the Pythagorean identity,

$$\sin x = -\sqrt{1 - \cos^2 x} = \sqrt{1 - \left(-\frac{12}{13}\right)^2} = -\sqrt{\frac{25}{169}} = -\frac{5}{13}.$$

Thus, $\tan x = \frac{\sin x}{\cos x} = \frac{-5/13}{-12/13} = \frac{5}{12}$. Finally,

$$\sin 2x = 2\sin x \cos x = 2\left(-\frac{5}{13}\right)\left(-\frac{12}{13}\right) = \frac{120}{169},$$

$$\cos 2x = 2\cos^2 x - 1 = 2\left(-\frac{12}{13}\right)^2 - 1 = \frac{119}{169}, \text{ and}$$

$$\tan 2x = \frac{2\tan x}{1 - \tan^2 x} = \frac{2 \cdot \frac{5}{12}}{1 - \left(\frac{5}{12}\right)^2} = \frac{120}{144 - 25} = \frac{120}{119}.$$

(b) $\sin\theta = \frac{8}{17}$ and θ is in quadrant II.

First find $\cos\theta$. Since θ is in quadrant II, $\cos\theta < 0$, so

$$\cos\theta = -\sqrt{1 - \sin^2\theta} = -\sqrt{1 - \left(\frac{8}{17}\right)^2} = -\sqrt{\frac{225}{289}} = -\frac{15}{17}.$$

Then $\tan\theta = \frac{\sin\theta}{\cos\theta} = \frac{8/17}{-15/17} = -\frac{8}{15}$. Finally,

$$\sin 2x = 2\sin x \cos x = 2\left(\frac{8}{17}\right)\left(-\frac{15}{17}\right) = -\frac{240}{289},$$

$$\cos 2x = 2\cos^2 x - 1 = 2\left(-\frac{15}{17}\right)^2 - 1 = \frac{161}{289}, \text{ and}$$

$$\tan 2x = \frac{2\tan x}{1 - \tan^2 x} = \frac{2\left(-\frac{8}{15}\right)}{1 - \left(-\frac{8}{15}\right)^2} = -\frac{240}{161}.$$

Exercise: Prove that the equation

$(\sin\alpha - \cos\alpha)^2 = 1 - \sin 2\alpha$ is an identity.

Answer: We expand the left-hand side:

$$(\sin\alpha - \cos\alpha)^2 = \sin^2\alpha - 2\sin\alpha\cos\alpha + \cos^2\alpha$$
$$= \sin^2\alpha + \cos^2\alpha - 2\sin\alpha\cos\alpha$$
$$= 1 - \sin\alpha$$

by a Pythagorean identity and the double-angle formula for sine.

Exercise: Prove that the equation

$$\cot x = \frac{\sin 2x}{1 - \cos 2x}$$ is an identity.

Answer: We use the double-angle formulas to expand the right-hand side. Since the denominator of $\cot x$ is $\sin x$, we choose the double-angle formula for cosine that is in terms of $\sin x$:

$$\frac{\sin 2x}{1 - \cos 2x} = \frac{2\sin x \cos x}{1 - (1 - 2\sin^2 x)} = \frac{2\sin x \cos x}{1 - 1 + 2\sin^2 x}$$
$$= \frac{2\sin x \cos x}{2\sin^2 x} = \frac{\cos x}{\sin x} = \cot x$$

(B) **Half-angle formulas.**

The double-angle formulas for $\cos 2x$ are used to establish the formulas for lowering powers:

$$\sin^2 x = \frac{1 - \cos 2x}{2} \qquad \cos^2 x = \frac{1 + \cos 2x}{2} \qquad \tan^2 x = \frac{1 - \cos 2x}{1 + \cos 2x}$$

By substituting u for $2x$, we get the half-angle formulas:

$$\sin \frac{u}{2} = \pm \sqrt{\frac{1 - \cos u}{2}} \qquad \cos \frac{u}{2} = \pm \sqrt{\frac{1 + \cos u}{2}}$$

$$\tan \frac{u}{2} = \pm \sqrt{\frac{1 - \cos u}{1 + \cos u}} = \frac{1 - \cos u}{\sin u} = \frac{\sin u}{1 + \cos u}$$

• **Using half-angle formulas to lower powers in a trigonometric expression.**

Exercise: Find an expression equivalent to $\sin^4 x$, reducing the powers as much as possible.

Answer:

$$\sin^4 x = \left(\sin^2 x\right)^2 = \left(\frac{1 - \cos 2x}{2}\right)^2$$

$$= \frac{1 - 2\cos 2x + \cos^2 2x}{4} = \frac{1}{4} - \frac{1}{2}\cos 2x + \frac{1}{4}\cos^2 2x$$

$$= \frac{1}{4} - \frac{1}{2}\cos 2x + \frac{1}{4}\left(\frac{1 + \cos 4x}{2}\right)$$

$$= \frac{3}{8} - \frac{1}{2}\cos 2x + \frac{1}{8}\cos 4x$$

This answer can also be written as $\frac{1}{8}\left(3 - 4\cos 2x + \cos 4x\right)$.

(C) **Product-to-sum and sum-to-product formulas.**

The **product-to-sum formulas** are useful in calculus. They are used to write products of sines and cosines as sums, and are derived using the addition and subtraction formulas.

$$\sin u \cos v = \frac{1}{2}\left[\sin\left(u + v\right) + \sin\left(u - v\right)\right] \qquad \cos u \sin v = \frac{1}{2}\left[\sin\left(u + v\right) - \sin\left(u - v\right)\right]$$

$$\cos u \cos v = \frac{1}{2}\left[\cos\left(u + v\right) + \cos\left(u - v\right)\right] \qquad \sin u \sin v = \frac{1}{2}\left[\cos\left(u - v\right) - \cos\left(u + v\right)\right]$$

The **sum-to-product formulas** are used to write sums of sines and cosines as products, which provides a very useful tool for solving equations by factoring.

$$\sin u + \sin v = 2 \sin \frac{u + v}{2} \cos \frac{u - v}{2} \qquad \sin u - \sin v = 2 \cos \frac{u + v}{2} \sin \frac{u - v}{2}$$

$$\cos u + \cos v = 2 \cos \frac{u + v}{2} \cos \frac{u - v}{2} \qquad \cos u - \cos v = -2 \sin \frac{u + v}{2} \sin \frac{u - v}{2}$$

- **Expressing a trigonometric product as a sum and vice versa.**

Exercise: Write the product as a sum. | **Answer:**

(a) $\sin 3x \sin x$

$$\sin 3x \sin x = \frac{1}{2} \left[\cos (3x - x) - \cos (3x + x) \right]$$
$$= \frac{1}{2} \left(\cos 2x - \cos 4x \right)$$

(b) $\cos 5x \sin 2x$

$$\cos 5x \sin 2x = \frac{1}{2} \left[\sin (5x + 2x) - \sin (5x - 2x) \right]$$
$$= \frac{1}{2} \left(\sin 7x - \sin 3x \right)$$

Exercise: Evaluate. | **Answer:**

(a) $\tan 82.5° \tan 52.5°$

$$\tan 82.5° \tan 52.5° = \frac{\sin 82.5°}{\cos 82.5°} \cdot \frac{\sin 52.5°}{\cos 52.5°}$$

$$= \frac{\sin 82.5° \cdot \sin 52.5°}{\cos 82.5° \cdot \cos 52.5°}$$

$$= \frac{\frac{1}{2} \left[\cos (82.5° - 52.5°) - \cos (82.5° + 52.5°) \right]}{\frac{1}{2} \left[\cos (82.5° + 52.5°) + \cos (82.5° - 52.5°) \right]}$$

$$= \frac{\cos 30° - \cos 135°}{\cos 135° + \cos 30°} = \frac{\frac{\sqrt{3}}{2} - \left(-\frac{\sqrt{2}}{2} \right)}{-\frac{\sqrt{2}}{2} + \frac{\sqrt{3}}{2}}$$

$$= \frac{\sqrt{3} + \sqrt{2}}{\sqrt{3} - \sqrt{2}} \cdot \frac{\sqrt{3} + \sqrt{2}}{\sqrt{3} + \sqrt{2}} = \frac{3 + 2 + 2\sqrt{6}}{1} = 5 + 2\sqrt{6}$$

(b) $\sin 97.5° \cos 37.5°$

$\sin 97.5° \cos 37.5°$

$$= \frac{1}{2} \left[\sin (97.5° + 37.5°) + \sin (97.5° - 37.5°) \right]$$
$$= \frac{1}{2} \left(\sin 135° + \sin 60° \right) = \frac{1}{2} \left(\frac{\sqrt{2}}{2} + \frac{\sqrt{3}}{2} \right)$$
$$= \frac{1}{4} \left(\sqrt{2} + \sqrt{3} \right)$$

Exercise: Write the sum as a product. | **Answer:**

(a) $\sin 3x - \sin x$

$$\sin 3x - \sin x = 2 \cos \frac{3x + x}{2} \sin \frac{3x - x}{2}$$
$$= 2 \cos 2x \sin x$$

(b) $\cos 4x + \cos 2x + \cos x$

We first express $\cos 4x + \cos 2x$ as a product:
$$\cos 4x + \cos 2x = 2 \cos \frac{4x + 2x}{2} \cos \frac{4x - 2x}{2}$$
$$= 2 \cos 3x \cos x$$

Thus,
$$\cos 4x + \cos 2x + \cos x = (2 \cos 3x \cos x) + \cos x$$
$$= \cos x (2 \cos 3x + 1)$$

◤ Core Exercises

5, 17, 29, 33, 65, 83

8.4 Inverse Trigonometric Functions

Concepts

(A) **The inverse sine and cosine functions.**

- Evaluating the inverse sine and cosine functions at special values.
- Using a calculator to evaluate inverse sines and cosines.
- Composing trigonometric functions and their inverses.

(B) **The inverse tangent function.**

- Evaluating the inverse tangent function at special values.
- Using a calculator to evaluate inverse tangents.

(C) **Applications.**

- Using inverse trigonometric functions to solve applied problems.

Definitions

- **Inverse sine, arcsine:** We say $\sin^{-1} x = y$ if $\sin y = x$ and y is in the range $\left[-\frac{\pi}{2}, \frac{\pi}{2}\right]$ of arcsine. The domain is $[-1, 1]$.

- **Inverse cosine, arccosine:** We say $\cos^{-1} x = y$ if $\cos y = x$ and y is in the range $[0, \pi]$ of arccosine. The domain is $[-1, 1]$.

- **Inverse tangent, arctangent:** We say $\tan^{-1} x = y$ if $\tan y = x$ and y is in the range $\left(-\frac{\pi}{2}, \frac{\pi}{2}\right)$ of arctangent. The domain is $(-\infty, \infty)$.

Hints and Tips

- This section introduces the worst notation in all of mathematics. Your textbook authors didn't invent it, I didn't invent it, your teacher didn't invent it, so don't blame us. But we are all stuck with it, and I want to draw your attention to it so you don't get confused under the pressure of a test. As we learned in previous sections, $2^{-1} = \frac{1}{2}$, $(x+5)^{-1} = \dfrac{1}{x+5}$, $\sin^3 x = (\sin x)^3$, and $\sin^5 x = (\sin x)^5$. So it is *natural* to misinterpret the notation $\sin^{-1} x$. As we've seen in this section, $\sin^{-1} x = \arcsin x$. It is *not* $\dfrac{1}{\sin x}$ — that is $\csc x$. It is unfortunate that the superscript means something totally different in $\sin^3 x$ and in $\sin^{-1} x$, but that is the way it is.

▌ Review

Ⓐ The inverse sine and cosine functions.

The properties of inverse functions are discussed in Section 3.7. Since sine is not a one-to-one function, we restrict the domain of sine to $\left[-\frac{\pi}{2}, \frac{\pi}{2}\right]$ so that it has an inverse.

> The inverse sine function $\sin^{-1}$ is the function with domain $[-1, 1]$ and range $\left[-\frac{\pi}{2}, \frac{\pi}{2}\right]$ defined by
>
> $$\sin^{-1} x = y \quad \Leftrightarrow \quad \sin y = x$$

Again, since cosine is not a one-to-one function, we restrict its domain to obtain a one-to-one function:

> The inverse cosine function $\cos^{-1}$ is the function with domain $[-1, 1]$ and range $[0, \pi]$ defined by
>
> $$\cos^{-1} x = y \quad \Leftrightarrow \quad \cos y = x$$

Note that cosine is positive throughout the range of arcsine, and sine is positive throughout the range of arccosine.

• **Evaluating the inverse sine and cosine functions at special values.**

Exercise: Find the exact value of $\sin^{-1}\left(-\frac{\sqrt{3}}{2}\right)$.

Answer: The number in $\left[-\frac{\pi}{2}, \frac{\pi}{2}\right]$ whose sine is $-\frac{\sqrt{3}}{2}$ is $-\frac{\pi}{3}$. Thus, $\sin^{-1}\left(-\frac{\sqrt{3}}{2}\right) = -\frac{\pi}{3}$.

Exercise: Find the exact value of $\cos^{-1}\left(-\frac{\sqrt{2}}{2}\right)$.

Answer: The number in $[0, \pi]$ whose sine is $-\frac{\sqrt{2}}{2}$ is $\frac{3\pi}{4}$. Thus, $\sin^{-1}\left(-\frac{\sqrt{2}}{2}\right) = \frac{3\pi}{4}$.

• **Using a calculator to evaluate inverse sine and cosine.**

Exercise: Find the approximate value of $\sin^{-1} 0.95$.

Answer: We use a calculator to approximate this value. Set the calculator to radian mode and then use either $\boxed{\text{INV}}$ $\boxed{\text{SIN}}$, $\boxed{\text{SIN}^{-1}}$, or $\boxed{\text{ARCSIN}}$ to find that $\sin^{-1} 0.95 \approx 1.2532$.

Exercise: Find the approximate value of $\arccos(-0.24)$

Answer: We use a calculator to approximate $\arccos(-0.24) \approx 1.8132$.

• **Composing trigonometric functions and their inverses.**

Exercise: Find $\cos\left(\sin^{-1}\frac{1}{3}\right)$ and $\tan\left(\sin^{-1}\frac{1}{3}\right)$.

Answer: Let $u = \sin^{-1}\frac{1}{3}$, so $\sin u = \frac{1}{3}$. Since $u \in \left[-\frac{\pi}{2}, \frac{\pi}{2}\right]$, we must have $\cos u > 0$. Using the Pythagorean identity, we get $\cos u = \sqrt{1 - \sin^2 u}$, so

$$\cos\left(\sin^{-1}\frac{1}{3}\right) = \sqrt{1 - \sin^2\left(\sin^{-1}\frac{1}{3}\right)}$$
$$= \sqrt{1 - \left(\frac{1}{3}\right)^2} = \sqrt{\frac{8}{9}} = \frac{2\sqrt{2}}{3}$$

Now $\tan\left(\sin^{-1}\frac{1}{3}\right) = \dfrac{\sin\left(\sin^{-1}\frac{1}{3}\right)}{\cos\left(\sin^{-1}\frac{1}{3}\right)} = \dfrac{\frac{1}{3}}{\frac{2\sqrt{2}}{3}} = \frac{\sqrt{2}}{4}$.

Exercise: Find $\tan\left(\cos^{-1}\left(-\frac{5}{13}\right)\right)$.

Answer: Let $u = \cos^{-1}\left(-\frac{5}{13}\right)$, so $\cos u = -\frac{5}{13}$. Since $u \in [0, \pi]$, $\sin u > 0$ and thus

$\sin u = \sqrt{1 - \cos^2 u} = \sqrt{1 - \left(-\frac{5}{13}\right)^2} = \frac{12}{13}$. Thus,

$$\tan\left(\cos^{-1}\left(-\frac{5}{13}\right)\right) = \frac{\sin\left(\cos^{-1}\left(-\frac{5}{13}\right)\right)}{\cos\left(\cos^{-1}\left(-\frac{5}{13}\right)\right)} = \frac{\frac{12}{13}}{-\frac{5}{13}} = -\frac{12}{5}.$$

Exercise: Rewrite the expression $\tan\left(\sin^{-1} x\right)$ as an algebraic function in x. Sketch a right triangle which shows the relationship between the trigonometric functions.

Answer: Let $\sin^{-1} x = \theta$. Then $\sin\theta = x$ and $\tan\left(\sin^{-1} x\right) = \tan\theta$. We draw the appropriate diagram:

Now $\cos\theta = \dfrac{\text{adjacent}}{\text{hypotenuse}} = \dfrac{\sqrt{1 - x^2}}{1}$ and

$\tan\left(\sin^{-1} x\right) = \tan\theta = \dfrac{x}{\sqrt{1 - x^2}}$.

(B) **The inverse tangent function.**

The inverse tangent function $\cos^{-1}$ is the function with domain $(-\infty, \infty)$ and range $\left(-\frac{\pi}{2}, \frac{\pi}{2}\right)$ defined by

$$\tan^{-1} x = y \quad \Leftrightarrow \quad \tan y = x$$

The inverse tangent function is also called **arctangent** and is denoted by arctan. Because secant and tangent are related by $\tan^2 x + 1 = \sec^2 x$, we choose to restrict the range of $\tan^{-1}$ to the interval where secant is positive.

• **Evaluating the inverse tangent function at special values.**

Exercise: Find $\tan^{-1}\left(-\frac{\sqrt{3}}{3}\right)$.

Answer: Since $\tan\left(-\frac{\pi}{6}\right) = -\frac{\sqrt{3}}{3}$ and $-\frac{\pi}{2} < -\frac{\pi}{6} < \frac{\pi}{2}$, we have $\tan^{-1}\left(-\frac{\sqrt{3}}{3}\right) = -\frac{\pi}{6}$.

Exercise: Find $\tan^{-1}\left(\tan\frac{7\pi}{4}\right)$.

Answer: Since $\tan\frac{7\pi}{4} = -1$, we have $\tan^{-1}\left(\tan\frac{7\pi}{4}\right) = \tan^{-1}(-1) = -\frac{\pi}{4}$. Note that $\tan^{-1}\left(\tan\frac{7\pi}{4}\right) \neq \frac{7\pi}{4}$ since $\frac{7\pi}{4}$ is not in the range of $\tan^{-1}$.

• **Using a calculator to evaluate inverse tangent.**

Exercise: Find the approximate value of $\tan^{-1} 50$.

Answer: We use a calculator to estimate $\tan^{-1} 50 \approx 1.55$.

(C) **Applications.**

- **Using inverse trigonometric functions to solve applied problems.**

Exercise:

A 85-ft rope is tied 10 ft from the base of a building. Find the measure of the angle θ that the rope makes with the ground.

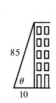

Answer: Since $\cos\theta = \frac{10}{85}$,

$\theta = \cos^{-1}\left(\frac{10}{85}\right) = 1.4529 \text{ rad} \approx 83.24°$.

Exercise:

A brace is made to support the frame of a gate, shown in the figure. Find the measure of θ.

Answer: Here we use the relation $\tan\theta = \dfrac{\text{opposite}}{\text{adjacent}} = \dfrac{41}{33}$

$\Leftrightarrow \quad \theta = \tan^{-1}\frac{41}{33} \approx 0.89308 \text{ rad} \approx 51.17°$.

◤ Core Exercises

3, 15, 23, 59

8.5   Trigonometric Equations

◤ Concepts

(A) **Trigonometric equations.**

- Solving trigonometric equations.
- Finding intersection points.

◤ Questions to Ask Your Teacher

- Given a trigonometric equation to solve, should we find all solutions in one period, or should we assume we need to find *all* solutions?

◤ Hints and Tips

- There are many techniques that we use to solve trigonometric equations. It can seem overwhelming at first. Realize that the goal is always the same: we start with an equation, and want to find a solution. Go through every example in the text — skip nary a one — and you will learn a whole menu of techniques. When you get a problem to solve, first think to yourself, "I want a solution." Now try something. If it looks like it is quadratic form, try using the factoring method. If you see something like $\sin x \cos x$, you may be inclined to try a trig identity. The first one or two things you try may not work — that happens to the best of us. Persevere. Even if you don't succeed, your teacher is sure to give you part marks if you try a few different techniques!

Review

(A) **Trigonometric equations.**

A **trigonometric equation** is an equation that contains trigonometric functions. To solve such an equation, first find all solutions in an appropriate interval (usually one period) and then use the period of the trigonometric function or functions to determine the other solutions.

- **Solving trigonometric equations.**

Exercise: Find all solutions of the equation.

Answer:

(a) $2 \sin x \cos x = \sin x$

As with polynomial equations, trigonometric equations can be solved by moving all terms to one side and factoring.
$2 \sin x \cos x = \sin x \quad \Leftrightarrow \quad 2 \sin x \cos x - \sin x = 0 \quad \Leftrightarrow$
$\sin x (2 \cos x - 1) = 0$. Now set each factor equal to zero and solve: Either $\sin x = 0$, in which case $x = 0$ or π, or
$2 \cos x - 1 = 0 \quad \Leftrightarrow \quad \cos x = \frac{1}{2}$, in which case $x = \frac{\pi}{3}$ or $\frac{5\pi}{3}$.
Because the periods of both sides of the equation are 2π, the solutions are $x = k\pi$, $\frac{\pi}{3} + 2k\pi$, or $\frac{5\pi}{3} + 2k\pi$, for any integer k.

(b) $\cos^2 x - 3 \sin x - 1 = 0$

Before we can factor, we first need to express this equation in terms of sine only or in terms of cosine only. Using
$\cos^2 x = 1 - \sin^2 x$, we get $\cos^2 x - 3 \sin x - 1 = 0 \quad \Leftrightarrow$
$\left(1 - \sin^2 x\right) - 3 \sin x - 1 = 0 \quad \Leftrightarrow \quad -\sin^2 x - 3 \sin x = 0$
$\Leftrightarrow \quad -\sin x (\sin x + 3) = 0 \quad \Leftrightarrow \quad x = 0$ or π, or
$\sin x = -3$ which has no solution. Thus, the solutions are
$x = 0 + 2k\pi$ or $x = \pi + 2k\pi$ for any integer k. That is,
$x = k\pi$ for any integer k.

Exercise: Solve the equation in the interval $[0, 2\pi)$.

Answer:

(a) $3 \tan^3 x - \tan x = 0$

$3 \tan^3 x - \tan x = 0 \quad \Leftrightarrow \quad \tan x \left(3 \tan^2 x - 1\right) = 0 \quad \Leftrightarrow$
$\tan x \left(\sqrt{3} \tan x - 1\right) \left(\sqrt{3} \tan x + 1\right) = 0 \quad \Leftrightarrow \quad \tan x = 0$,
$\frac{\sqrt{3}}{3}$, or $-\frac{\sqrt{3}}{3} \quad \Leftrightarrow \quad x = 0, \pi, \frac{\pi}{6}, \frac{7\pi}{6}, \frac{5\pi}{6}$, or $\frac{11\pi}{6}$.

(b) $\sec x \tan x = 4 \sin x$

$\sec x \tan x = 4 \sin x \quad \Leftrightarrow \quad \sec x \tan x - 4 \sin x = 0 \quad \Leftrightarrow$
$\frac{1}{\cos x} \frac{\sin x}{\cos x} - 4 \sin x = 0 \quad \Leftrightarrow$
$\sin x \left(\frac{1}{\cos^2 x} - 4\right) = \sin x \left(\sec^2 x - 4\right) = 0$. Thus, either
$\sin x = 0$ (in which case $x = 0$ or π) or $\sec x = \pm 2$. If
$\sec x = 2$, then $x = \frac{\pi}{3}$ or $x = \frac{5\pi}{3}$. If $\sec x = -2$, then $x = \frac{2\pi}{3}$
or $x = \frac{4\pi}{3}$. Thus, the solutions are $0, \frac{\pi}{3}, \frac{2\pi}{3}, \pi, \frac{4\pi}{3}$, and $\frac{5\pi}{3}$.

(c) $3\cos x + 1 = \cos 2x$

$3\cos x + 1 = \cos 2x \quad \Leftrightarrow \quad \cos 2x - 3\cos x - 1 = 0$. We use the double-angle formula for $\cos 2x$ that includes only terms of $\cos x$, since there is already a $\cos x$ term. Continuing:
$(2\cos^2 x - 1) - 3\cos x - 1 = 0 \quad \Leftrightarrow$
$2\cos^2 x - 3\cos x - 2 = 0 \quad \Leftrightarrow$
$(2\cos x + 1)(\cos x - 2) = 0$. Since $\cos x \neq 2$, the only solutions occur when $2\cos x + 1 = 0 \quad \Leftrightarrow \quad 2\cos x = -1$
$\Leftrightarrow \quad \cos x = -\frac{1}{2} \quad \Leftrightarrow \quad x = \frac{2\pi}{3}$ or $\frac{4\pi}{3}$. Recall that we are only concerned with the interval $[0, 2\pi)$. So the only solutions are $x = \frac{2\pi}{3}$ and $\frac{4\pi}{3}$.

Exercise: Solve the equation $\cos 5x = \cos 3x$.

Answer: $\cos 5x = \cos 3x \quad \Leftrightarrow \quad \cos 5x - \cos 3x = 0 \quad \Leftrightarrow$
$-2\sin \dfrac{5x + 3x}{2} \sin \dfrac{5x - 3x}{2} = 0 \quad \Leftrightarrow \quad -2\sin 4x \sin x = 0$
$\Leftrightarrow \quad \sin 4x = 0$ or $\sin x = 0 \quad \Leftrightarrow \quad x = \frac{1}{4}k\pi$ or $x = k\pi$.
Note that the first case includes all solutions, $x = \frac{1}{4}k\pi$.

• **Finding intersection points.**

Exercise: Sketch $f(x) = 2\sin x$ and $g(x) = \tan x$ on the same axes and find their points of intersection.

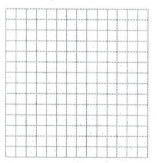

Answer:

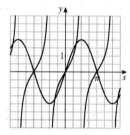

$2\sin x - \tan x = 0 \quad \Leftrightarrow \quad 2\sin x - \dfrac{\sin x}{\cos x} = 0 \quad \Leftrightarrow$
$\sin x \left(2 - \dfrac{1}{\cos x}\right) = 0 \quad \Leftrightarrow \quad \sin x = 0$ or $2 = \dfrac{1}{\cos x} \quad \Leftrightarrow$
$x = 0, \pi$ or $\cos x = \frac{1}{2}$, in which case $x = \frac{\pi}{3}, \frac{5\pi}{3}$. Thus, the solutions are $x = k\pi, \frac{\pi}{3} + 2k\pi$, or $\frac{5\pi}{3} + 2k\pi$, k any integer.

In most cases, we cannot find exact solutions to equations involving both algebraic and trigonometric expressions. However, with the aid of a graphing device we can approximate the solutions by graphing each side of the equation. Note that many graphing calculators have either an equation solver or an "intersection" feature that gives you one solution quickly. These tools are great to use, but you should still look at the graphs, because the equation solver will give you only one solution, and there are many.

Exercise: Use a graphing device to find the solutions of each equation, correct to two decimal places.

(a) $\sin x = \frac{1}{5}x$

Answer:

Since $-1 \le \sin x \le 1$, we must have $-1 \le \frac{1}{5}x \le 1$ $\Leftrightarrow$ $-5 \le x \le 5$.

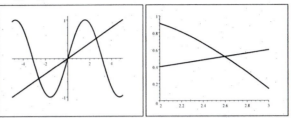

We see from the first graph that there are three solutions: one at $x = 0$, and one each between ± 2 and ± 3. By zooming in, we see that one solution is $x \approx 2.60$, and by symmetry the third solution is approximately -2.60. The three solutions are $x = 0$ and $x \approx \pm 2.60$.

(b) $\sin^2 x = x^2 - 1$

Since $\sin^2 x \le 1$, we must have $x^2 - 1 \le 1$ $\Leftrightarrow$ $x^2 \le 2$, so $-\sqrt{2} \le x \le \sqrt{2}$. Since $\sqrt{2} < 1.5$, we choose the viewing rectangle $[-1.5, 1.5] \times [-1.25, 1.25]$.

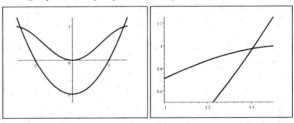

We see that there are two solutions: one between -1.5 and -1, and another between 1 and 1.5. Zooming in, we see that the positive solution is $x \approx 1.40$. By symmetry, the other solution is $x \approx -1.40$. The solutions are $x \approx \pm 1.40$.

▰ Core Exercises

31, 45, 65, 71

9 Polar Coordinates and Vectors

9.1 Polar Coordinates

▶ Concepts

(A) Polar coordinates.

- Plotting polar points.
- Converting between polar and rectangular coordinates.

(B) Polar equations.

- Converting between polar and rectangular equations.

▶ Definitions

- **Polar coordinates:** An alternate method of specifying points in a plane. The two coordinates are r, the distance from a given point P to the origin, and θ, the angle between the polar axis and $\overrightarrow{OP}$.

- **Polar equations:** Equations involving polar r and θ instead of (for example) rectangular x and y.

▶ Review

(A) Polar coordinates.

The **polar coordinate system** uses distances and directions to specify the location of points in the plane. Choose a fixed point O called the **origin** (or **pole**), then draw a ray starting at O, called the **polar axis**. This axis coincides with the x-axis in rectangular coordinates. Now let P be any point in the plane. Let r be the distance from P to the origin, and let θ be the angle between the polar axis and $\overrightarrow{OP}$. The ordered pair (r, θ) uniquely specifies the location of P; r and θ are referred to as the **polar coordinates of** P. Although the ordered pair (r, θ) specifies a point, there are infinitely many other ordered pairs that give the same point; in particular, $(r, \theta) = (r, \theta + 2k\pi) = (-r, \theta + (2k + 1)\pi)$ for any integer k. When we go from rectangular to polar coordinates, we lose unique representation of points, but we gain simplification in expressing the equations of *some* graphs.

● **Plotting polar points.**

Exercise: Plot the points $\left(3, \frac{\pi}{4}\right)$, $\left(1, \frac{\pi}{2}\right)$, $\left(3, \frac{9\pi}{4}\right)$, $\left(-2, \frac{\pi}{3}\right)$, and $\left(1, -\frac{\pi}{8}\right)$ in polar coordinates.

Answer:

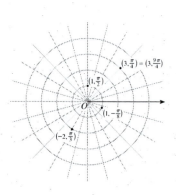

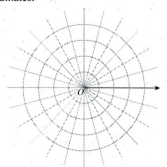

● **Converting between polar and rectangular coordinates.**

Polar and rectangular coordinates are related by the following formulas:

Polar to Rectangular:	$x = r\cos\theta$	$y = r\sin\theta$
Rectangular to Polar:	$\tan\theta = y/x \ (x \neq 0)$	$r^2 = x^2 + y^2$

Exercise: Find rectangular coordinates for the point with the given polar coordinates.

Answer:

(a) $\left(3, \frac{\pi}{4}\right)$

Apply the formulas: $x = r\cos\theta = 3\cos\frac{\pi}{4} = 3 \cdot \frac{\sqrt{2}}{2} = \frac{3\sqrt{2}}{2}$, $y = r\sin\theta = 3\sin\frac{\pi}{4} = 3 \cdot \frac{\sqrt{2}}{2} = \frac{3\sqrt{2}}{2}$. Thus, the rectangular coordinates are $\left(\frac{3\sqrt{2}}{2}, \frac{3\sqrt{2}}{2}\right)$.

(b) $\left(-4, \frac{7\pi}{6}\right)$

Apply the formulas:
$x = r\cos\theta = -4\cos\frac{7\pi}{6} = -4\left(-\frac{\sqrt{3}}{2}\right) = 2\sqrt{3}$,
$y = r\sin\theta = -4\sin\frac{7\pi}{6} = -4\left(-\frac{1}{2}\right) = 2$. Thus, the rectangular coordinates are $\left(2\sqrt{3}, 2\right)$.

Exercise: Find polar coordinates (with $r > 0$ and $0 \le \theta < 2\pi$) for the point with the given rectangular coordinates.

Answer:

(a) $(4, 4\sqrt{3})$

Apply the formulas:

$$r = \sqrt{x^2 + y^2} = \sqrt{4^2 + \left(4\sqrt{3}\right)^2} = \sqrt{16 + 48} = \sqrt{64} = 8.$$

Since $(4, 4\sqrt{3})$ is in quadrant I,

$\theta = \tan^{-1}\left(\frac{y}{x}\right) = \tan^{-1}\left(\frac{4\sqrt{3}}{4}\right) = \tan^{-1}\sqrt{3} = \frac{\pi}{3}$. Thus,

polar coordinates are $\left(8, \frac{\pi}{3}\right)$.

(b) $(-5, 12)$

Apply the formulas: $r = \sqrt{x^2 + y^2} = \sqrt{(-5)^2 + 12^2} = 13$.

Since $(-5, 12)$ is in quadrant II,

$\theta = \pi + \tan^{-1}\left(\frac{y}{x}\right) = \pi + \tan^{-1}\left(-\frac{12}{5}\right) \approx 1.97$. Thus, polar

coordinates are approximately $(13, 1.97)$.

(B) Polar equations.

- **Converting between polar and rectangular equations.**

Exercise: Convert the equation $4x^2 + y^2 = 16$ to polar form.

Answer: Recall that this equation represents an ellipse. Using the formulas, we substitute $x = r \cos\theta$ and $y = r \sin\theta$:

$4\left(r \cos\theta\right)^2 + \left(r \sin\theta\right)^2 = 16 \quad \Leftrightarrow$

$4r^2 \cos^2\theta + r^2 \sin^2\theta = 16 \quad \Leftrightarrow$

$r^2 \left(4\cos^2\theta + \sin^2\theta\right) = 16 \quad \Leftrightarrow \quad r^2 = \dfrac{16}{3\cos^2\theta + 1} \quad \Leftrightarrow$

$r = \pm\dfrac{4}{\sqrt{3\cos^2 +1}}.$

Exercise: Convert the polar equation $r = 2\sin\theta - \cos\theta$ to rectangular form.

Answer: We try to write this equation in terms of r^2, $r\cos\theta$, and $r\sin\theta$, since these terms are easily converted to rectangular coordinates. Multiplying both sides by r, we get

$r^2 = 2r\sin\theta - r\cos\theta$. Substituting, this is equivalent to $x^2 + y^2 = 2y - x$. This is the equation of a circle. We complete the square: $x^2 + x + \frac{1}{4} + y^2 - 2y + 1 = \frac{1}{4} + 1 = \frac{5}{4}$

$\Leftrightarrow \left(x + \frac{1}{2}\right)^2 + (y - 1)^2 = \left(\frac{\sqrt{5}}{2}\right)^2$. Thus, the center of the

circle is $\left(-\frac{1}{2}, 1\right)$ and its radius is $\frac{\sqrt{5}}{2}$.

Core Exercises

17, 21, 41, 61

9.2   Graphs of Polar Equations

▶ Concepts

(A) **Sketching the graph of a polar equation.**

- Sketching by converting to rectangular coordinates.

- Sketching by plotting points.

- Sketching using technology.

(B) **Symmetry.**

- Determining the symmetry of a polar graph.

▶ Definitions

- **Symmetry about the polar axis:** Corresponds to symmetry about the x-axis in rectangular coordinates.

- **Symmetry about the pole:** Corresponds to symmetry about the origin in rectangular coordinates.

- **Symmetry about the line $\theta = \frac{\pi}{2}$:** Corresponds to symmetry about the y-axis in rectangular coordinates.

▶ Hints and Tips

- Once you've figured out how to use your calculator to draw polar graphs, it is a good idea to take a few minutes to just goof around with it. Put various functions $r = f(\theta)$ into the calculator and see what comes out. If your graphs are looking jagged, make sure that your t-step is sufficiently small to get an accurate picture. The idea is to get a feel for how polar equations work.

▶ Review

(A) **Sketching the graph of a polar equation.**

The graph of a polar equation $r = f(\theta)$ consists of all points P that have at least one polar representation (r, θ) whose coordinates satisfy the equation. When graphing a polar equation it is important to find the intervals where r is positive and the intervals where r is negative. For negative r, the point (r, θ) will lie in the quadrant opposite angle θ.

Several polar graphs have names:

limaçon	(lima bean-shaped or heart-shaped): $r = a \pm b\cos\theta$ or $r = a \pm b\sin\theta$
cardioid	(heart-shaped; a special type of limaçon): $r = a(1 \pm \cos\theta)$ or $r = a(1 \pm \sin\theta)$
rose	(has $2n$ leaves for n even; n leaves for n odd): $r = a\cos n\theta$

• **Sketching by converting to rectangular coordinates.**

Exercise: Sketch the curve $r = 4$.

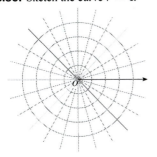

Answer: We convert to rectangular coordinates: $x^2 + y^2 = 4$, and this is a circle with radius 4.

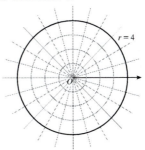

• **Sketching by plotting points.**

Exercise: Sketch the curve with the given polar equation.

(a) $r = -4\cos\theta$

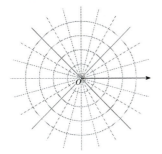

Answer:

The tables below show values of r for some convenient values of θ.

θ	r	θ	r	θ	r	θ	r
0	-4	$\frac{\pi}{2}$	0	π	4	$\frac{3\pi}{2}$	0
$\frac{\pi}{6}$	$-2\sqrt{3}$	$\frac{2\pi}{3}$	2	$\frac{7\pi}{6}$	$2\sqrt{3}$	$\frac{5\pi}{3}$	-2
$\frac{\pi}{4}$	$-2\sqrt{2}$	$\frac{3\pi}{4}$	$2\sqrt{2}$	$\frac{5\pi}{4}$	$2\sqrt{2}$	$\frac{7\pi}{4}$	$-2\sqrt{2}$
$\frac{\pi}{3}$	-2	$\frac{5\pi}{6}$	$2\sqrt{3}$	$\frac{4\pi}{3}$	2	$\frac{11\pi}{6}$	$-2\sqrt{3}$

We can see how r changes as θ changes: As θ ranges from 0 to $\frac{\pi}{2}$, r ranges from -4 to 0. As θ ranges from $\frac{\pi}{2}$ to π, r ranges from 0 to 4. As θ ranges from π to $\frac{3\pi}{2}$, r ranges from 4 back to 0, and as θ ranges from $\frac{3\pi}{2}$ to 2π, r ranges from 0 to -4. Since $r = -4\cos\theta \quad \Rightarrow \quad r^2 = -4r\cos\theta \quad \Rightarrow \quad x^2 + y^2 = -4x$, the equation is that of the circle $x^2 + 4x + y^2 = 0 \quad \Leftrightarrow \quad (x+2)^2 + y^2 = 4$.

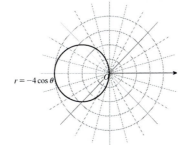

(b) $r = 1 + 3\sin\theta$

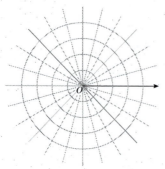

$r = 1 + 3\sin\theta = 0$ $\Rightarrow$ $\sin\theta = -\frac{1}{3}$. Since $\sin^{-1}\left(-\frac{1}{3}\right) \approx -19.5°$, $r = 0$ when $\theta = -19.5°$ and when $\theta = 199.5°$. The tables below show values of r for some convenient values of θ.

θ	r
0	1
$\frac{\pi}{4}$	3.12
$\frac{\pi}{2}$	4
$\frac{3\pi}{4}$	3.12

θ	r
π	1
$\frac{5\pi}{4}$	-1.12
$\frac{3\pi}{2}$	-2
$\frac{7\pi}{4}$	-1.12

Thus, $r > 0$ for $-19.5° < \theta < 199.5°$ and $r < 0$ for $199.5° < \theta < 340.5°$. This graph is called a **limaçon** with an inner loop. The graph below shows the angles at which r changes sign.

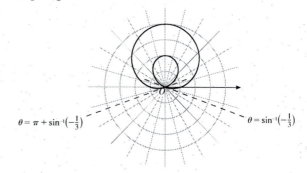

$\theta = \pi + \sin^{-1}\left(-\frac{1}{3}\right)$ $\theta = \sin^{-1}\left(-\frac{1}{3}\right)$

• **Sketching using technology.**

Just as when we use a calculator to graph rectangular functions, the toughest part of using technology to sketch polar graphs is selecting a good viewing rectangle. It is also important to select an appropriate range of θ-values, and to ensure that the step size of θ is sufficiently small to avoid a jagged graph.

Exercise: Use a calculator to graph the polar curve $r = \theta$ for $\theta > 0$.

Answer: We graph the curve for $0 \leq \theta \leq 10$ in the viewing rectangle $[-10, 7]$ by $[-6, 11]$.

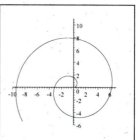

Exercise: Use a calculator to graph the polar curve $r = 2 - 4 \sin \theta$.

Answer: We graph the curve for $-\pi \le \theta \le \pi$ in the viewing rectangle $[-4, 4]$ by $[-6.5, 1.5]$.

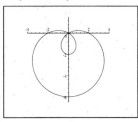

(B) **Symmetry.**

When sketching the graph of a polar equation, it is often helpful to take advantage of symmetry.

> 1. A graph is **symmetric about the polar axis** if its polar equation is unchanged when we replace θ by $-\theta$.
>
> 2. A graph is **symmetric about the pole** if its polar equation is unchanged when we replace r by $-r$.
>
> 3. A graph is **symmetric about the vertical line** $\theta = \frac{\pi}{2}$ if its polar equation is unchanged when we replace θ by $\pi - \theta$.

• **Determining the symmetry of a polar graph.**

Exercise: Determine the symmetry of the graph of the given polar equation.

Answer:

(a) $r = -4 \cos \theta$

$-4 \cos(-\theta) = -4 \cos \theta = r$, so the graph is symmetric about the polar axis.

$-r = -(-4 \cos \theta) = 4 \cos \theta \ne r$, so the graph is not symmetric about the pole.

$-4 \cos(\pi - \theta) = -4 (\cos \pi \cos \theta + \sin \pi \sin \theta) = 4 \cos \theta \ne r$

so the graph is not symmetric about the vertical line $\theta = \frac{\pi}{2}$. See the graph in the first exercise in this section.

(b) $r = 1 + 3 \sin \theta$

$1 + 3 \sin(-\theta) = 1 - 3 \sin \theta \ne r$, so the graph is not symmetric about the polar axis.

$-r = -(1 + 3 \sin \theta) = -1 - 3 \sin \theta \ne r$, so the graph is not symmetric about the pole.

$1 + 3 \sin(\pi - \theta) = 1 + 3 (\sin \pi \cos \theta - \cos \pi \sin \theta)$
$$= 1 + 3 \sin \theta = r$$

so the graph is symmetric about the vertical line $\theta = \frac{\pi}{2}$. See the graph in the first exercise in this section.

▱ Core Exercises

15, 39, 41, 49, 57

9.3 Polar Form of Complex Numbers; DeMoivre's Theorem

Concepts

(A) Complex numbers in polar form.

- Putting complex numbers in polar form.
- Multiplying and dividing complex numbers.
- Graphing complex numbers.

(B) DeMoivre's Theorem.

- Raising complex numbers to powers.
- Taking nth roots of complex numbers.

Definitions

- **The polar** or **trigonometric form of a complex number:** Given a complex number $a + bi$, we can write it in the **polar form** $r(\cos\theta + i\sin\theta)$, where $r = |z| = \sqrt{a^2 + b^2}$ and $\tan\theta = b/a$. This can also be written r cis θ.

Hints and Tips

- When I first learned this material, I thought of a complex number as $a + bi$ and the "polar form" as a whimsical, useless thing that I had to learn. As I got my engineering degree, things changed, and now I think of complex numbers in their polar form, and as points on a plane, and I think of $a + bi$ as a form which is only occasionally useful.

  The obvious advantage of $a + bi$ (rectangular) form is that it makes it easy to add and subtract complex numbers. The advantage of polar form is it makes it much easier to multiply and to take roots. Which is better overall depends on the applications you will be using. Engineers often deal with quantities like voltage that have a magnitude and a phase angle. The magnitude is r, the phase angle is θ, and it is easy to tell from the polar form of two signals whether they reinforce each other or cancel each other out.

  I would advise getting comfortable with both forms of a complex number, because both forms wind up being useful.

Review

(A) Complex numbers in polar form.

Let $z = a+bi$ be a complex number. Then z has the **polar form** $z = r(\cos\theta + i\sin\theta)$, where $r = |z| = \sqrt{a^2 + b^2}$ and $\tan\theta = b/a$. θ is the **argument** of z and r is the **modulus**.

• **Putting complex numbers in polar form.**

Exercise: Write each complex number in polar form.

Answer:

(a) $-4 + 4i$

Since $\tan\theta = 4/(-4) = -1$, an argument is $\theta = \frac{3\pi}{4}$ and

$$r = \sqrt{(-4)^2 + 4^2} = \sqrt{32} = 4\sqrt{2}.\text{ Thus,}$$

$$-4 + 4i = 4\sqrt{2}\left(\cos\tfrac{3\pi}{4} + i\sin\tfrac{3\pi}{4}\right).$$

(b) $7 - i$

$r = \sqrt{7^2 + (-1)^2} = \sqrt{50} = 5\sqrt{2}$. To find an argument θ, we note that $7 - i$ is in quadrant IV, as shown.

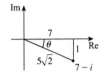

We see from the diagram that

$$\theta = \sin^{-1}\tfrac{-1}{5\sqrt{2}} = \sin^{-1}\left(-\tfrac{\sqrt{2}}{10}\right) \approx -0.142.\text{ Thus,}$$

$$7 - i \approx 5\sqrt{2}\left(\cos\left(-0.142\right) + i\sin\left(-0.142\right)\right).$$

• **Multiplying and dividing complex numbers.**

If $z_1 = r_1\left(\cos\theta_1 + i\sin\theta_1\right)$ and $z_2 = r_2\left(\cos\theta_2 + i\sin\theta_2\right)$, then their product and quotient are

$$z_1 z_2 = r_1 r_2\left[\cos\left(\theta_1 + \theta_2\right) + i\sin\left(\theta_1 + \theta_2\right)\right]$$

$$\frac{z_1}{z_2} = \frac{r_1}{r_2}\left[\cos\left(\theta_1 - \theta_2\right) + i\sin\left(\theta_1 - \theta_2\right)\right]$$

Exercise: Let $z_1 = 2 + 2\sqrt{3}i$ and $z_2 = -i$.

Answer: We convert these numbers to polar form:

$$r_1 = \sqrt{2^2 + \left(2\sqrt{3}\right)^2} = \sqrt{4 + 12} = 4\text{ and}$$

$$\tan\theta_1 = \tfrac{2\sqrt{3}}{2} = \sqrt{3} \quad\Rightarrow\quad \theta_1 = \tfrac{\pi}{3}.$$

$$r_2 = \sqrt{0^2 + (-1)^2} = 1.\text{ Since }\tan\theta_2\text{ is undefined and}$$

$z_2 = -i$, we see that $\theta_2 = \frac{3\pi}{2}$. So $z_1 = 4\left(\cos\frac{\pi}{3} + i\sin\frac{\pi}{3}\right)$

and $z_2 = \cos\frac{3\pi}{2} + i\sin\frac{3\pi}{2}$.

(a) Find $z_1 z_2$.

$$z_1 z_2 = 4 \cdot 1\left[\cos\left(\tfrac{\pi}{3} + \tfrac{3\pi}{2}\right) + i\sin\left(\tfrac{\pi}{3} + \tfrac{3\pi}{2}\right)\right]$$
$$= 4\left(\cos\tfrac{11\pi}{6} + i\sin\tfrac{11\pi}{6}\right)$$

(b) Find $\dfrac{z_1}{z_2}$.

$$\frac{z_1}{z_2} = \frac{4}{1}\left[\cos\left(\tfrac{\pi}{3} - \tfrac{3\pi}{2}\right) + i\sin\left(\tfrac{\pi}{3} - \tfrac{3\pi}{2}\right)\right]$$
$$= 4\left[\cos\left(-\tfrac{7\pi}{6}\right) + i\sin\left(-\tfrac{7\pi}{6}\right)\right]$$
$$= 4\left(\cos\tfrac{5\pi}{6} + i\sin\tfrac{5\pi}{6}\right)$$

Exercise: Let $z_1 = 2 + 2i$ and $z_2 = -\sqrt{2} - \sqrt{2}i$.

Answer: We convert these numbers to polar form:

$r_1 = \sqrt{2^2 + 2^2} = 2\sqrt{2}$ and $\tan \theta_1 = \frac{2}{2} = 1$ with $a, b > 0$, so

$\theta_1 = \frac{\pi}{4}$. $r_2 = \sqrt{\left(-\sqrt{2}\right)^2 + \left(-\sqrt{2}\right)^2} = 2$ and $\tan \theta_2 = \frac{-\sqrt{2}}{-\sqrt{2}}$

with $a, b < 0$, so $\theta_2 = \frac{5\pi}{4}$. Thus, $z_1 = 2\sqrt{2}\left(\cos \frac{\pi}{4} + i \sin \frac{\pi}{4}\right)$

and $z_2 = 2\left(\cos \frac{5\pi}{4} + i \sin \frac{5\pi}{4}\right)$.

(a) Find $z_1 z_2$.

$$z_1 z_2 = 2\sqrt{2} \cdot 2 \left[\cos\left(\frac{\pi}{4} + \frac{5\pi}{4}\right) + i \sin\left(\frac{\pi}{4} + \frac{5\pi}{4}\right)\right]$$
$$= 4\sqrt{2}\left(\cos \frac{3\pi}{2} + i \sin \frac{3\pi}{2}\right) = -4\sqrt{2}i$$

(b) Find $\dfrac{z_1}{z_2}$.

$$\frac{z_1}{z_2} = \frac{2\sqrt{2}}{2}\left[\cos\left(\frac{\pi}{4} - \frac{5\pi}{4}\right) + i \sin\left(\frac{\pi}{4} - \frac{5\pi}{4}\right)\right]$$
$$= \sqrt{2}\left[\cos\left(-\pi\right) + i \sin\left(-\pi\right)\right] = -\sqrt{2}$$

• **Graphing complex numbers.**

To graph complex numbers we need two axes, one for the real part and one for the imaginary part. These are called the **real axis** and the **imaginary axis**. The plane determined by these two axes is called the **complex plane**. To graph the complex number $a + bi$, we plot the ordered pair (a, b) in this plane.

Exercise: Graph the complex numbers $z_1 = -2 - 3i$, $z_2 = 3 - i$, $z_1 + z_2$, and $z_1 z_2$.

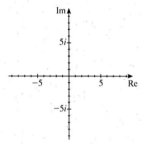

Answer: We calculate

$z_1 + z_2 = (-2 - 3i) + (3 - i) = 1 - 4i$ and

$z_1 z_2 = (-2 - 3i)(3 - i) = -6 + 2i - 9i + 3i^2$
$\qquad\quad = -6 - 7i - 3 = -9 - 7i$

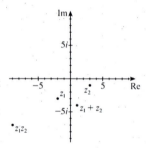

Exercise: Graph the set of complex numbers $\{z = a + bi \mid 0 \leq a \leq 5, |b| < 3\}$.

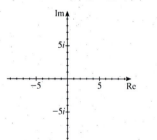

Answer: $|b| < 3 \quad \Leftrightarrow \quad -3 < b < 3$

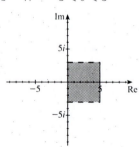

(B) DeMoivre's Theorem.

Repeated use of the multiplication formula gives

> **DeMoivre's Theorem**
>
> If $z = r \left(\cos \theta + i \sin \theta \right)$, then for any integer n, $z^n = r^n \left(\cos n\theta + i \sin n\theta \right)$.

- **Raising complex numbers to powers.**

Exercise: Find the indicated power using DeMoivre's Theorem.

Answer: In each case, we find r and θ and the apply DeMoivre's Theorem.

(a) $\left(4 + 4\sqrt{3}i \right)^5$

$r = \sqrt{4^2 + \left(4\sqrt{3} \right)^2} = \sqrt{16 + 48} = 8$ and

$\tan \theta = \frac{4\sqrt{3}}{4} = \sqrt{3}$ with $a, b > 0$, so $\theta = \frac{\pi}{3}$. Therefore,

$z = 8 \left(\cos \frac{\pi}{3} + i \sin \frac{\pi}{3} \right)$ and by DeMoivre's Theorem,

$$\left(4 + 4\sqrt{3}i \right)^5 = 8^5 \left[\cos \left(5 \cdot \frac{\pi}{3} \right) + i \sin \left(5 \cdot \frac{\pi}{3} \right) \right]$$
$$= 32{,}768 \left(\cos \frac{5\pi}{3} + i \sin \frac{5\pi}{3} \right)$$
$$= 32{,}768 \left(\frac{1}{2} - i \frac{\sqrt{3}}{2} \right) = 16{,}384 - 16{,}384\sqrt{3}i$$

(b) $(-2 - 2i)^6$

$r = \sqrt{(-2)^2 + (-2)^2} = 2\sqrt{2}$ and $\tan \theta = 1$ with $a, b < 0$, so

$\theta = \frac{5\pi}{4}$. Therefore, $z = 2\sqrt{2} \left(\cos \frac{5\pi}{4} + i \sin \frac{5\pi}{4} \right)$ and by DeMoivre's Theorem,

$$(-2 - 2i)^6 = \left(2\sqrt{2} \right)^6 \left[\cos \left(6 \cdot \frac{5\pi}{4} \right) + i \sin \left(6 \cdot \frac{5\pi}{4} \right) \right]$$
$$= 512 \left(\cos \frac{15\pi}{2} + i \sin \frac{15\pi}{2} \right) = 512 \left(0 - i \right)$$
$$= -512i$$

(c) $(3 - 4i)^3$

$r = \sqrt{3^2 + (-4)^2} = 5$ and $\theta = \tan^{-1} \left(-\frac{4}{3} \right) \approx -0.927$, so

$z \approx 5 \left(\cos \left(-0.927 \right) + i \sin \left(-0.927 \right) \right)$ and by DeMoivre's Theorem,

$$(3 - 4i)^3 \approx 5^3 \left[\cos \left(3 \left(-0.927 \right) \right) + i \sin \left(3 \left(-0.927 \right) \right) \right]$$
$$\approx 125 \left(\cos \left(-2.781 \right) + i \sin \left(-2.781 \right) \right)$$
$$\approx -117 - 44i$$

- **Taking nth roots of complex numbers.**

> For n a positive integer, the n distinct nth roots of the complex number $z = r \left(\cos \theta + i \sin \theta \right)$ are given by the formula
>
> $$w_k = r^{1/n} \left[\cos \left(\frac{\theta + 2k\pi}{n} \right) + i \sin \left(\frac{\theta + 2k\pi}{n} \right) \right] \quad \text{for } k = 0, 1, 2, \ldots, n - 1$$

The main concepts are:

1. The modulus of each nth root is $r^{1/n}$.

2. The argument of the first root is $\dfrac{\theta}{n}$.

3. We repeatedly add $\dfrac{2\pi}{n}$ to get the argument of each successive root.

Exercise: Find the indicated roots and sketch them in the complex plane.

Answer:

(a) The cube roots of 2

Find r and θ: $r = \sqrt{4+0} = 2$ and $\theta = 0$. Thus, $\theta_0 = 0$ and $w_0 = \sqrt[3]{2}\,(\cos 0 + i\sin 0) \approx 1.260$, $\theta_1 = \frac{2\pi}{3}$ and $w_1 = \sqrt[3]{2}\left(\cos\frac{2\pi}{3} + i\sin\frac{2\pi}{3}\right) \approx -0.630 + 1.091i$, $\theta_2 = \frac{4\pi}{3}$ and $w_2 = \sqrt[3]{2}\left(\cos\frac{4\pi}{3} + i\sin\frac{4\pi}{3}\right) \approx -0.630 - 1.091i$.

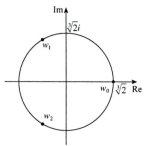

(b) The square roots of $5\sqrt{3} - 5i$

$r = \sqrt{\left(5\sqrt{3}\right)^2 + (-5)^2} = \sqrt{75+25} = 10$ and $\theta = \frac{11\pi}{6}$ (this is $\arctan\frac{-5}{5\sqrt{3}} + 2\pi$). Thus, $\theta_0 = \frac{1}{2}\left(\frac{11\pi}{6}\right) = \frac{11\pi}{12}$ and $w_0 = \sqrt{10}\left(\cos\frac{11\pi}{12} + i\sin\frac{11\pi}{12}\right) \approx -3.055 + 0.818i$, $\theta_1 = \frac{11\pi}{12} + \frac{2\pi}{2} = \frac{23\pi}{12}$ and $w_1 = \sqrt{10}\left(\cos\frac{23\pi}{12} + i\sin\frac{23\pi}{12}\right) \approx 3.055 - 0.818i$.

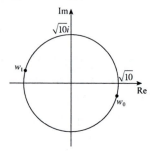

Exercise: Solve the equation.	**Answer:**
(a) $z^3 - i = 0$	$z^3 - i = 0 \iff z^3 = i$. Thus, the solutions are the cube roots of i. Because $i = \cos\frac{\pi}{2} + i\sin\frac{\pi}{2}$, we have $r = 1$ and $\theta = \frac{\pi}{2}$. So $\theta_0 = \frac{1}{3}\left(\frac{\pi}{2}\right) = \frac{\pi}{6}$ and $z_0 = \cos\frac{\pi}{6} + i\sin\frac{\pi}{6} = \frac{\sqrt{3}}{2} + \frac{1}{2}i$, $\theta_1 = \frac{\pi}{6} + \frac{2\pi}{3} = \frac{5\pi}{6}$ and $z_1 = \cos\frac{5\pi}{6} + i\sin\frac{5\pi}{6} = -\frac{\sqrt{3}}{2} + \frac{1}{2}i$, $\theta_2 = \frac{\pi}{6} + \frac{4\pi}{3}$ and $z_2 = \cos\frac{3\pi}{2} + i\sin\frac{3\pi}{2} = -i$.
(b) $z^2 + 4i = 0$	$z^2 + 4i = 0 \iff z^2 = -4i \iff z = \pm\sqrt{-4i} = \pm 2\sqrt{-i}$. We find the square roots of $-i$, which has $r = 1$ and $\theta = \frac{3\pi}{2}$. Thus, $\theta_0 = \frac{1}{2}\left(\frac{3\pi}{2}\right) = \frac{3\pi}{4}$ and $w_0 = \sqrt{1}\left(\cos\frac{3\pi}{4} + i\sin\frac{3\pi}{4}\right) = -\frac{\sqrt{2}}{2} + \frac{\sqrt{2}}{2}i$, $\theta_1 = \frac{3\pi}{4} + \frac{1}{2}(2\pi) = \frac{7\pi}{4}$ and $w_1 = \sqrt{1}\left(\cos\frac{7\pi}{4} + i\sin\frac{7\pi}{4}\right) = \frac{\sqrt{2}}{2} - \frac{\sqrt{2}}{2}i$. Since $w_1 = -w_0$, there are only two solutions to $z^2 + 4i = 0$. They are $z = \pm 2\sqrt{-i} = \pm 2\left(-\frac{\sqrt{2}}{2} + \frac{\sqrt{2}}{2}i\right) = \pm\sqrt{2} \mp \sqrt{2}i$.

Core Exercises

11, 29, 67

9.4   Vectors

Concepts

(A) Descriptions of vectors.

- Describing vectors in component form.
- Finding the magnitude of a given vector.

(B) Operations on vectors.

- Adding and subtracting vectors; scalar multiplication.
- Resolving vectors into horizontal and vertical components.

(C) Applications of vectors.

- Using vectors to model velocity and force.

Definitions

- **Vector:** A quantity with both magnitude and direction. It can be represented as a line segment with an arrowhead.
- **Scalar:** A quantity with a magnitude but no direction. The real numbers are all examples of scalars.

- **Speed**: If an object is moving, its speed is how fast it is going, and is a positive scalar quantity.

- **Velocity**: If an object is moving, its velocity is a vector whose magnitude is the speed, and whose direction is the direction of motion.

- **Initial point, terminal point:** If a vector is represented as an arrow, the arrow goes from the initial point to the terminal point.

- **Zero vector**: The vector whose magnitude is zero, and which therefore represents no displacement.

- **Horizontal** and **vertical components**: If we write a vector **v** as $\langle a, b \rangle$, the scalars a and b are the horizontal and vertical components of the vector.

- **i, j**: The vectors $\mathbf{i} = \langle 1, 0 \rangle$ and $\mathbf{j} = \langle 0, 1 \rangle$. Do not confuse the vector **i** with the imaginary number i.

Questions to Ask Your Teacher

- What is the preferred notation for a specific vector such as $\langle 1, 2 \rangle$, or a general vector such as **v**?

Hints and Tips

- Vectors are mind-bogglingly useful. Think about all the things you know about that have magnitudes and directions. Gravity, wind, a poke in the belly, the velocity of a car, the sound from a trombone; all of these quantities can be modeled by vectors.

Remember: You can add two vectors. You can subtract two vectors. You can multiply a vector by a scalar. In all these cases, your answer is a vector. *You can never add a vector to a scalar.* Multiplication of a vector by a vector is a little less straightforward, and is dealt with in the next section.

Review

(A) **Descriptions of vectors.**

We can think of vectors as quantities with a magnitude and a direction. We can represent them as line segments in a plane, with an assigned direction. If a vector goes from point A to point B, we can denote it $\overrightarrow{AB}$. We can also represent a vector as an ordered pair of real numbers $\mathbf{v} = \langle a, b \rangle$, where a is the horizontal component of **v** and b is the vertical component. We define special vectors $\mathbf{i} = \langle 1, 0 \rangle$ and $\mathbf{j} = \langle 0, 1 \rangle$, and we can represent a vector $\langle a, b \rangle$ as $a\mathbf{i} + b\mathbf{j}$.

- **Describing vectors in component form.**

If a vector $\mathbf{v}$ is represented in the plane with initial point $P(x_1, y_1)$ and terminal point $Q(x_2, y_2)$, then $\mathbf{v} = \langle x_2 - x_1, y_2 - y_1 \rangle$. Vectors $\mathbf{u} = \langle x_1, y_1 \rangle$ and $\mathbf{v} = \langle x_2, y_2 \rangle$ are equal if and only if $x_1 = x_2$ and $y_1 = y_2$.

Exercise: Write the vector with initial point P and terminal point Q in component form.

Answer:

(a) $P = (1, 2), Q = (-3, 3)$

$\mathbf{v} = \langle -3 - 1, 3 - 2 \rangle = \langle -4, 1 \rangle$

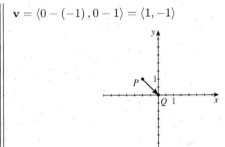

(b) $P = (-1, 1), Q = (0, 0)$

$\mathbf{v} = \langle 0 - (-1), 0 - 1 \rangle = \langle 1, -1 \rangle$

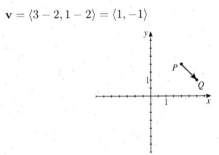

(c) $P = (2, 2), Q = (3, 1)$

$\mathbf{v} = \langle 3 - 2, 1 - 2 \rangle = \langle 1, -1 \rangle$

Notice that the vectors in parts (b) and (c) are *the same vector*. They represent the same displacement, and are therefore equal. Vector equality is independent of where the vector is drawn in the plane.

- **Finding the magnitude of a given vector.**

  The **magnitude** or **length** of the vector $\mathbf{v} = \langle a, b \rangle$ is $|\mathbf{v}| = \sqrt{a^2 + b^2}$.

Exercise: Find the magnitude of the vector. ‖ **Answer:**

(a) $\mathbf{v} = \langle 6, 8 \rangle$

$\quad\quad |\mathbf{v}| = \sqrt{6^2 + 8^2} = 10$

(b) $\mathbf{v} = 2\mathbf{i} - 6\mathbf{j}$

$\quad\quad |\mathbf{v}| = \sqrt{2^2 + (-6)^2} = 2\sqrt{10}$

(c) $\mathbf{v} = \langle 0, -10 \rangle$

$\quad\quad |\mathbf{v}| = \sqrt{0^2 + (-10)^2} = 10$

(B) **Operations on vectors.**

- **Adding and subtracting vectors; scalar multiplication.**

  Geometrically, the **sum** of two vectors is the vector found by placing the initial point of one vector at the terminal point of the other. The **difference** of two vectors $\mathbf{u}$ and $\mathbf{v}$ is defined by $\mathbf{u} - \mathbf{v} = \mathbf{u} + (-\mathbf{v})$. Each of $\mathbf{u} + \mathbf{v}$ and $\mathbf{u} - \mathbf{v}$ is a diagonal of the parallelogram generated by $\mathbf{u}$ and $\mathbf{v}$.

  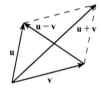

  If a is a **scalar** (that is, a number) and $\mathbf{v}$ is a vector, then the vector $a\mathbf{v}$ is defined to have magnitude $|a|\,|\mathbf{v}|$ and to have the same direction as $\mathbf{v}$ (if $a > 0$) or the opposite direction (if $a < 0$). If $a = 0$, then $a\mathbf{v} = \mathbf{0} = \langle 0, 0 \rangle$, the zero vector.

  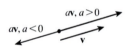

Exercise: The vectors $\mathbf{u}$ and $\mathbf{v}$ are shown. Sketch the vectors $2\mathbf{u}$, $\mathbf{u} + \mathbf{v}$, $\mathbf{v} - \mathbf{u}$, and $\frac{1}{2}\mathbf{u} - 3\mathbf{v}$.

Answer:

$2\mathbf{u}$ has twice the magnitude of $\mathbf{u}$ and the same direction.

To draw $\mathbf{u} + \mathbf{v}$, we put the initial point of $\mathbf{v}$ at the terminal point of $\mathbf{u}$, or make a parallelogram with $\mathbf{u}$ and $\mathbf{v}$ and draw the diagonal.

To draw $\mathbf{v} - \mathbf{u}$, use $\mathbf{v} + (-\mathbf{u})$ or the other diagonal in the parallelogram above.

To draw $\frac{1}{2}\mathbf{u} - 3\mathbf{v}$, first find $\frac{1}{2}\mathbf{u}$ and $-3\mathbf{v}$, and then add the two.

Algebraic operations on vectors are defined as follows:

If $\mathbf{u} = \langle a_1, b_1 \rangle$, $\mathbf{v} = \langle a_2, b_2 \rangle$, and c is any constant, then

$$\mathbf{u} + \mathbf{v} = \langle a_1 + a_2, b_1 + b_2 \rangle \qquad \mathbf{u} - \mathbf{v} = \langle a_1 - a_2, b_1 - b_2 \rangle \qquad c\mathbf{u} = \langle ca_1, cb_1 \rangle$$

Exercise: Let $\mathbf{u} = \langle -12, 18 \rangle$ and $\mathbf{v} = \langle 10, 8 \rangle$. Find $2\mathbf{u}$, $\mathbf{u} + \mathbf{v}$, $\mathbf{v} - \mathbf{u}$, and $\frac{1}{2}\mathbf{u} - 3\mathbf{v}$.

Answer:
$$2\mathbf{u} = 2\langle -12, 18 \rangle = \langle 2(-12), 2(18) \rangle = \langle -24, 36 \rangle,$$
$$\mathbf{u} + \mathbf{v} = \langle -12, 18 \rangle + \langle 10, 8 \rangle = \langle -12 + 10, 18 + 8 \rangle$$
$$= \langle -2, 26 \rangle,$$
$$\mathbf{v} - \mathbf{u} = \langle 10, 8 \rangle - \langle -12, 18 \rangle = \langle 10 + 12, 8 - 18 \rangle = \langle 22, -10 \rangle,$$
and
$$\frac{1}{2}\mathbf{u} - 3\mathbf{v} = \frac{1}{2}\langle -12, 18 \rangle - 3\langle 10, 8 \rangle = \langle -6, 9 \rangle + \langle -30, -24 \rangle$$
$$= \langle -36, -15 \rangle.$$

Exercise: Let $\mathbf{u} = \langle -12, 5 \rangle$ and $\mathbf{v} = \langle 4, 3 \rangle$. Write $\mathbf{u}$, $\mathbf{u} + \mathbf{v}$, and $2\mathbf{u} - 3\mathbf{v}$ in terms of $\mathbf{i}$ and $\mathbf{j}$, and find their lengths.

Answer: $\mathbf{u} = \langle -12, 5 \rangle = -12\mathbf{i} + 5\mathbf{j}$ has length
$$|\mathbf{u}| = \sqrt{(-12)^2 + 5^2} = 13.$$
$\mathbf{u} + \mathbf{v} = \langle -12, 5 \rangle + \langle 4, 3 \rangle = \langle -8, 8 \rangle = -8\mathbf{i} + 8\mathbf{j}$ has length
$$|\mathbf{u} + \mathbf{v}| = \sqrt{(-8)^2 + 8^2} = 8\sqrt{2}.$$
$2\mathbf{u} - 3\mathbf{v} = 2\langle -12, 5 \rangle - 3\langle 4, 3 \rangle = \langle -24, 10 \rangle + \langle -12, -9 \rangle =$
$-36\mathbf{i} + \mathbf{j}$ has length $|2\mathbf{u} - 3\mathbf{v}| = \sqrt{(-36)^2 + 1^2} = \sqrt{1297}$.

The properties of vector addition and multiplication by a scalar are summarized below.

Vector addition	Multiplication of a vector by a scalar
$\mathbf{u} + \mathbf{v} = \mathbf{v} + \mathbf{u}$	$c(\mathbf{u} + \mathbf{v}) = c\mathbf{u} + c\mathbf{v}$
$\mathbf{u} + (\mathbf{v} + \mathbf{w}) = (\mathbf{u} + \mathbf{v}) + \mathbf{w}$	$(c + d)\mathbf{u} = c\mathbf{u} + d\mathbf{u}$
$\mathbf{u} + \mathbf{0} = \mathbf{u}$	$(cd)\mathbf{u} = c(d\mathbf{u}) = d(c\mathbf{u})$
$\mathbf{u} + (-\mathbf{u}) = \mathbf{0}$	$1\mathbf{u} = \mathbf{u}$
	$0\mathbf{u} = \mathbf{0}$
Length of a vector	$c\mathbf{0} = \mathbf{0}$
$|c\mathbf{v}| = |c|\,|\mathbf{v}|$	

- **Resolving vectors into horizontal and vertical components.**

  If $\mathbf{v}$ is a vector with magnitude $|\mathbf{v}|$ and direction θ, then the horizontal component of $\mathbf{v}$ is $a = |\mathbf{v}| \cos \theta$ and the vertical component is $b = |\mathbf{v}| \sin \theta$.

Exercise: Find the direction of the vector $\mathbf{v} = -2\mathbf{i} + 2\sqrt{3}\mathbf{j}$, and sketch the vector.

Answer: We sketch $\mathbf{v}$ first, placing the initial point at $(0,0)$ and the terminal point at $(-2, 2\sqrt{3})$.

The length of $\mathbf{v}$ is $|\mathbf{v}| = \sqrt{(-2)^2 + (2\sqrt{3})^2} = \sqrt{4 + 12} = 4$, and since θ is in the second quadrant,

$\theta = \cos^{-1}\left(\frac{-2}{4}\right) = \cos^{-1}\left(-\frac{1}{2}\right) = \frac{2\pi}{3}$.

Exercise: A vector $\mathbf{u}$ has length $6\sqrt{3}$ and direction $\frac{7\pi}{6}$. Find the horizontal and vertical components and write $\mathbf{v}$ in terms of $\mathbf{i}$ and $\mathbf{j}$.

Answer: The horizontal component of $\mathbf{v}$ is $a = |\mathbf{v}| \cos \theta = 6\sqrt{3} \cos \frac{7\pi}{6} = -9$, and its vertical component is $b = |\mathbf{v}| = 6\sqrt{3} \sin \frac{7\pi}{6} = -3\sqrt{3}$. Thus, $\mathbf{v} = -9\mathbf{i} - 3\sqrt{3}\mathbf{j}$.

(C) **Applications of vectors.**

We can use vectors to model any quantity possessing both magnitude and direction.

- **Using vectors to model velocity and force.**

  The **velocity** of a moving object is described by a vector whose direction is the direction of motion and whose magnitude is the speed of the object. **Force** is another quantity that can be represented by a vector.

Exercise: A plane flies due north at 550 mi/h against a 75 mi/h headwind blowing from N 45° W. Find the true speed and direction of the plane.

Answer: Let $\mathbf{p}$ represent the velocity of the plane and $\mathbf{w}$ the velocity of the wind. Since the plane is flying north at 550 mi/h, $\mathbf{p} = \langle 0, 550 \rangle$. Since the wind is coming *from* N 45° W, its direction is S 45° E and its components resolve as

$\mathbf{w} = \langle -75 \cos(-45°), 75 \sin(-45°) \rangle = \left\langle \frac{75\sqrt{2}}{2}, -\frac{75\sqrt{2}}{2} \right\rangle$.

The sum of these two velocities is the vector

$\mathbf{p} + \mathbf{w} = \langle 0, 550 \rangle + \left\langle \frac{75\sqrt{2}}{2}, -\frac{75\sqrt{2}}{2} \right\rangle$

$\approx \langle 53.033, 496.967 \rangle$

The true speed of the plane is

$|\mathbf{p} + \mathbf{w}| \approx \sqrt{53.033^2 + 496.967^2} \approx 499.8$ mi/h, and the direction is given by

$\theta \approx \tan^{-1} \frac{496.967}{53.033} \approx \tan^{-1}(9.371) \approx 83.9°$. So the plane's true direction is N 83.9° E.

Note that for problems of this type, directions are often given as bearings. A direction such as S 15° W means 15° to the west of south. The equivalent angle is $270° - 15° = 255°$ or $\frac{17\pi}{12}$.

Exercise: A beam weighing 600 pounds is supported by two ropes, as shown in the figure. Find the magnitudes of the tensions $|\mathbf{T}_1|$ and $|\mathbf{T}_2|$.

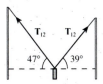

Answer: The other force involved is $|\mathbf{T}_3|$, the weight of the beam.

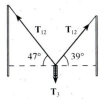

Because the beam is not accelerating, $\mathbf{T}_1 + \mathbf{T}_2 + \mathbf{T}_3 = 0$. Now $\mathbf{T}_1 = \langle a \cos 133°, a \sin 133° \rangle \approx \langle -0.68a, 0.73a \rangle$, $\mathbf{T}_2 = \langle b \cos 39°, b \sin 39° \rangle \approx \langle 0.78b, 0.63b \rangle$, and $\mathbf{T}_3 = \langle 0, -600 \rangle$, where a and b are the magnitudes of the tensions. We set the components equal to 0 and solve: $-0.68a + 0.78b = 0$ and $0.73a + 0.63b - 600 = 0$. Solving the first equation for a, we find $a = 1.15b$. Substituting in the second equation, $0.73 (1.15b) + 0.63b = 600 \Leftrightarrow b \approx 408.2$ pounds. Thus, $a = 1.15b \approx 469.4$ pounds.

Core Exercises

5, 13, 39, 57, 59

9.5   The Dot Product

Concepts

(A) **The dot product of vectors.**

- Calculating dot products.

(B) **The Dot Product Theorem.**

- Finding the angle between two vectors.
- Determining if two vectors are orthogonal.

(C) **Resolving vectors.**

- Finding the component of a vector $\mathbf{u}$ along another vector $\mathbf{v}$.
- Calculating the projection of a vector $\mathbf{u}$ along another vector $\mathbf{v}$.
- Resolving a vector into orthogonal vectors.

(D) **Applications.**

- Solving problems involving work.

Definitions

- **Dot** or **Scalar product**: If $\mathbf{u} = \langle u_1, u_2 \rangle$ and $\mathbf{v} = \langle v_1, v_2 \rangle$ are vectors, their dot product $\mathbf{u} \cdot \mathbf{v}$ is the scalar $u_1 v_1 + u_2 v_2$.

- **Angle between two vectors**: Recall that vectors can be drawn anywhere in the plane. The angle between two vectors is the angle between the line segments if we draw both vectors with initial points at the origin.

- **Orthogonal**: Perpendicular. We use the word "orthogonal" because it is a more general term that is used for various types of mathematical objects you will be exposed to in the future.

- **The component of u along v**: The magnitude of the portion of $\mathbf{u}$ that points in the direction of $\mathbf{v}$, as illustrated in the text.

- **The projection of u onto v**: The component of $\mathbf{u}$ along $\mathbf{v}$ is a scalar quantity. The projection of $\mathbf{u}$ onto $\mathbf{v}$ is the corresponding vector quantity, pointing in the direction of $\mathbf{v}$.

- **Work:** A scalar quantity that measures total effort, loosely speaking.

Hints and Tips

- We know how to add and subtract vectors, and in this section we learn how to multiply a vector by a vector. We call this a "dot product" to distinguish it from scalar multiplication, and from the "cross product" which will be covered in a future course. *Remember that the dot product of two vectors is a scalar.* If they point in the same direction, the dot product is the product of their magnitudes — its largest possible value. If they are orthogonal, the dot product is zero.

The Dot Product Theorem is used so often that I often forget it is a theorem — I think of the dot product as $u_1 v_1 + u_2 v_2$ or as $|\mathbf{u}|\,|\mathbf{v}| \cos\theta$ interchangeably. If I know the components, I use the first expression; if I know the vector magnitudes and the angle, I use the second.

Review

(A) **The dot product of vectors.**

If $\mathbf{u} = \langle a_1, b_1 \rangle$ and $\mathbf{v} = \langle a_2, b_2 \rangle$, then their **dot product**, denoted $\mathbf{u} \cdot \mathbf{v}$, is defined by

$$\mathbf{u} \cdot \mathbf{v} = a_1 a_2 + b_1 b_2$$

The dot product has the following properties:

$$\mathbf{u} \cdot \mathbf{v} = \mathbf{v} \cdot \mathbf{u} \qquad\qquad (a\mathbf{u}) \cdot \mathbf{v} = a\,(\mathbf{u} \cdot \mathbf{v}) = \mathbf{u} \cdot (a\mathbf{v})$$

$$(\mathbf{u} + \mathbf{v}) \cdot \mathbf{w} = \mathbf{u} \cdot \mathbf{w} + \mathbf{v} \cdot \mathbf{w} \qquad\qquad |\mathbf{u}|^2 = \mathbf{u} \cdot \mathbf{u}$$

- **Calculating dot products.**

   Exercise: Let $\mathbf{u} = 3\mathbf{i} + 2\mathbf{j}$ and $\mathbf{v} = -\mathbf{i} - 2\mathbf{j}$. Find $\mathbf{u} \cdot \mathbf{v}$.

   Answer: $\mathbf{u} \cdot \mathbf{v} = (3)(-1) + (2)(-2) = -7$

(B) The Dot Product Theorem

Let **u** and **v** be vectors with initial points at the origin. Then the smaller of the angles formed by these representations of **u** and **v** is called the **angle between u and v**. If θ is the angle between two nonzero vectors **u** and **v**, then $\mathbf{u} \cdot \mathbf{v} = |\mathbf{u}| \, |\mathbf{v}| \cos \theta$. This fact can also be expressed as $\cos \theta = \dfrac{\mathbf{u} \cdot \mathbf{v}}{|\mathbf{u}| \, |\mathbf{v}|}$. This allows us to use the dot product to test whether two vectors are perpendicular; in particular, two nonzero vectors **u** and v are perpendicular if and only if $\mathbf{u} \cdot \mathbf{v} = 0$.

• **Finding the angle between two vectors.**

Exercise: Find $\mathbf{u} \cdot \mathbf{v}$ and the angle between **u** and **v**, to the nearest degree.

Answer:

(a) $\mathbf{u} = \langle -2, 5 \rangle$, $\mathbf{v} = \langle 4, 3 \rangle$

$\mathbf{u} \cdot \mathbf{v} = -2\,(4) + 5\,(3) = 7$. $|\mathbf{u}| = \sqrt{(-2)^2 + 5^2} = \sqrt{29}$ and $|\mathbf{v}| = \sqrt{4^2 + 3^2} = 5$, so $\cos \theta = \dfrac{\mathbf{u} \cdot \mathbf{v}}{|\mathbf{u}| \, |\mathbf{v}|} = \dfrac{7}{5\sqrt{29}} \approx 0.260$, so $\theta \approx \arccos 0.260 \approx 75°$.

(b) $\mathbf{u} = \langle 4, -3 \rangle$, $\mathbf{v} = \langle 6, -8 \rangle$

$\mathbf{u} \cdot \mathbf{v} = 4\,(6) - 3\,(-8) = 48$. $|\mathbf{u}| = \sqrt{4^2 + (-3)^2} = 5$ and $|\mathbf{v}| = \sqrt{6^2 + (-8)^2} = 10$, so $\cos \theta = \dfrac{\mathbf{u} \cdot \mathbf{v}}{|\mathbf{u}| \, |\mathbf{v}|} = \dfrac{48}{5 \cdot 10} = 0.96$ and $\theta = \arccos 0.96 \approx 16°$.

• **Determining if two vectors are orthogonal.**

Exercise: Determine whether the given vectors are orthogonal.

Answer:

(a) $\mathbf{u} = \langle -2, 4 \rangle$, $\mathbf{v} = \langle 7, 3 \rangle$

$\mathbf{u} \cdot \mathbf{v} = -2\,(7) + 4\,(3) = -2 \neq 0$, so **u** and **v** are not orthogonal.

(b) $\mathbf{u} = \langle -4, 14 \rangle$, $\mathbf{v} = \langle -7, -2 \rangle$

$\mathbf{u} \cdot \mathbf{v} = -4\,(-7) + 14\,(-2) = 0$, so **u** and **v** are orthogonal.

(c) $\mathbf{u} = 9\mathbf{i} - 7\mathbf{j}$, $\mathbf{v} = 21\mathbf{i} + 27\mathbf{j}$

$\mathbf{u} \cdot \mathbf{v} = 9\,(21) - 7\,(27) = 0$, so **u** and **v** are orthogonal.

(C) Resolving vectors.

The **component of u along v** (or the **component of u in the direction of v**) is defined as $|\mathbf{u}| \cos \theta = \dfrac{\mathbf{u} \cdot \mathbf{v}}{|\mathbf{v}|}$. The **projection of u onto v**, denoted $\text{proj}_{\mathbf{v}} \mathbf{u}$, is the vector whose *direction* is the same as **v** and whose *length* is the component of **u** along **v**. The projection of **u** onto **v** is given by

$$\text{proj}_{\mathbf{v}} \mathbf{u} = (\text{component of } \mathbf{u} \text{ along } \mathbf{v})\,(\text{unit vector in the direction of } \mathbf{v})$$
$$= \left(\frac{\mathbf{u} \cdot \mathbf{v}}{|\mathbf{v}|} \right) \frac{\mathbf{v}}{|\mathbf{v}|} = \left(\frac{\mathbf{u} \cdot \mathbf{v}}{|\mathbf{v}|^2} \right) \mathbf{v}$$

• **Finding the component of a vector u along another vector v.**

Exercise: Find the component of $u = 3i + 2j$ along $v = -i - 2j$.

Answer: The component of u along v is $|u| \cos \theta = \dfrac{u \cdot v}{|v|}$.

$u \cdot v = -7$ and $|v| = \sqrt{(-1)^2 + (-2)^2} = \sqrt{5}$, so the

component of u along v is therefore $\dfrac{-7}{\sqrt{5}} = -\dfrac{7\sqrt{5}}{5}$.

• **Calculating the projection of a vector u along another vector v.**

Exercise: Let $u = \langle -1, 5 \rangle$ and $v = \langle 4, 3 \rangle$. Find the component of u along v, then find $\text{proj}_v \, u$. Sketch u, v, and $\text{proj}_v \, u$.

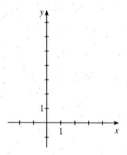

Answer: The component of u along v is

$\dfrac{u \cdot v}{|v|} = \dfrac{-1(4) + 5(3)}{\sqrt{4^2 + 3^2}} = \dfrac{11}{5}$ and the projection is

$\text{proj}_v \, u = \left(\dfrac{u \cdot v}{|v|^2} \right) v = \dfrac{11}{5^2} \langle 4, 3 \rangle = \left\langle \dfrac{44}{25}, \dfrac{33}{25} \right\rangle$.

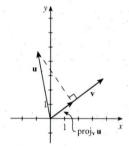

• **Resolving a vector into orthogonal vectors.**

Exercise: Let $u = \langle -1, 5 \rangle$ and $v = \langle 4, 3 \rangle$. Resolve u into u_1 and u_2, where u_1 is parallel to v and u_2 is orthogonal (perpendicular) to v.

Answer: $u_1 = \text{proj}_v \, u = \left\langle \dfrac{44}{25}, \dfrac{33}{25} \right\rangle$ (see the previous exercise). $u_2 = u - \text{proj}_v \, u = \left\langle -\dfrac{69}{25}, \dfrac{92}{25} \right\rangle$.

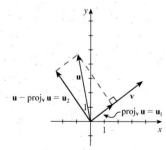

Ⓓ **Applications.**

- **Solving problems involving work.**

Work is defined as force times distance. So the work W done by a force $\mathbf{F}$ in the direction of a vector $\mathbf{D}$ is

$$\boxed{W = \mathbf{F} \cdot \mathbf{D}}$$

Exercise: Find the work done by a force $\mathbf{F} = 2\mathbf{i} + 3\mathbf{j}$ in moving an object from $P(4,7)$ to $Q(5,10)$.

Answer: We find $\mathbf{D}$ and apply the formula: $\mathbf{D} = (5-4)\,\mathbf{i} + (10-7)\,\mathbf{j} = \mathbf{i} + 3\mathbf{j}$, so the work done is $W = \mathbf{F} \cdot \mathbf{D} = 2(1) + 3(3) = 11$. Since we are given no units, we leave the answer without units.

Exercise: A man pulls a wagon up a 15° incline by exerting a force of 40 lb on the handle. If the handle makes an angle of 45° with the incline, find the work done in moving the wagon 20 feet up this incline.

Answer: Of course we start with a sketch.

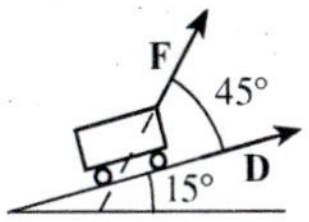

The handle makes an angle of $45^\circ + 15^\circ = 60^\circ$ with the horizontal. Thus,
$\mathbf{F} = 40\cos 60^\circ\,\mathbf{i} + 40\sin 60^\circ\,\mathbf{j} = 20\mathbf{i} + 20\sqrt{3}\mathbf{j}$. Also,
$\mathbf{D} = 20\cos 15^\circ\,\mathbf{i} + 20\sin 15^\circ\,\mathbf{j} \approx 19.32\mathbf{i} + 5.18\mathbf{j}$, so
$W = \mathbf{F} \cdot \mathbf{D} \approx (20)(19.32) + \left(20\sqrt{3}\right)(5.18) \approx 566$ ft-lb.
Alternately, we could use the Dot Product Theorem. We know that the angle between $\mathbf{F}$ and $\mathbf{D}$ is 45°, so
$W = \mathbf{F} \cdot \mathbf{D} = |\mathbf{F}|\,|\mathbf{D}|\cos 45^\circ = 40 \cdot 20 \cdot \frac{\sqrt{2}}{2} = 400\sqrt{2}$ ft-lb.

Core Exercises

3, 21, 31, 41, 47

10 Systems of Equations and Inequalities

10.1 Systems of Equations

◤ Concepts

Ⓐ **Solving systems of equations.**

- Solving systems of equations using the substitution method.
- Solving systems of equations using the elimination method.
- Solving systems of equations using the graphical method.

◤ Definitions

- **System of equations:** Two or more equations with a common solution set.

◤ Questions to Ask Your Teacher

- When solving a system, should we always check the answers explicitly, or is it okay to just do this step on a calculator without writing everything down?

◤ Hints and Tips

- Many calculators will solve a system of equations automatically (giving an approximation as an answer). Others will allow you to automatically find the intersection point between two graphs, which is tantamount to solving a system of two equations with two unknowns (assuming both equations can be solved for the same variable).

When dealing with a system of *linear* equations (of the form $ax + by = c$) it is best to use the elimination method whenever possible. When the equations are not linear, it is usually easier to use the substitution method. After you've practiced, you will find your preferences, although you should be able to use all the methods presented in the section.

◤ Review

Ⓐ **Solving systems of equations.**

A **solution** to a system of equations is a point that simultaneously makes each equation of the system a true statement. A system of equation can have one solution, many solutions, or no solutions. This section provides several methods of finding the solution or solutions to a system of equations.

- **Solving systems of equations using the substitution method.**

The **substitution method** is used when an equation in the system can be solved for one variable in terms of the other variables. The steps of the substitution method are as follows:

1. **Solve one equation** for one variable in terms of the other variables.

2. **Substitute** this value into *another equation*. (It is very important that you substitute this value into a *different* equation.) Solve for the second variable.

3. **Back-substitute:** Substitute the value you found in Step 2 back into the equation used in Step 1 to solve for the remaining variable.

Exercise: Find all real solutions (x, y) of the system of equations
$$\begin{cases} x^2 - xy - y^2 = 1 \\ 2x + y = 1 \end{cases}$$

Answer: Solve the second equation for y: $y = 1 - 2x$. Now substitute this value into the first equation:
$$x^2 - x(1 - 2x) - (1 - 2x)^2 = 1 \quad \Leftrightarrow$$
$$x^2 - x + 2x^2 - (1 - 4x + 4x^2) = 1 \quad \Leftrightarrow$$
$$x^2 - x + 2x^2 - 1 + 4x - 4x^2 = 1 \quad \Leftrightarrow \quad -x^2 + 3x - 1 = 1$$
$$\Leftrightarrow \quad -x^2 + 3x - 2 = 0 \quad \Leftrightarrow \quad x^2 - 3x + 2 = 0 \quad \Leftrightarrow$$
$$(x - 2)(x - 1) = 0 \quad \Leftrightarrow \quad x = 2 \text{ or } x = 1 \text{ Now}$$
back-substitute: if $x = 2$, then $y = 1 - 2(2) = -3$. If $x = 1$, then $y = 1 - 2(1) = -1$. So the possible solutions are $(2, -3)$ and $(1, -1)$.

We check $(2, -3)$: $(2)^2 - (2)(-3) - (-3)^2 = 4 + 6 - 9 = 1$ and $2(2) + (-3) = 4 - 3 = 1$, so $(2, -3)$ is indeed a solution. Now check $(1, -1)$: $(1)^2 - (1)(-1) - (-1)^2 = 1 + 1 - 1 = 1$ and $2(1) + (-1) = 2 - 1 = 1$. So $(1, -1)$ is also a solution.

Exercise: Find all real solutions (x, y) of the system of equations
$$\begin{cases} 3x + y = -1 \\ -2x + 3y = 8 \end{cases}$$

Answer: Solve $3x + y = -1$ for y (since it is easier): $y = -3x - 1$. Substitute for y in the second equation. $-2x + 3(-3x - 1) = 8 \quad \Leftrightarrow \quad -2x - 9x - 3 = 8$ $\Leftrightarrow \quad -11x = 11 \quad \Leftrightarrow \quad x = -1$. Substituting this value of x back into $y = -3x - 1$ gives $y = -3(-1) - 1 = 3 - 1 = 2$. So the only possible solution is $(-1, 2)$. Check: $3(-1) + (2) = -3 + 2 = -1$ and $-2(-1) + 3(2) = 2 + 6 = 8$. So the solution is $(-1, 2)$.

- **Solving systems of equations using the elimination method.**

Use the **elimination method** when a variable can be eliminated by the addition of a multiple of one equation to a multiple of another equation. The steps of this method are as follows:

---

1. **Pick the variable to eliminate. Adjust the coefficients** of this variable by multiplying *each* equation by the appropriate *nonzero* numbers so that the coefficients of this chosen variable add to zero.

2. **Add the two equations** to eliminate the variable you chose in Step 1. Solve this new equation for the other variable.

3. **Back-substitute:** Substitute the value you found in Step 2 back into the equation used in Step 1 to solve for the remaining variable.

Exercise: Find all real solutions (x, y) of the system of equations $\begin{cases} 3x + 2y = -5 \\ x + 4y = 5 \end{cases}$

Answer: We first eliminate x by multiplying the second equation by -3 and adding it to the first:

$$\begin{array}{r} 3x + 2y = -5 \\ -3\,(x + 4y) = -3\,(5) \\ \hline -10y = -20 \end{array}$$

Thus $y = 2$. Substituting into the original second equation and solving for x, we get $x + 4\,(2) = 5 \iff x + 8 = 5 \iff x = -3$.

Another method: Eliminate y by multiplying the first equation by -2 and add it to the second:

$$\begin{array}{r} -2\,(3x + 2y) = -2\,(-5) \\ x + 4y = 5 \\ \hline -5x = 15 \end{array}$$

Thus $x = 2$, and substitution gives $y = 2$ as before.

Exercise: Find all real solutions (x, y) of the system of equations

$$\begin{cases} x^2 + y^2 + 2y = 10 \\ x^2 + 4y = 7 \end{cases}$$

Answer: The x^2 term can be eliminated by subtracting the second equation from the first equation to give $y^2 - 2y = 3$ $\iff y^2 - 2y - 3 = 0 \iff (y - 3)(y + 1) = 0 \iff y = 3$ or $y = -1$. If $y = 3$, then the second equation gives $x^2 + 12 = 7 \iff x^2 = -5$ which has no real solution. If $y = -1$, then the second equation gives $x^2 - 4 = 7 \iff x = \pm\sqrt{11}$. The solutions are $\left(\pm\sqrt{11}, -1\right)$. Be sure to check that each is actually a solution to each of the original equations.

- **Solving systems of equations using the graphical method.**

Graphing devices are sometimes useful in solving systems of equations in two variables. However, with most graphing devices, an equation must first be expressed in terms of one or more functions of the form $y = f(x)$ before it can be graphed. Not all equations can be expressed in this way, so not all systems can be solved this way.

Exercise: Find all real solutions to the system of equations

$$\begin{cases} x^2y + 3xy + 4y - 4 = 0 \\ x^2 + 4x - y = 6 \end{cases}$$

correct to two decimal places.

Answer: We try to isolate y in the first equation: $x^2y + 3xy + 4y - 4 = 0 \quad \Leftrightarrow \quad x^2y + 3xy + 4y = 4 \quad \Leftrightarrow$ $y(x^2 + 3x + 4) = 4 \quad \Leftrightarrow \quad y = \dfrac{4}{x^2 + 3x + 4}$. In the second equation, $x^2 + 4x - y = 6 \quad \Leftrightarrow \quad y = x^2 + 4x - 6$. We plot both of these expressions in the viewing rectangle $[-8, 5]$ by $[-2, 3]$:

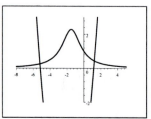

We see that the solutions are approximately $(-5.20, 0.26)$ and $(1.23, 0.43)$.

Exercise: Find all real solutions to the system of equations

$$\begin{cases} e^x y = x^2 \\ (x+2)^2 + (y-2)^2 = 9 \end{cases}$$

correct to two decimal places.

Answer: We isolate y in each equation: $e^x y = x^2 \quad \Leftrightarrow$ $y = \dfrac{x^2}{e^x}$ and $(x+2)^2 + (y-2)^2 = 9 \quad \Leftrightarrow$ $(y-2)^2 = 9 - (x+2)^2 \quad \Leftrightarrow \quad y - 2 = \pm\sqrt{9 - (x+2)^2}$ $\Leftrightarrow \quad y = 2 \pm \sqrt{9 - (x+2)^2}$. We graph $y = 2 \pm \sqrt{9 - (x+2)^2}$ and find any points of intersection with $y = \dfrac{x^2}{e^x}$:

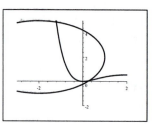

The solutions are approximately $(-1.21, 4.90)$ and $(0.29, 0.06)$.

Exercise: Find all real solutions to the system of equations

$$\begin{cases} xy = x^2 - 3x + 2 \\ x = 3y \end{cases}$$

correct to three decimal places.

Answer: We express each equation as a function of x:

$$xy = x^2 - 3x + 2 \quad \Leftrightarrow \quad y = \frac{x^2 - 3x + 2}{x} \text{ and } x = 3y$$

$$\Leftrightarrow \quad y = \tfrac{1}{3}x. \text{ We graph each expression and find the points of intersection.}$$

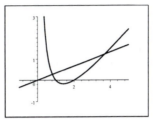

The solutions are approximately $(0.814, 0.271)$ and $(3.686, 1.229)$.

Core Exercises

19, 25, 31, 51, 55

10.2   Systems of Linear Equations in Two Variables

Concepts

(A) **Linear systems of two equations with two variables.**

- Solving these systems.
- Modeling with these systems.

Definitions

- **Linear equation:** An equation that can be put into the form $ax + by = c$ where a, b, and c are constants.
- **Inconsistent system:** A system of equations that has no solution.
- **Dependent system**: A system of equations that has infinitely many solutions.

Hints and Tips

- When you solve a system, you should be hoping to get $x =$ something, $y =$ something else. One solution.
- If instead you get something that is false, like $1 = 2$, then you know there is no solution. (No values of x and y will magically make one equal to two.)
- If you get something that is true but useless, like $1 = 1$ then you know there are infinitely many solutions. (Yes, yes, one is equal to one. What does that tell me about the values of x and y? Nothing!)

▸ Review

(A) **Linear systems of two equations with two variables.**

A system of two linear equations in two unknowns can have *infinitely many* solutions, *exactly one* solution, or *no* solution. A system that has no solution is said to be **inconsistent**, while a system with infinitely many solutions is called **dependent**.

- **Solving these systems.**

  We solve systems of linear equations using the methods from the previous section.

Exercise: Solve the system

$$\begin{cases} 3x + 2y = 3 \\ x - 2y = 3 \end{cases}$$

If it has infinitely many solutions, express them in terms of x.

Answer: We add the two equations to eliminate y:

$(3x + 2y) + (x - 2y) = 3 + 3 \quad \Leftrightarrow \quad 4x = 6 \quad \Leftrightarrow$

$x = \frac{2}{3}$. Back-substitution gives $\frac{3}{2} - 2y = 3 \quad \Leftrightarrow \quad y = -\frac{3}{4}$.

The solution is thus $\left(\frac{3}{2}, -\frac{3}{4}\right)$.

Exercise: Solve the system

$$\begin{cases} 2x - 3y = 9 \\ -4x + 6y = 4 \end{cases}$$

If it has infinitely many solutions, express them in terms of x.

Answer: To eliminate x, we multiply the first equation by 2 and add it to the second equation:

$2(2x - 3y) + (-4x + 6y) = 2(9) + 4 \quad \Leftrightarrow \quad 0 = 22$. Thus, the system is inconsistent and there is no solution. Further examination of this system shows that the lines $2x - 3y = 9$ and $-4x + 6y = 4$ are parallel, and thus do not intersect.

Exercise: Solve the system

$$\begin{cases} \frac{3}{4}x + y = 1 \\ 3x + 2y = 1 \end{cases}$$

If it has infinitely many solutions, express them in terms of x.

Answer: To eliminate x, we multiply the first equation by -4 and add it to the second equation:

$-4\left(\frac{3}{4}x + y\right) + (3x + 2y) = -4(1) + 1 \quad \Leftrightarrow \quad -2y = -3$

$\Leftrightarrow \quad y = \frac{3}{2}$. Back-substituting gives $3x + 2\left(\frac{3}{2}\right) = 1 \quad \Leftrightarrow$

$3x = -2 \quad \Leftrightarrow \quad x = -\frac{2}{3}$. The solution is $\left(-\frac{2}{3}, \frac{3}{2}\right)$.

Exercise: Solve the system

$$\begin{cases} \frac{3}{4}x - \frac{1}{2}y = 1 \\ -3x + 2y = -4 \end{cases}$$

If it has infinitely many solutions, express them in terms of x.

Answer: To eliminate x, we multiply the first equation by 4 and add it to the second equation:

$4\left(\frac{3}{4}x - \frac{1}{2}y\right) + (-3x + 2y) = 4(1) - 4 \quad \Leftrightarrow \quad 0 = 0$. So the two equations in the original system are two different ways of expressing the same equation. We solve the second equation for y: $-3x + 2y = -4 \quad \Leftrightarrow \quad 2y = 3x - 4 \quad \Leftrightarrow$

$y = \frac{3}{2}x - 2$. Thus, the solutions are $\left(x, \frac{3}{2}x - 2\right)$, where x is any real number.

● **Modeling with these systems.**

Systems of linear equations appear frequently in science applications and models in other areas. Use these guidelines when modeling with systems of equations:

1. **Identify the variables.** Identify the quantities the problem asks you to find. This is done by **reading the question carefully**. Assign letters to these quantities in the problem and define what each letter represents. This helps you and others to understand what values you are looking for.

2. **Express all unknown quantities in terms of the variables.** Read the problem again and express all the quantities mentioned in the problem in terms of the variables you defined in Step 1.

3. **Set up a system of equations.** Find the crucial facts in the problem that give the relationships between the expressions you found in Step 2. Set up a system of equations that expresses these relationships.

4. **Solve the system and interpret the results.** Solve the system found in Step 3, check your solutions, and state your final answer as a sentence that answers the question posed.

Exercise: The sum of two numbers is 43 and their difference is 11. Find the two numbers.

Answer: Let x be the larger number and y the smaller number. Then "the sum of two numbers is 43" means that $x + y = 43$, and "their difference is 11" means that $x - y = 11$. So we get

the system $\begin{cases} x + y = 43 \\ x - y = 11 \end{cases}$ Adding these two equations gives

$2x = 54$, so $x = 27$. Substituting into the first equation gives $27 + y = 43 \iff y = 16$. The numbers are 27 and 16.

Exercise: A movie theater charges $4.50 for children and $8.50 for adults. On a certain day, 625 people saw a movie and the theater collected $3632.50 in receipts. How many children and how many adults were admitted?

Answer: Let x be the number of children admitted and y the number of adults admitted. Then "625 people saw a movie" means that $x + y = 625$ and "the theater collected $3632.50 in receipts" means that $4.50x + 8.50y = 3632.50$. So we get the

system $\begin{cases} x + y = 625 \\ 4.5x + 8.5y = 3632.5 \end{cases}$ To solve, multiply the

first equation by -4.5 and add it to the second equation:
$-4.5(x + y) + (4.5x + 8.5y) = -4.5(625) + 3632.5 \iff$
$4y = 820 \iff y = 205$. Substituting into the first equation gives $x + 205 = 625 \iff x = 420$, so the theater admitted 420 children and 205 adults.

Exercise: A city plans to plant 50 new trees in a subdivision. If birch trees cost $40 each and oak trees cost $55 each, how many of each kind of tree can they plant for $2300?

Answer: Let x be the number of birch trees and y the number of oak trees. There are to be 50 new trees, so $x + y = 50$. The total cost is $2300, so $40x + 55y = 2300$. We get the system

$$\begin{cases} x + y = 50 \\ 40x + 55y = 2300 \end{cases}$$ Multiply the first equation by -40

and add it to the second:

$$-40(x+y) + (40x + 55y) = -40(50) + 2300 \quad \Leftrightarrow$$

$15y = 300 \quad \Leftrightarrow \quad y = 20$. Substituting into the first equation gives $x = 30$, so the city should plant 30 birch trees and 20 oak trees in the subdivision.

Core Exercises

9, 23, 39, 55

10.3 Systems of Linear Equations in Several Variables

Concepts

(A) **Systems of linear equations.**

- Solving these systems.
- Modeling with these systems.

Definitions

- **Triangular form:** A linear system is in triangular form if it looks like this:

$$v + 4w - x - 4y - z = -1$$
$$2w + x - 2y + 4z = 5$$
$$x + y + z = 3$$
$$-2y + 3z = 1$$
$$z = 1$$

We like triangular systems because they are easy to solve.

- **Gaussian elimination:** The process of transforming a general system of equations into an equivalent system in triangular form.

Review

(A) Systems of linear equations.

A **linear equation in n variables** is an equation of the form $a_1x_1 + a_2x_2 + a_3x_3 + \cdots + a_nx_n = c$, where $a_1, a_2, a_3, \ldots, a_n, c$ are constants and $x_1, x_2, x_3, \ldots, x_n$ are variables. When $n \leq 4$, the letters x, y, z, and w are usually used for the variables. Each term in a linear equation is either a constant or a constant multiple of *one* of the variables raised to the first power.

A system is in **triangular form** when the first variable fails to appear in the second equation, the first two variables fail to appear in the third equation, and so on. It is easy to solve a system that is in triangular form using back-substitution. For a system of three equations, the idea is to eliminate all but one variable from one equation, then all but two variables from another equation.

• Solving these systems.

Two systems of linear equations are **equivalent** if they have the same solution set. The only allowable algebraic operations for transforming one system into an equivalent system are:

1. Add a multiple of one equation to another equation.
2. Multiply an equation by a nonzero constant.
3. Rearrange the order of the equations.

Gaussian elimination is a systematic process to transform a system of linear equations using elimination into a triangular system:

1. Use two equations to create a new equation where one of the variables is eliminated. Then use a different pair of equations to create another new equation where the same variable is eliminated.
2. Use these two new equations to eliminate another variable.
3. Back-substitute to find the solution. Remember to check the resulting solution in all three original equations.

Exercise: Solve the system

$$\begin{cases} x + 3y + 3z = -1 \\ y - z = -7 \\ z = 2 \end{cases}$$

using back-substitution.

Answer: This system is in triangular form. Back-substituting $z = 2$ into the second equation gives $y - (2) = -7 \Leftrightarrow y = -5$, and back-substituting $y = -5$ and $z = 2$ into the first equation gives $x + 3(-5) + 3(2) = -1 \Leftrightarrow x = 8$. Thus, the solution is $(8, -5, 2)$.

Exercise: Solve $\begin{cases} x + y + 3z = 1 \\ 3x + 2y + 3z = 1 \\ 2x + y + 2z = -3 \end{cases}$

using Gaussian elimination.

Answer: We use the first equation to eliminate x from the second and third equations:

$$\begin{cases} x + y + 3z = 1 \\ y + 6z = 2 \quad \text{3 × Equation 1 − Equation 2} \\ y + 4z = 5 \quad \text{2 × Equation 1 − Equation 3} \end{cases}$$

Now subtract the third equation from the second:

$$\begin{cases} x + y + 3z = 1 \\ y + 6z = 2 \\ 2z = -3 \end{cases}$$

This is a triangular system. Back-substituting $z = -\frac{3}{2}$ in the second equation gives $y + 6\left(-\frac{3}{2}\right) = 2 \quad \Leftrightarrow \quad y - 9 = 2 \Leftrightarrow \quad y = 11$, and back-substituting for y and z in the first equation gives $x + (11) + 3\left(-\frac{3}{2}\right) = 1 \quad \Leftrightarrow \quad x + 11 - \frac{9}{2} = 1 \Leftrightarrow \quad x = -\frac{11}{2}$. So the solution is $\left(-\frac{11}{2}, 11, -\frac{3}{2}\right)$.

Check: $-\frac{11}{2} + (11) + 3\left(-\frac{3}{2}\right) = 1$,

$3\left(-\frac{11}{2}\right) + 2(11) + 3\left(-\frac{3}{2}\right) = 1$, and

$2\left(-\frac{11}{2}\right) + (11) + 2\left(-\frac{3}{2}\right) = -3$.

Exercise: Solve the system

$$\begin{cases} x - 2z + w = 2 \\ x + y + 5z + w = 7 \\ 2x + y + 3z + 2w = 9 \\ 3x + y + z + 3w = 11 \end{cases}$$

or show

that it has no solution.

Answer: We use the first equation to eliminate x from the second, third, and fourth equations.

$$\begin{cases} x - 2z + w = 2 \\ y + 7z = 5 \quad \text{Equation 2 − Equation 1} \\ y + 7z = 5 \quad \text{Equation 3 − 2 × Equation 1} \\ y + 7z = 5 \quad \text{Equation 4 − 3 × Equation 1} \end{cases}$$

Since the last three equations are the same, this system

equivalent to the system $\begin{cases} x - 2z + w = 2 \\ y + 7z = 5 \end{cases}$

This system is dependent since this it has only 2 equations in 4 variables. We solve the first equation for x in terms of z and w, and then we solve the second equation for y in terms of z:

$x - 2z + w = 2 \quad \Leftrightarrow \quad x = 2 + 2z - w$ and $y + 7z = 5 \quad \Leftrightarrow$

$y = 5 - 7z$. Thus, the solutions are $(2 + 2z - w, 5 - 7z, z, w)$ where z and w are any real numbers.

Exercise: Solve the system

$$\begin{cases} x - 7y + 5z = 2 \\ 2x + 3y - z = 5 \\ 4x - 11y + 9z = 10 \end{cases}$$ or show that it

has no solution.

Answer: We use the first equation to eliminate x from the second and third equations:

$$\begin{cases} x - 7y + 5z = 2 \\ 17y - 11z = 1 \quad \text{Equation 2} - 2 \times \text{Equation 1} \\ 17y - 11z = 0 \quad \text{Equation 3} - 4 \times \text{Equation 1} \end{cases}$$

But this is equivalent to

$$\begin{cases} x - 7y + 5z = 2 \\ 17y - 11z = 1 \\ 0 = 1 \quad \text{Equation 2} - \text{Equation 3} \end{cases}$$

Since the last equation is false, the system has no solution.

• **Modeling with these systems**

Two points in the coordinate plane determine a unique line (first degree polynomial) and three non-collinear points determine a unique quadratic (second-degree polynomial). Of course, the x-coordinates must be distinct, otherwise the graph fails the Vertical Line Test and cannot represent a function.

Exercise: Find an equation of a function whose graph passes through the points $(3, 7)$ and $(-8, 2)$.

Answer: Since we are given two points, we are looking for a first-degree equation of the form $y = Ax + B$. Substituting $(3, 7)$ for (x, y), we get $3A + B = 7$, and substituting $(-8, 2)$, we get $-8A + B = 2$. Thus we must solve the

system $\begin{cases} 3A + B = 7 \\ -8A + B = 2 \end{cases} \Leftrightarrow$

$$\begin{cases} 3A + B = 7 \\ -11A = -5 \end{cases} \quad \text{Thus, } -11A = -5 \quad \Leftrightarrow \quad A = \tfrac{5}{11},$$

and back-substituting gives $3\left(\tfrac{5}{11}\right) + B = 7 \quad \Leftrightarrow$

$B = \tfrac{77}{11} - \tfrac{15}{11} = \tfrac{62}{11}$. Thus, one such function is $y = \tfrac{5}{11}x + \tfrac{62}{11}$.

We check that both points do indeed lie on the curve:

$\tfrac{5}{11}(3) + \tfrac{62}{11} = 7$ and $\tfrac{5}{11}(-8) + \tfrac{62}{11} = 2$.

Exercise: Find the equation of a curve passing through the points $(-1, -2)$, $(2, -5)$, and $(3, 2)$.

Answer: Since we are given three points, we are looking for a second-degree equation of the form $y = Ax^2 + Bx + C$. Substituting $(x, y) = (-1, -2)$ gives $A - B + C = -2$, $(2, -5)$ gives $4A + 2B + C = -5$, and $(3, 2)$ gives $9A + 3B + C = 2$. The system is thus

$$\begin{cases} A - B + C = -2 \\ 4A + 2B + C = -5 \\ 9A + 3B + C = 2 \end{cases} \Leftrightarrow$$

$$\begin{cases} A - B + C = -2 \\ 6B - 3C = 3 \quad \text{Equation } 2 - 4 \times \text{Equation } 1 \\ 12B - 8C = 20 \quad \text{Equation } 3 - 9 \times \text{Equation } 1 \end{cases} \Leftrightarrow$$

$$\begin{cases} A - B + C = -2 \\ 2B - C = 1 \quad \frac{1}{3} \times \text{Equation } 2 \\ -2C = 14 \quad \text{Equation } 3 - 2 \times \text{Equation } 2 \end{cases}$$

Thus, $C = -7$ and back-substituting gives $2B - (-7) = 1$ $\Leftrightarrow$ $B = -3$ and $A - (-3) + (-7) = -2$ $\Leftrightarrow$ $A = 2$. The parabola $y = 2x^2 - 3x - 7$ passes through the desired points.

Check: $2(-1)^2 - 3(-1) - 7 = -2$, $2(2)^2 - 3(2) - 7 = -5$, and $2(3)^2 - 3(3) - 7 = 2$.

◢ Core Exercises

1, 11, 17, 21, 35, 37, 39

◢ 10.4   Systems of Linear Equations: Matrices

◢ Concepts

(A) **Solving linear systems using matrices.**

- Converting a linear system into an augmented matrix.
- Solving a system using Gaussian elimination.
- Solving a system using Gauss-Jordan elimination.
- Solving inconsistent systems and dependent systems.

Definitions

- $m \times n$ **matrix:** A rectangular array of numbers with m rows and n columns.

- **Augmented matrix of a linear system:** The matrix that is obtained by writing the coefficients and constants of a linear system in matrix form.

- **Row-echelon form / Reduced row-echelon form:** A matrix is in echelon form if it satisfies the following conditions:
  1. The first nonzero number in each row (reading from left to right) is 1. This is called the **leading entry**.
  2. The leading entry in each row is to the right of the of the leading entry in the row immediately above it.
  3. All rows consisting entirely of zeros are at the bottom of the matrix.

  A matrix in echelon form is in **reduced row-echelon form** if it also satisfies the condition that every entry above and below each leading entry is a 0.

Hints and Tips

- You may notice that the Gaussian elimination described in this section is very similar to the Gaussian elimination in the previous section. You are right! The only difference is that when we use matrices, we don't have to mess around with writing out variables, plus signs, equal signs, and the like. We only have to write out the coefficients.

Review

(A) **Solving linear systems using matrices**

An $m \times n$ **matrix** is a rectangular array of numbers with m rows and n columns.

$$\begin{bmatrix} a_{11} & a_{12} & a_{13} & \cdots & a_{1n} \\ a_{21} & a_{22} & a_{23} & \cdots & a_{2n} \\ a_{31} & a_{32} & a_{33} & \cdots & a_{3n} \\ \vdots & \vdots & \vdots & \ddots & \vdots \\ a_{m1} & a_{m2} & a_{m3} & \cdots & a_{mn} \end{bmatrix}$$

The **dimension** of the matrix is $m \times n$, which is read "m by n". The numbers are called the **entries** of the matrix. The entry a_{ij} indicates that it is in the ith row and the jth column. Instead of writing the equations of a system in full, we may write only the coefficients and constants in a matrix called an **augmented matrix**. Each equation corresponds to a row of the matrix and each variable to a column of the matrix. The constants also contribute a column.

- **Converting a linear system into an augmented matrix.**

Exercise: Write the augmented matrix of the system of equations

$$\begin{cases} x - 7y + 3z = 5 \\ 2x + 3y - 2z = 11 \\ -x + 5y + 4z = -2 \end{cases}$$

Answer: Since the system has three equations in three variables, the augmented matrix is the 3×4 matrix

$$\begin{bmatrix} 1 & -7 & 3 & 5 \\ 2 & 3 & -2 & 11 \\ -1 & 5 & 4 & -2 \end{bmatrix}.$$

Exercise: Write the system of equations that corresponds to the augmented matrix

$$\begin{bmatrix} 1 & 6 & 0 & 8 \\ 0 & 1 & -5 & -2 \\ 0 & 0 & 1 & 3 \end{bmatrix}$$

and solve.

Answer: This is a 3×4 matrix, so we are trying to solve a system of three equations in three unknowns, namely

$$\begin{cases} x + 6y & = & 8 \\ & y - 5z = -2 \\ & z = & 3 \end{cases}$$

This is a triangular system. Using back-substitution, we find $y - 5(3) = -2 \Leftrightarrow y = 13$ and $x + 6(13) = 8 \Leftrightarrow x = -70$. Thus, the solution is $(-70, 13, 3)$.

● **Solving a system using Gaussian elimination.**

One matrix is transformed into an equivalent matrix by these **elementary row operations**:

> 1. Add a multiple of one row to another row.
>
> 2. Multiply a row by a nonzero constant.
>
> 2. Interchange two rows.

There is no *one way* to solve a system of equations using matrices, so it is helpful to yourself (and others who read your work) to leave a trail of the row operations you perform. We use the following notation to do this.

$$R_i + kR_j \rightarrow R_i \quad \text{Replace row } i \text{ with row } i \text{ plus } k \text{ times row } j.$$
$$kR_i \qquad\qquad \text{Multiply row } i \text{ by } k.$$
$$R_i \leftrightarrow R_j \qquad \text{Interchange rows } i \text{ and } j.$$

The systematic technique of using elementary row operations to arrive at a matrix in row-echelon form is called **Gaussian elimination**.

Exercise: Write the system

$$\begin{cases} x + 2y + 3z = & 1 \\ 2x - y - 3z = & 1 \\ x + 3y + 4z = -2 \end{cases}$$

in matrix form and use Gaussian elimination to solve.

Answer: In matrix form, we have

$$\begin{bmatrix} 1 & 2 & 3 & 1 \\ 2 & -1 & -3 & 1 \\ 1 & 3 & 4 & -2 \end{bmatrix} \xrightarrow[R_3 - R_1 \rightarrow R_3]{R_2 - 2R_1 \rightarrow R_2} \begin{bmatrix} 1 & 2 & 3 & 1 \\ 0 & -5 & -9 & -1 \\ 0 & 1 & 1 & -3 \end{bmatrix}$$

$$\xrightarrow{R_2 \leftrightarrow R_3} \begin{bmatrix} 1 & 2 & 3 & 1 \\ 0 & 1 & 1 & -3 \\ 0 & -5 & -9 & -1 \end{bmatrix} \xrightarrow{R_3 + 5R_2 \rightarrow R_3}$$

$$\begin{bmatrix} 1 & 2 & 3 & 1 \\ 0 & 1 & 1 & -3 \\ 0 & 0 & -4 & -16 \end{bmatrix} \xrightarrow{-\frac{1}{4}R_3} \begin{bmatrix} 1 & 2 & 3 & 1 \\ 0 & 1 & 1 & -3 \\ 0 & 0 & 1 & 4 \end{bmatrix}.$$

From the last row we have $z = 4$. Back-substituting into $y + z = -3$, we have $y + (4) = -3 \Leftrightarrow y = -7$. Back-substituting into $x + 2y + 3z = 1$, we have $x + 2(-7) + 3(4) = 1 \Leftrightarrow x - 14 + 12 = 1 \Leftrightarrow x = 3$. So the solution is $(3, -7, 4)$. Check the solution to make sure it satisfies all three equations in the original system.

Exercise: Write the system

$$\begin{cases} x + 2y + 5z & = 6 \\ -x + 2y - 5z - 4w & = -2 \\ x - 2y + w & = 1 \\ 2x + 10z + w & = 2 \end{cases}$$

in matrix form and use Gaussian elimination to solve.

Answer: Don't forget to insert 0 for the coefficients of missing terms.

$$\begin{bmatrix} 1 & 2 & 5 & 0 & 6 \\ -1 & 2 & -5 & -4 & -2 \\ 1 & -2 & 0 & 1 & 1 \\ 2 & 0 & 10 & 1 & 2 \end{bmatrix} \quad \begin{array}{c} R_2 + R_1 \to R_1 \\ R_3 - R_1 \to R_3 \\ \hline R_4 - 2R_1 \to R_4 \end{array}$$

$$\begin{bmatrix} 1 & 2 & 5 & 0 & 6 \\ 0 & 4 & 0 & -4 & 4 \\ 0 & -4 & -5 & 1 & -5 \\ 0 & -4 & 0 & 1 & -10 \end{bmatrix}$$. We divide the second row by 4 in

order to more easily eliminate the entries below the $(2, 2)$ entry.

$$\begin{bmatrix} 1 & 2 & 5 & 0 & 6 \\ 0 & 1 & 0 & -1 & 1 \\ 0 & -4 & -5 & 1 & -5 \\ 0 & -4 & 0 & 1 & -10 \end{bmatrix} \quad \begin{array}{c} R_3 + 4R_2 \to R_3 \\ \hline R_4 + 4R_2 \to R_4 \end{array}$$

$$\begin{bmatrix} 1 & 2 & 5 & 0 & 6 \\ 0 & 1 & 0 & -1 & 1 \\ 0 & 0 & -5 & -3 & -1 \\ 0 & 0 & 0 & -3 & -6 \end{bmatrix} \quad \xrightarrow{-\frac{1}{3}R_4} \quad \begin{bmatrix} 1 & 2 & 5 & 0 & 6 \\ 0 & 1 & 0 & -1 & 1 \\ 0 & 0 & -5 & -3 & -1 \\ 0 & 0 & 0 & 1 & 2 \end{bmatrix} .$$

Thus $w = 2$. Back-substituting gives $-5z - 3(2) = -1 \Leftrightarrow z = -1$, $y - (2) = 1 \Leftrightarrow y = 3$, and $x + 2(3) + 5(-1) = 6 \Leftrightarrow x = 5$. So the solution is $(5, 3, -1, 2)$. Check the solution to make sure it satisfies all four equations in the original system.

● **Solving a system using Gauss-Jordan elimination.**

The systematic technique of using elementary row operations to arrive at a matrix in reduced row-echelon form is called **Gauss-Jordan elimination**.

Exercise: Find the unique solution to

$$\begin{cases} x + 3y = 2 \\ 2x + 5y = 7 \end{cases}$$

using Gauss-Jordan elimination.

Answer: The augmented matrix is

$$\begin{bmatrix} 1 & 3 & 2 \\ 2 & 5 & 7 \end{bmatrix} \quad \xrightarrow{R_2 - 2R_1 \to R_2} \quad \begin{bmatrix} 1 & 3 & 2 \\ 0 & -1 & 3 \end{bmatrix} \quad \xrightarrow{-R_2}$$

$$\begin{bmatrix} 1 & 3 & 2 \\ 0 & 1 & -3 \end{bmatrix} \quad \xrightarrow{R_1 - 3R_2 \to R_1} \quad \begin{bmatrix} 1 & 0 & 11 \\ 0 & 1 & -3 \end{bmatrix} . \text{ So } x = 11$$

and $y = -3$. The solution of the system is $(11, -3)$.

Exercise: Find the unique solution to

$$\begin{cases} x - 3y + 3z = -2 \\ 2x + 3y - z = 1 \\ 3x - 10y + 6z = 1 \end{cases}$$

using Gauss-Jordan elimination.

Answer:

$$\begin{bmatrix} 1 & -3 & 3 & -2 \\ 2 & 3 & -1 & 1 \\ 3 & -10 & 6 & 1 \end{bmatrix} \xrightarrow[R_3 - 3R_1 \rightarrow R_3]{R_2 - 2R_1 \rightarrow R_2} \begin{bmatrix} 1 & -3 & 3 & -2 \\ 0 & 9 & -7 & 5 \\ 0 & -1 & -3 & 7 \end{bmatrix}$$

$$\xrightarrow{R_2 \leftrightarrow -R_3} \begin{bmatrix} 1 & -3 & 3 & -2 \\ 0 & 1 & 3 & -7 \\ 0 & 9 & -7 & 5 \end{bmatrix} \xrightarrow{R_3 - 9R_2 \rightarrow R_3}$$

$$\begin{bmatrix} 1 & -3 & 3 & -2 \\ 0 & 1 & 3 & -7 \\ 0 & 0 & -34 & 68 \end{bmatrix} \xrightarrow{-\frac{1}{34}R_3} \begin{bmatrix} 1 & -3 & 3 & -2 \\ 0 & 1 & 3 & -7 \\ 0 & 0 & 1 & -2 \end{bmatrix}$$

$$\xrightarrow[R_2 - 3R_3 \rightarrow R_2]{R_1 - 3R_3 \rightarrow R_1} \begin{bmatrix} 1 & -3 & 0 & 4 \\ 0 & 1 & 0 & -1 \\ 0 & 0 & 1 & -2 \end{bmatrix} \xrightarrow{R_1 + 3R_2 \rightarrow R_1}$$

$$\begin{bmatrix} 1 & 0 & 0 & 1 \\ 0 & 1 & 0 & -1 \\ 0 & 0 & 1 & -2 \end{bmatrix}. \text{ Thus, the solution is } (1, -1, -2).$$

• **Solving inconsistent systems and dependent systems.**

A system that has no solution is called an **inconsistent system**. A system is inconsistent when we try to use Gaussian elimination to get a triangular system and we end up with an equation that is always false. The false equation always has the form $0 = N$, where N is nonzero number.

A system that has infinitely many solutions is called a **dependent system**. The complete solution to such a system has one or more arbitrary variables called **parameters** on which the other variables depend.

Exercise: Find the complete solution to the system

$$\begin{cases} x - 5y + 3z = 3 \\ 2x + 3y - z = 5 \\ 4x - 7y + 5z = 8 \end{cases}$$

or show that none exists.

Answer:

$$\begin{bmatrix} 1 & -5 & 3 & 3 \\ 2 & 3 & -1 & 5 \\ 4 & -7 & 5 & 8 \end{bmatrix} \xrightarrow[R_3 - 4R_1 \rightarrow R_3]{R_3 - 2R_1 \rightarrow R_2} \begin{bmatrix} 1 & -5 & 3 & 3 \\ 0 & 13 & -7 & -1 \\ 0 & 13 & -7 & -4 \end{bmatrix}$$

$$\xrightarrow{R_3 - R_2 \rightarrow R_3} \begin{bmatrix} 1 & -5 & 3 & 3 \\ 0 & 13 & -7 & -1 \\ 0 & 0 & 0 & -3 \end{bmatrix}.$$

Since the last row corresponds to the equation $0 = -3$, which is always false, the system is inconsistent and there is no solution.

Exercise: Find the complete solution to the system

$$\begin{cases} x \quad\;\; -2z +\; w = 2 \\ x + y + 5z +\;\; w = 7 \\ 2x + y + 3z + 2w = 9 \\ 3x + y + \;\; z + 3w = 13 \end{cases}$$

or show that none exists.

Answer:

$$\begin{bmatrix} 1 & 0 & -2 & 1 & 2 \\ 1 & 1 & 5 & 1 & 7 \\ 2 & 1 & 3 & 2 & 9 \\ 3 & 1 & 1 & 3 & 13 \end{bmatrix} \xrightarrow[\substack{R_2 - R_1 \to R_2 \\ R_3 - 2R_1 \to R_3 \\ R_4 - 3R_1 \to R_4}]{} \begin{bmatrix} 1 & 0 & -2 & 1 & 6 \\ 0 & 1 & 7 & 0 & 5 \\ 0 & 1 & 7 & 0 & 5 \\ 0 & 1 & 7 & 0 & 7 \end{bmatrix}$$

$$\xrightarrow[\substack{R_3 - R_2 \to R_3 \\ R_4 - R_2 \to R_4}]{} \begin{bmatrix} 1 & 0 & -2 & 1 & 6 \\ 0 & 1 & 7 & 0 & 5 \\ 0 & 0 & 0 & 0 & 0 \\ 0 & 0 & 0 & 0 & 2 \end{bmatrix}.$$

Since the last row corresponds to the equation $0 = 2$, which is always false, the system is inconsistent and there is no solution.

Exercise: Find the complete solution to the system

$$\begin{cases} x - 2y - 2z + 2w = -3 \\ x + 2y + 2z + 4w = \;\;\; 5 \\ x - 6y - 6z \quad\quad = -11 \\ 2x - 2y - 2z + 5w = -2 \end{cases}$$

or show that none exists.

Answer:

$$\begin{bmatrix} 1 & -2 & -2 & 2 & -3 \\ 1 & 2 & 2 & 4 & 5 \\ 1 & -6 & -6 & 0 & -11 \\ 2 & -2 & -2 & 5 & -2 \end{bmatrix} \xrightarrow[\substack{R_2 - R_1 \to R_2 \\ R_3 - R_1 \to R_3 \\ R_4 - 2R_1 \to R_4}]{} \begin{bmatrix} 1 & -2 & -2 & 2 & -3 \\ 0 & 4 & 4 & 2 & 8 \\ 0 & -4 & -4 & -2 & -8 \\ 0 & 2 & 2 & 1 & 4 \end{bmatrix}$$

$$\xrightarrow[R_2 \leftrightarrow R_4]{} \begin{bmatrix} 1 & -2 & -2 & 2 & -3 \\ 0 & 2 & 2 & 1 & 4 \\ 0 & -4 & -4 & -2 & -8 \\ 0 & 4 & 4 & 2 & 8 \end{bmatrix} \xrightarrow[\substack{R_3 + 2R_2 \to R_3 \\ R_4 - 2R_2 \to R_4}]{}$$

$$\begin{bmatrix} 1 & -2 & -2 & 2 & -3 \\ 0 & 2 & 2 & 1 & 4 \\ 0 & 0 & 0 & 0 & 0 \\ 0 & 0 & 0 & 0 & 0 \end{bmatrix} \xrightarrow[\frac{1}{2}R_2]{} \begin{bmatrix} 1 & -2 & -2 & 2 & -3 \\ 0 & 1 & 1 & \frac{1}{2} & 2 \\ 0 & 0 & 0 & 0 & 0 \\ 0 & 0 & 0 & 0 & 0 \end{bmatrix}$$

$$\xrightarrow[R_1 + 2R_2 \to R_1]{} \begin{bmatrix} 1 & 0 & 0 & 3 & 1 \\ 0 & 1 & 1 & \frac{1}{2} & 2 \\ 0 & 0 & 0 & 0 & 0 \\ 0 & 0 & 0 & 0 & 0 \end{bmatrix}.$$

This system is dependent because its matrix has 2 equations in 4 variables. We solve the first equation for x in terms of w, and then we solve the second equation for y in terms of z and w:

$x + 3w = 1 \;\;\Leftrightarrow\;\; x = 1 - 3w$ and $y + z + \frac{1}{2}w = 2 \;\;\Leftrightarrow\;\;$

$y = 2 - z - \frac{1}{2}w$. Thus, the solutions are

$\left(1 - 3t, 2 - s - \frac{1}{2}t, s, t\right)$ where s and t are any real numbers.

Exercise: Find the complete solution to the system

$$\begin{cases} x + 2y \quad\quad - w = 0 \\ x + y + z + w = 8 \\ 3x + 4y - z + w = -5 \\ 2x + 3y - z \quad\quad = -6 \end{cases}$$

or show that none exists.

Answer:

$$\begin{bmatrix} 1 & 2 & 0 & -1 & 0 \\ 1 & 1 & 1 & 1 & 8 \\ 3 & 4 & -1 & 1 & -5 \\ 2 & 3 & -1 & 0 & -6 \end{bmatrix} \begin{array}{l} R_2 - R_1 \to R_2 \\ R_3 - 3R_1 \to R_3 \\ \xrightarrow{} \\ R_4 - 2R_1 \to R_4 \end{array}$$

$$\begin{bmatrix} 1 & 2 & 0 & -1 & 0 \\ 0 & -1 & 1 & 2 & 8 \\ 0 & -2 & -1 & 4 & -5 \\ 0 & -1 & -1 & 2 & -6 \end{bmatrix} \xrightarrow{-R_2} \begin{bmatrix} 1 & 2 & 0 & -1 & 0 \\ 0 & 1 & -1 & -2 & -8 \\ 0 & -2 & -1 & 4 & -5 \\ 0 & -1 & -1 & 2 & -6 \end{bmatrix}$$

$$\xrightarrow[R_4 + R_2 \to R_4]{R_3 + 2R_2 \to R_3} \begin{bmatrix} 1 & 2 & 0 & -1 & 0 \\ 0 & 1 & -1 & -2 & -8 \\ 0 & 0 & -3 & 0 & -21 \\ 0 & 0 & -2 & 0 & -14 \end{bmatrix} \xrightarrow{-\frac{1}{3}R_3}$$

$$\begin{bmatrix} 1 & 2 & 0 & -1 & 0 \\ 0 & 1 & -1 & -2 & -8 \\ 0 & 0 & 1 & 0 & 7 \\ 0 & 0 & -2 & 0 & -14 \end{bmatrix} \xrightarrow{R_4 + 2R_3 \to R_4} \begin{bmatrix} 1 & 2 & 0 & -1 & 0 \\ 0 & 1 & -1 & -2 & -8 \\ 0 & 0 & 1 & 0 & 7 \\ 0 & 0 & 0 & 0 & 0 \end{bmatrix}$$

$$\xrightarrow{R_2 + R_3 \to R_2} \begin{bmatrix} 1 & 2 & 0 & -1 & 0 \\ 0 & 1 & 0 & -2 & -1 \\ 0 & 0 & 1 & 0 & 7 \\ 0 & 0 & 0 & 0 & 0 \end{bmatrix} \xrightarrow{R_1 - 2R_2 \to R_1}$$

$$\begin{bmatrix} 1 & 0 & 0 & 3 & 2 \\ 0 & 1 & 0 & -2 & -1 \\ 0 & 0 & 1 & 0 & 7 \\ 0 & 0 & 0 & 0 & 0 \end{bmatrix}.$$

Now $z = 7 \iff y - 2w = -1 \iff y = -1 + 2w$ and $x + 3w = 2 \iff x = 2 - 3w$, so the solutions are $(2 - 3t, -1 + 2t, 7, t)$ where t is any real number.

Core Exercises

11, 29, 37

10.5   The Algebra of Matrices

Concepts

(A) **Basic matrix arithmetic.**

- Adding and subtracting matrices.
- Calculating the scalar product of a matrix and a scalar.
- Solving a matrix equation.

(B) **Matrix multiplication.**

- Computing the product of two matrices.

(C) **Properties of matrix arithmetic.**

(D) **Applications of matrices.**

- Using matrices to solve modeling problems.

Definitions

- **Inner product:** The inner product of a row of matrix row and matrix column is a real number determined as follows:

$$
\begin{bmatrix} a_1 & a_2 & a_3 & a_4 \end{bmatrix} \begin{bmatrix} b_1 \\ b_2 \\ b_3 \\ b_4 \end{bmatrix} = a_1 b_1 + a_2 b_2 + a_3 b_3 + a_4 b_4.
$$ Notice the similarity between the inner product (between a matrix row and matrix column) and the dot product (of two vectors).

Hints and Tips

- Once upon a time, the quantities you knew about were real numbers. Since then, you have learned about complex numbers and vectors — new types of quantities with their own rules for addition, subtraction, and so on. Matrices are another new kind of quantity. They are useful because a given matrix can hold quite a lot of numerical information about an object. Just as with vectors, matrix addition, subtraction, and scalar multiplication are fairly straightforward, while it takes a bit more work to multiply a matrix by another matrix.

Review

(A) **Basic matrix arithmetic.**

The **dimension** of a matrix is the number of rows by the number of columns, expressed as $R \times C$ and read as "R by C". The numbers in a matrix are called **entries**, and are referred to by their row and column position. In the matrix A, the quantity in the ith row and the jth column is called the (i, j) entry and denoted by a_{ij}. Two matrices are **equal** if and only if they have the same dimensions and the corresponding entries are equal.

- **Adding and subtracting matrices.**

If A and B are matrices of the same dimension and c is any real number, then

---

1. The **sum** $A + B$ is a matrix of the same dimension as A and B whose (i, j) entry is $a_{ij} + b_{ij}$.

2. The **difference** $A - B$ is a matrix of the same dimension as A and B whose (i, j) entry is $a_{ij} - b_{ij}$.

3. The **scalar product** cA is a matrix of the same dimension as A whose (i, j) entry is ca_{ij}.

Exercise: Calculate

$$\begin{bmatrix} 2 & -5 \\ 0 & 7 \end{bmatrix} + \begin{bmatrix} 8 & -5 \\ -5 & 4 \end{bmatrix}$$

or explain why it is undefined.

Answer: Adding the corresponding entries, we get

$$\begin{bmatrix} 2 & -5 \\ 0 & 7 \end{bmatrix} + \begin{bmatrix} 8 & -5 \\ -5 & 4 \end{bmatrix} = \begin{bmatrix} 2+8 & -5-5 \\ 0-5 & 7+4 \end{bmatrix}$$

$$= \begin{bmatrix} 10 & -10 \\ -5 & 11 \end{bmatrix}.$$

Exercise: Calculate

$$\begin{bmatrix} 2 & 1 & 8 \\ 11 & -3 & 0 \\ 7 & 0 & 9 \\ -2 & 1 & 8 \end{bmatrix} - \begin{bmatrix} 8 & -2 & -1 \\ 0 & -5 & 4 \\ 1 & 3 & -2 \\ 1 & 0 & 3 \end{bmatrix}$$

or explain why it is undefined.

Answer: Subtracting the corresponding entries, we get

$$\begin{bmatrix} 2 & 1 & 8 \\ 11 & -3 & 0 \\ 7 & 0 & 9 \\ -2 & 1 & 8 \end{bmatrix} - \begin{bmatrix} 8 & -2 & -1 \\ 0 & -5 & 4 \\ 1 & 3 & -2 \\ 1 & 0 & 3 \end{bmatrix}$$

$$= \begin{bmatrix} 2-8 & 1-(-2) & 8-(-1) \\ 11-0 & -3-(-5) & 0-4 \\ 7-1 & 0-3 & 9-(-2) \\ -2-1 & 1-0 & 8-3 \end{bmatrix} = \begin{bmatrix} -6 & 3 & 9 \\ 11 & 2 & -4 \\ 6 & -3 & 11 \\ -3 & 1 & 5 \end{bmatrix}$$

- **Calculating the scalar product of a matrix and a scalar.**

Exercise: Calculate

$$2 \begin{bmatrix} 4 & 0 & -2 & 3 & 4 \\ -3 & 2 & -6 & 0 & -5 \\ 6 & 2 & -5 & 4 & 1 \\ 3 & -2 & 1 & 0 & 3 \end{bmatrix}$$

or explain why it is undefined.

Answer: Multiply each entry by 2 to get

$$2 \begin{bmatrix} 4 & 0 & -2 & 3 & 4 \\ -3 & 2 & -6 & 0 & -5 \\ 6 & 2 & -5 & 4 & 1 \\ 3 & -2 & 1 & 0 & 3 \end{bmatrix}$$

$$= \begin{bmatrix} 2(4) & 2(0) & 2(-2) & 2(3) & 2(4) \\ 2(-3) & 2(2) & 2(-6) & 2(0) & 2(-5) \\ 2(6) & 2(2) & 2(-5) & 2(4) & 2(1) \\ 2(3) & 2(-2) & 2(1) & 2(0) & 2(3) \end{bmatrix}$$

$$= \begin{bmatrix} 8 & 0 & -4 & 6 & 8 \\ -6 & 4 & -12 & 0 & -10 \\ 12 & 4 & -10 & 8 & 2 \\ 6 & -4 & 2 & 0 & 6 \end{bmatrix}$$

Exercise: Calculate

$$2 \begin{bmatrix} 4 & 0 & 7 \\ 3 & 4 & -11 \\ 2 & -6 & 0 \end{bmatrix} - \begin{bmatrix} 6 & 2 \\ -5 & 4 \\ 1 & 3 \\ -2 & 1 \\ 0 & -9 \end{bmatrix}$$

or explain why it is undefined.

Answer: Although the scalar product

$$2 \begin{bmatrix} 4 & 0 & 7 \\ 3 & 4 & -11 \\ 2 & -6 & 0 \end{bmatrix}$$

is defined, the result is a 3×3 matrix and that cannot be added to a 5×2 matrix. So this matrix subtraction is not possible.

• **Solving a matrix equation.**

Exercise: Find the matrix A such that

$$2A + \begin{bmatrix} -4 & 2 & 2 \\ 1 & 0 & -1 \end{bmatrix} = \begin{bmatrix} 2 & 6 & 8 \\ 9 & 10 & 1 \end{bmatrix}.$$

Answer: Let $A = \begin{bmatrix} a_{11} & a_{12} & a_{13} \\ a_{21} & a_{22} & a_{23} \end{bmatrix}$. This now becomes six

tiny problems. For example, $2a_{11} + -4 = 2$, so $a_{11} = 3$. $2a_{12} + 2 = 6$, so $a_{12} = 2$. The final answer is

$$A = \begin{bmatrix} 3 & 2 & 3 \\ 4 & 5 & 1 \end{bmatrix}.$$

(B) **Matrix multiplication.**

If A is a **row matrix** (dimension $1 \times n$) and B is a **column matrix** (dimension $n \times 1$) then the **inner product**, AB is the sum of the products of the corresponding entries:

$$\begin{bmatrix} a_1 & a_2 & \cdots & a_n \end{bmatrix} \cdot \begin{bmatrix} b_1 \\ b_2 \\ \vdots \\ b_n \end{bmatrix} = a_1 b_1 + a_2 b_2 + \cdots + a_n b_n$$

Suppose that A is an $m \times n$ matrix and B is an $n \times k$ matrix; then the **product** $C = AB$ of two matrices is an $m \times k$ matrix, where the (i, j) entry, c_{ij}, is the inner product of the ith row of A and the jth column of B. Always check the dimension of the matrices first to make sure that the product is possible and to find the dimension of the resulting matrix.

• **Computing the product of two matrices.**

Exercise: Find

$$\begin{bmatrix} -3 & 1 \\ -4 & 5 \\ -1 & 2 \end{bmatrix} \cdot \begin{bmatrix} -7 & 2 \\ 3 & 1 \end{bmatrix}$$

or explain why it is undefined.

Answer: This is the product of a 3×2 matrix times a 2×2 matrix. The product exists and is a 3×2 matrix:

$$\begin{bmatrix} -3 & 1 \\ -4 & 5 \\ -1 & 2 \end{bmatrix} \cdot \begin{bmatrix} -7 & 2 \\ 3 & 1 \end{bmatrix}$$

$$= \begin{bmatrix} -3(-7) + 1(3) & -3(2) + 1(1) \\ -4(-7) + 5(3) & -4(2) + 5(1) \\ -1(-7) + 2(3) & -1(2) + 2(1) \end{bmatrix} = \begin{bmatrix} 24 & -5 \\ 43 & -3 \\ 13 & 0 \end{bmatrix}$$

Exercise: Find

$$\begin{bmatrix} 2 & 1 & 0 & 4 & 2 \\ 0 & 0 & -3 & -1 & 2 \end{bmatrix} \cdot \begin{bmatrix} -7 & 2 \\ 2 & 5 \\ 0 & -2 \\ 4 & 3 \end{bmatrix}$$

or explain why it is undefined.

Answer: The product of a 2×5 matrix and a 4×2 matrix is undefined because $5 \neq 4$.

Exercise: Find

$$\begin{bmatrix} 0 & 9 & -3 & 2 \\ 1 & 0 & 3 & -4 \\ 5 & 0 & -1 & 2 \end{bmatrix} \cdot \begin{bmatrix} 1 & 0 \\ 2 & 1 \\ 3 & 0 \\ 0 & 5 \end{bmatrix}$$

or explain why it is undefined.

Answer: This is the product of a 3×4 matrix times a 4×2 matrix, namely a 3×2 matrix:

$$\begin{bmatrix} 0 & 9 & -3 & 2 \\ 1 & 0 & 3 & -4 \\ 5 & 0 & -1 & 2 \end{bmatrix} \cdot \begin{bmatrix} 1 & 0 \\ 2 & 1 \\ 3 & 0 \\ 0 & 5 \end{bmatrix}$$

$$= \begin{bmatrix} 0 + 18 - 9 + 0 & 0 + 9 - 0 + 10 \\ 1 + 0 + 9 - 0 & 0 + 0 + 0 - 20 \\ 5 + 0 - 3 + 0 & 0 + 0 - 0 + 10 \end{bmatrix}$$

$$= \begin{bmatrix} 9 & 19 \\ 10 & -20 \\ 2 & 10 \end{bmatrix}$$

Exercise: Find

$$\begin{bmatrix} 4 & 0 & 7 \\ 3 & 4 & -11 \\ 2 & -6 & 0 \end{bmatrix} \begin{bmatrix} 6 & 2 & -5 \\ 4 & 1 & 3 \\ -2 & 1 & 3 \end{bmatrix}$$

or explain why it is undefined.

Answer: This is the product of a 3×3 matrix times a 3×3 matrix, namely a 3×3 matrix:

$$\begin{bmatrix} 4 & 0 & 7 \\ 3 & 4 & -11 \\ 2 & -6 & 0 \end{bmatrix} \cdot \begin{bmatrix} 6 & 2 & -5 \\ 4 & 1 & 3 \\ -2 & 1 & 3 \end{bmatrix}$$

$$= \begin{bmatrix} 24 + 0 - 14 & 8 + 0 + 7 & -20 + 0 + 21 \\ 18 + 16 + 22 & 6 + 4 - 11 & -15 + 12 - 33 \\ 12 - 24 + 0 & 4 - 6 + 0 & -10 - 18 + 0 \end{bmatrix}$$

$$= \begin{bmatrix} 10 & 15 & 1 \\ 56 & -1 & -36 \\ -12 & -2 & -28 \end{bmatrix}$$

(C) **Properties of matrix arithmetic.**

If A, B, C, and D are matrices for which the following products are defined, then

$(AB)C = A(BC)$	Associate Property
$A(B+C) = AB + AC$ and $(B+C)D = BD + CD$	Distributive Property

The matrix product is not commutative, that is, order is important when multiplying matrices.

(D) **Applications of matrices.**

• **Using matrices to solve modeling problems.**

Exercise: Uncle Nutenius owns three restaurants in Chicago: The Ordinary Pan, Brushfire, and Mon Ennemi Gabi. Even though they all look different, they sell mostly steak dinners, buffalo wings, and desserts. On a certain day, sales were distributed as follows:

	Ordinary Pan	Brushfire	Mon Ennemi Gabi
Steak Dinners	100	150	200
Wings	0	300	300
Desserts	400	50	200

We can put this information in the matrix $A = \begin{bmatrix} 100 & 150 & 200 \\ 0 & 300 & 300 \\ 400 & 50 & 200 \end{bmatrix}$.

The price of each item is the same at each restaurant. Steaks are $15, wings are $8, and desserts are $4. We can put this information in a matrix as well:

$$B = \begin{bmatrix} 15 & 8 & 4 \end{bmatrix}$$

‖ **Answer:**

(a) Calculate the product BA.

‖ $\begin{bmatrix} 15 & 8 & 4 \end{bmatrix} \begin{bmatrix} 100 & 150 & 200 \\ 0 & 300 & 300 \\ 400 & 50 & 200 \end{bmatrix} = \begin{bmatrix} 3100 & 4850 & 6200 \end{bmatrix}$

(b) Interpret the entries in the product matrix BA in real-world terms.

‖ The entries of BA are the gross incomes of the three restaurants.

▰ **Core Exercises**

13, 33, 47, 55

10.6   Inverses of Matrices and Matrix Equations

▼ Concepts

(A) **Inverse matrices.**

- Computing the inverse of a 2×2 matrix.
- Computing the inverse of an $n \times n$ matrix.

(B) **Matrix equations.**

- Solving systems of linear equations by solving the corresponding matrix equation.

▼ Definitions

- **Identity matrix:** An $n \times n$ matrix with 1s along the diagonal and 0s everywhere else. It is denoted I_n.
- **Inverse of a square matrix:** If A is an $n \times n$ square matrix, then A^{-1} is the unique matrix such that
  $(A)\left(A^{-1}\right) = \left(A^{-1}\right)(A) = I_n$.
- **Singular matrix:** A matrix that does not have an inverse.

▼ Hints and Tips

- You might think that this all is just another way to solve systems of equations. Matrix equations turn out to be very useful things in their own right. If you wind up taking a course on linear algebra you will find many uses for matrices beyond just solving equations.

- I find this section easier to understand if I think in analogies. If I have a number like 3, there is a number called 3^{-1}, namely $\frac{1}{3}$. The number $\frac{1}{3}$ has the property that $3 \cdot \frac{1}{3} = 1$. Now let's say I want to solve an equation like $3x = 6$. I know that $x = \left(\frac{1}{3}\right)6$. In other words, I multiply both sides of the equation $ax = b$ by a^{-1} to obtain $x = a^{-1}b$. This is exactly what we are doing in this section, with matrices instead of numbers.

▼ Review

(A) **Inverse matrices.**

The **identity matrix** I_n is the $n \times n$ matrix whose main diagonal consists entirely of 1s and whose other entries are 0. Identity matrices behave like the number 1 does, in the sense that $A \cdot I_n = A$ and $I_n \cdot A = A$ whenever these products are defined. If A and B are $n \times n$ matrices where $A \cdot B = B \cdot A = I_n$, then we say that B is the **inverse** of A, and we write $B = A^{-1}$. The concept of the inverse of a matrix is analogous to that of a reciprocal of a real number, but *not every square matrix has an inverse*.

- **Computing the inverse of a 2×2 matrix.**

  If $A = \begin{bmatrix} a & b \\ c & d \end{bmatrix}$ is a 2×2 matrix, then the inverse of A is $A^{-1} = \dfrac{1}{ad - bc} \begin{bmatrix} d & -b \\ -c & a \end{bmatrix}$ provided $ad - bc \neq 0$.

  The quantity $ad - bc$ is called the **determinant** of the matrix A. A has an inverse if and only if its determinant is not 0.

Exercise: Are the following pairs of matrices inverses of each other?

Answer: This question asks whether or not $AB = I_n$. Check by finding the product.

(a) $\begin{bmatrix} 4 & 7 \\ 3 & 3 \end{bmatrix}$ and $\begin{bmatrix} -\frac{1}{3} & \frac{7}{9} \\ \frac{1}{3} & -\frac{4}{9} \end{bmatrix}$

$\begin{bmatrix} 4 & 7 \\ 3 & 3 \end{bmatrix} \cdot \begin{bmatrix} -\frac{1}{3} & \frac{7}{9} \\ \frac{1}{3} & -\frac{4}{9} \end{bmatrix} = \begin{bmatrix} 1 & 0 \\ 0 & 1 \end{bmatrix}$, so yes, these two matrices are inverses of each other.

(b) $\begin{bmatrix} 1 & 1 & -1 \\ 1 & 2 & -1 \\ -1 & -1 & 2 \end{bmatrix}$ and $\begin{bmatrix} 3 & -1 & 1 \\ -1 & 1 & 0 \\ 1 & 0 & 1 \end{bmatrix}$

$\begin{bmatrix} 1 & 1 & -1 \\ 1 & 2 & -1 \\ -1 & -1 & 2 \end{bmatrix} \cdot \begin{bmatrix} 3 & -1 & 1 \\ -1 & 1 & 0 \\ 1 & 0 & 1 \end{bmatrix} = \begin{bmatrix} 1 & 0 & 0 \\ 0 & 1 & 0 \\ 0 & 0 & 1 \end{bmatrix}$, so yes, these two matrices are inverses of each other.

Exercise: Find the inverse of each 2×2 matrix, if it exists.

Answer:

(a) $\begin{bmatrix} 4 & 7 \\ 3 & 5 \end{bmatrix}$

Use the formula: $A = \begin{bmatrix} a & b \\ c & d \end{bmatrix} = \begin{bmatrix} 4 & 7 \\ 3 & 5 \end{bmatrix}$, so

$A^{-1} = \frac{1}{ad - bc} \begin{bmatrix} d & -b \\ -c & a \end{bmatrix} = \frac{1}{4 \cdot 5 - 7 \cdot 3} \begin{bmatrix} 5 & -7 \\ -3 & 4 \end{bmatrix}$

$= \begin{bmatrix} -5 & 7 \\ 3 & -4 \end{bmatrix}$

(b) $\begin{bmatrix} 9 & 12 \\ 6 & 8 \end{bmatrix}$

Here $ad - bc = 9 \cdot 8 - 12 \cdot 6 = 0$, so this matrix has no inverse.

(c) $\begin{bmatrix} 2 & 3 \\ -3 & 5 \end{bmatrix}$

$A^{-1} = \frac{1}{2 \cdot 5 - 3 \cdot (-3)} \begin{bmatrix} 5 & -3 \\ 3 & 2 \end{bmatrix} = \begin{bmatrix} \frac{5}{19} & -\frac{3}{19} \\ \frac{3}{19} & \frac{2}{19} \end{bmatrix}$

- **Computing the inverse of a $n \times n$ matrix.**
One technique used to find the inverse of an $n \times n$ matrix A is to augment A with I_n, the identity matrix. Then use row operations to change the left side to the identity matrix. That is, we start with $[A \mid I_n]$ and use row operations to get the matrix $[I_n \mid A^{-1}]$.

Exercise: Find the inverse of each matrix. ‖ **Answer:**

(a) $A = \begin{bmatrix} 1 & 0 & 2 \\ 6 & 1 & 16 \\ 2 & 0 & 3 \end{bmatrix}$

Augment A with I_3: $\left[\begin{array}{ccc|ccc} 1 & 0 & 2 & 1 & 0 & 0 \\ 6 & 1 & 16 & 0 & 1 & 0 \\ 2 & 0 & 3 & 0 & 0 & 1 \end{array}\right]$ $\begin{array}{c} R_2 - 6R_1 \rightarrow R_2 \\ R_3 - 2R_1 \rightarrow R_3 \end{array}$

$\left[\begin{array}{ccc|ccc} 1 & 0 & 2 & 1 & 0 & 0 \\ 0 & 1 & 4 & -6 & 1 & 0 \\ 0 & 0 & -1 & -2 & 0 & 1 \end{array}\right]$ $\xrightarrow{R_3}$ $\left[\begin{array}{ccc|ccc} 1 & 0 & 2 & 1 & 0 & 0 \\ 0 & 1 & 4 & -6 & 1 & 0 \\ 0 & 0 & 1 & 2 & 0 & -1 \end{array}\right]$

$\begin{array}{c} R_1 - 2R_3 \rightarrow R_1 \\ R_2 - 4R_3 \rightarrow R_2 \end{array}$ $\left[\begin{array}{ccc|ccc} 1 & 0 & 0 & -3 & 0 & 2 \\ 0 & 1 & 0 & -14 & 1 & 4 \\ 0 & 0 & 1 & 2 & 0 & -1 \end{array}\right]$.

So $A^{-1} = \begin{bmatrix} -3 & 0 & 2 \\ -14 & 1 & 4 \\ 2 & 0 & -1 \end{bmatrix}$.

Check: $\begin{bmatrix} 1 & 0 & 2 \\ 6 & 1 & 16 \\ 2 & 0 & 3 \end{bmatrix} \cdot \begin{bmatrix} -3 & 0 & 2 \\ -14 & 1 & 4 \\ 2 & 0 & -1 \end{bmatrix} = \begin{bmatrix} 1 & 0 & 0 \\ 0 & 1 & 0 \\ 0 & 0 & 1 \end{bmatrix}$.

(b) $B = \begin{bmatrix} 1 & 2 & 1 \\ 1 & 1 & -1 \\ 3 & 4 & -1 \end{bmatrix}$

Augment B with I_3: $\left[\begin{array}{ccc|ccc} 1 & 2 & 1 & 1 & 0 & 0 \\ 1 & 1 & -1 & 0 & 1 & 0 \\ 3 & 4 & -1 & 0 & 0 & 1 \end{array}\right]$ $\begin{array}{c} R_2 - R_1 \rightarrow R_2 \\ R_3 - 3R_1 \rightarrow R_3 \end{array}$

$\left[\begin{array}{ccc|ccc} 1 & 2 & 1 & 1 & 0 & 0 \\ 0 & -1 & -2 & -1 & 1 & 0 \\ 0 & -2 & -4 & -3 & 0 & 1 \end{array}\right]$ $\begin{array}{c} -R_2 \\ R_3 - 2R_2 \rightarrow R_3 \end{array}$

$\left[\begin{array}{ccc|ccc} 1 & 2 & 1 & 1 & 0 & 0 \\ 0 & 1 & 2 & 1 & -1 & 0 \\ 0 & 0 & 0 & -1 & -2 & 1 \end{array}\right]$.

Since the last row of the left half of the matrix consists entirely of 0s, B has no inverse.

(c) $C = \begin{bmatrix} 7 & -3 & -3 \\ -1 & 1 & 0 \\ -1 & 0 & 1 \end{bmatrix}$

Augment C with I_3:

$$\left[\begin{array}{ccc|ccc} 7 & -3 & -3 & 1 & 0 & 0 \\ -1 & 1 & 0 & 0 & 1 & 0 \\ -1 & 0 & 1 & 0 & 0 & 1 \end{array}\right] \quad \xrightarrow{R_1 \leftrightarrow -R_2}$$

$$\left[\begin{array}{ccc|ccc} 1 & -1 & 0 & 0 & -1 & 0 \\ 7 & -3 & -3 & 1 & 0 & 0 \\ -1 & 0 & 1 & 0 & 0 & 1 \end{array}\right] \quad \begin{array}{c} R_2 - 7R_1 \to R_2 \\ \xrightarrow{} \\ R_3 + R_1 \to R_3 \end{array}$$

$$\left[\begin{array}{ccc|ccc} 1 & -1 & 0 & 0 & -1 & 0 \\ 0 & 4 & -3 & 1 & 7 & 0 \\ 0 & -1 & 1 & 0 & -1 & 1 \end{array}\right] \quad \xrightarrow{R_2 \leftrightarrow -R_3}$$

$$\left[\begin{array}{ccc|ccc} 1 & -1 & 0 & 0 & -1 & 0 \\ 0 & 1 & -1 & 0 & 1 & -1 \\ 0 & 4 & -3 & 1 & 7 & 0 \end{array}\right] \quad \xrightarrow{R_3 - 4R_2 \to R_3}$$

$$\left[\begin{array}{ccc|ccc} 1 & -1 & 0 & 0 & -1 & 0 \\ 0 & 1 & -1 & 0 & 1 & -1 \\ 0 & 0 & 1 & 1 & 3 & 4 \end{array}\right] \quad \xrightarrow{R_2 + R_3 \to R_2}$$

$$\left[\begin{array}{ccc|ccc} 1 & -1 & 0 & 0 & -1 & 0 \\ 0 & 1 & 0 & 1 & 4 & 3 \\ 0 & 0 & 1 & 1 & 3 & 4 \end{array}\right] \quad \xrightarrow{R_1 + R_2 \to R_1}$$

$$\left[\begin{array}{ccc|ccc} 1 & 0 & 0 & 1 & 3 & 3 \\ 0 & 1 & 0 & 1 & 4 & 3 \\ 0 & 0 & 1 & 1 & 3 & 4 \end{array}\right].$$

So $C^{-1} = \begin{bmatrix} 1 & 3 & 3 \\ 1 & 4 & 3 \\ 1 & 3 & 4 \end{bmatrix}$.

Check: $\begin{bmatrix} 7 & -3 & -3 \\ -1 & 1 & 0 \\ -1 & 0 & 1 \end{bmatrix} \cdot \begin{bmatrix} 1 & 3 & 3 \\ 1 & 4 & 3 \\ 1 & 3 & 4 \end{bmatrix} = \begin{bmatrix} 1 & 0 & 0 \\ 0 & 1 & 0 \\ 0 & 0 & 1 \end{bmatrix}$.

B **Matrix equations.**

A system of linear equations can be written as a **matrix equation** $A \cdot X = B$, where A is the **coefficient matrix**, X is the **matrix of unknowns**, and B is the **constant matrix**. The solution to the matrix equation is found by multiplying both sides by A^{-1}, provided the inverse exists. Then $A \cdot X = B \Leftrightarrow A^{-1}(A \cdot X) = A^{-1}B \Leftrightarrow (A^{-1}A) \cdot X = A^{-1}B \Leftrightarrow I_n \cdot X = A^{-1}B \Leftrightarrow X = A^{-1}B$. *Remember that the matrix product is associative but it is not commutative*, so it is important to multiply A^{-1} from the correct (left) side.

• **Solving systems of linear equations by solving the corresponding matrix equation.**

Exercise: Solve $\begin{cases} 4x + 7y = 7 \\ 3x + 3y = -3 \end{cases}$ using the corresponding matrix equation.

Answer: The system can be represented by the matrix equation $\begin{bmatrix} 4 & 7 \\ 3 & 3 \end{bmatrix} \cdot \begin{bmatrix} x \\ y \end{bmatrix} = \begin{bmatrix} 7 \\ -3 \end{bmatrix}$. The inverse of $\begin{bmatrix} 4 & 7 \\ 3 & 3 \end{bmatrix}$

was found in a previous exercise; it is $\begin{bmatrix} -\frac{1}{3} & \frac{7}{9} \\ \frac{1}{3} & -\frac{4}{9} \end{bmatrix}$. Since

$X = A^{-1}B$, we have

$$\begin{bmatrix} x \\ y \end{bmatrix} = \begin{bmatrix} -\frac{1}{3} & \frac{7}{9} \\ \frac{1}{3} & -\frac{4}{9} \end{bmatrix} \cdot \begin{bmatrix} 7 \\ -3 \end{bmatrix} = \begin{bmatrix} -\frac{1}{3}(7) + \frac{7}{9}(-3) \\ \frac{1}{3}(7) - \frac{4}{9}(-3) \end{bmatrix}$$

$$= \begin{bmatrix} -\frac{14}{3} \\ \frac{11}{3} \end{bmatrix}$$

So $x = -\frac{14}{3}$ and $y = \frac{11}{3}$. Check:

$4\left(-\frac{14}{3}\right) + 7\left(\frac{11}{3}\right) = -\frac{56}{3} + \frac{77}{3} = \frac{21}{3} = 7$ and

$3\left(-\frac{14}{3}\right) + 3\left(\frac{11}{3}\right) = -14 + 11 = -3.$

Exercise: Solve $\begin{cases} 4x + 7y = 2 \\ 3x + 5y = 1 \end{cases}$ using the corresponding matrix equation.

Answer: The system can be represented by the matrix equation $\begin{bmatrix} 4 & 7 \\ 3 & 5 \end{bmatrix} \cdot \begin{bmatrix} x \\ y \end{bmatrix} = \begin{bmatrix} 2 \\ 1 \end{bmatrix}$. In a previous exercise we

found that $\begin{bmatrix} 4 & 7 \\ 3 & 5 \end{bmatrix}^{-1} = \begin{bmatrix} -5 & 7 \\ 3 & -4 \end{bmatrix}$, so $X = A^{-1}B \Leftrightarrow$

$$\begin{bmatrix} x \\ y \end{bmatrix} = \begin{bmatrix} -5 & 7 \\ 3 & -4 \end{bmatrix} \cdot \begin{bmatrix} 2 \\ 1 \end{bmatrix} = \begin{bmatrix} -3 \\ 2 \end{bmatrix}.$$

Thus $x = -3$ and $y = 2$. Check:

$4(-3) + 7(2) = -12 + 14 = 2$ and

$3(-3) + 5(2) = -9 + 10 = 1.$

Exercise: Solve $\begin{cases} 2x + 3y = -95 \\ -3x + 5y = 76 \end{cases}$

using the corresponding matrix equation.

Answer: The system can be represented by the matrix equation

$$\begin{bmatrix} 2 & 3 \\ -3 & 5 \end{bmatrix} \cdot \begin{bmatrix} x \\ y \end{bmatrix} = \begin{bmatrix} -95 \\ 76 \end{bmatrix}.$$ In a previous exercise we

calculated $\begin{bmatrix} 2 & 3 \\ -3 & 5 \end{bmatrix}^{-1} = \begin{bmatrix} \frac{5}{19} & -\frac{3}{19} \\ \frac{3}{19} & \frac{2}{19} \end{bmatrix}$, so $X = A^{-1}B \iff$

$$\begin{bmatrix} x \\ y \end{bmatrix} = \begin{bmatrix} \frac{5}{19} & -\frac{3}{19} \\ \frac{3}{19} & \frac{2}{19} \end{bmatrix} \cdot \begin{bmatrix} -95 \\ 76 \end{bmatrix} = \begin{bmatrix} -37 \\ -7 \end{bmatrix}.$$

So $x = -37$ and $y = -7$. Check:

$2(-37) + 3(-7) = -74 - 21 = -95$ and

$-3(-37) + 5(-7) = 111 - 35 = 76.$

Exercise: Solve $\begin{cases} x + y - z = -3 \\ x + 2y - z = 6 \\ -x - y + 2z = 4 \end{cases}$ using

the corresponding matrix equation.

Answer: The system can be represented by the matrix

equation $\begin{bmatrix} 1 & 1 & -1 \\ 1 & 2 & -1 \\ -1 & -1 & 2 \end{bmatrix} \cdot \begin{bmatrix} x \\ y \\ z \end{bmatrix} = \begin{bmatrix} -3 \\ 6 \\ 4 \end{bmatrix}.$ In a previous

exercise we calculated

$$\begin{bmatrix} 1 & 1 & -1 \\ 1 & 2 & -1 \\ -1 & -1 & 2 \end{bmatrix}^{-1} = \begin{bmatrix} 3 & -1 & 1 \\ -1 & 1 & 0 \\ 1 & 0 & 1 \end{bmatrix},$$ so $X = A^{-1}B \iff$

$$\begin{bmatrix} x \\ y \\ z \end{bmatrix} = \begin{bmatrix} 3 & -1 & 1 \\ -1 & 1 & 0 \\ 1 & 0 & 1 \end{bmatrix} \cdot \begin{bmatrix} -3 \\ 6 \\ 4 \end{bmatrix} = \begin{bmatrix} -11 \\ 9 \\ 1 \end{bmatrix}.$$

Thus $x = -11$, $y = 9$, and $z = 1$.

Check:

$(-11) + (9) - (1) = -11 + 9 - 1 = -3 \checkmark$

$(-11) + 2(9) - (1) = -11 + 18 - 1 = 6 \checkmark$

$-(-11) - (9) + 2(1) = 11 - 9 + 2 = 4 \checkmark$

Exercise: Solve $\begin{cases} x + 2z = 5 \\ 6x + y + 16z = 3 \\ 2x + 3z = -5 \end{cases}$ using the corresponding matrix equation.

Answer: The system can be represented by the matrix

equation $\begin{bmatrix} 1 & 0 & 2 \\ 6 & 1 & 16 \\ 2 & 0 & 3 \end{bmatrix} \cdot \begin{bmatrix} x \\ y \\ z \end{bmatrix} = \begin{bmatrix} 5 \\ 3 \\ -5 \end{bmatrix}$. In a previous

exercise we calculated $\begin{bmatrix} 1 & 0 & 2 \\ 6 & 1 & 16 \\ 2 & 0 & 3 \end{bmatrix}^{-1} = \begin{bmatrix} -3 & 0 & 2 \\ -14 & 1 & 4 \\ 2 & 0 & -1 \end{bmatrix}$, so

$X = A^{-1}B \iff$

$\begin{bmatrix} x \\ y \\ z \end{bmatrix} = \begin{bmatrix} -3 & 0 & 2 \\ -14 & 1 & 4 \\ 2 & 0 & -1 \end{bmatrix} \cdot \begin{bmatrix} 5 \\ 3 \\ -5 \end{bmatrix} = \begin{bmatrix} -25 \\ -87 \\ 15 \end{bmatrix}$.

Thus $x = -25$, $y = -87$, and $z = 15$. Check:

$(-25) + 2(15) = -25 + 30 = 5$,

$6(-25) + (-87) + 16(15) = -150 - 87 + 240 = 3$, and

$2(-25) + 3(15) = -50 + 45 = -5$.

Exercise: Solve $\begin{cases} 7x - 3y - 3z = 4 \\ -x + y = -2 \\ -x + z = 1 \end{cases}$ using the corresponding matrix equation.

Answer: The system can be represented by the matrix

equation $\begin{bmatrix} 7 & -3 & -3 \\ -1 & 1 & 0 \\ -1 & 0 & 1 \end{bmatrix} \cdot \begin{bmatrix} x \\ y \\ z \end{bmatrix} = \begin{bmatrix} 4 \\ -2 \\ 1 \end{bmatrix}$. In a previous

exercise we calculated $\begin{bmatrix} 7 & -3 & -3 \\ -1 & 1 & 0 \\ -1 & 0 & 1 \end{bmatrix}^{-1} = \begin{bmatrix} 1 & 3 & 3 \\ 1 & 4 & 3 \\ 1 & 3 & 4 \end{bmatrix}$, so

$X = A^{-1}B \iff$

$\begin{bmatrix} x \\ y \\ z \end{bmatrix} = \begin{bmatrix} 1 & 3 & 3 \\ 1 & 4 & 3 \\ 1 & 3 & 4 \end{bmatrix} \cdot \begin{bmatrix} 4 \\ -2 \\ 1 \end{bmatrix} = \begin{bmatrix} 1 \\ -1 \\ 2 \end{bmatrix}$.

Thus $x = 1$, $y = -1$, and $z = 2$. Check:

$7(1) - 3(-1) - 3(2) = 7 + 3 - 6 = 4$,

$-(1) + (-1) = -1 - 1 = -2$, and

$-(1) + (2) = -1 + 2 = 1$.

▰ **Core Exercises**

3, 17, 33, 37, 39, 47

10.7   Determinants and Cramer's Rule

Concepts

(A) **Computing determinants.**
- Computing the determinant of a 2×2 matrix.
- Computing the determinant of a $n \times n$ matrix.

(B) **Applications of determinants.**
- Testing for invertibility.
- Solving systems of equations with Cramer's Rule.
- Finding areas of triangles.

Definitions

- **Determinant:** A useful number that is associated to a square matrix. Every square matrix has a determinant.
- **Minor:** Every element a_{ij} in a square matrix A has a minor M_{ij} The minor of the element a_{ij} is the determinant of the matrix obtained by deleting the ith row and jth column of A. The minor helps us to compute determinants of large matrices by reducing it to a problem of computing determinants of smaller matrices.
- **Cofactor:** Every element a_{ij} in a square matrix A has a cofactor A_{ij} The cofactor of the element a_{ij} is $(-1)^{i+j} M_{ij}$

Hints and Tips

- The determinant of a matrix can be used to tell if a given matrix has an inverse (check to see if $\det A = 0$), to solve equations (using Cramer's Rule), and to find the area of a triangle. As the study of matrices goes on, determinants keep coming up. A matrix and its determinant are as closely associated as a person is to his or her name.

Review

(A) **Computing determinants.**

- **Computing the determinant of a 2×2 matrix.**
  The **determinant** of the square matrix A is a value denoted by the symbol $|A|$. For a 1×1 matrix, the determinant is its only entry. When A is the 2×2 matrix $\begin{bmatrix} a & b \\ c & d \end{bmatrix}$, the determinant of A is $|A| = \begin{vmatrix} a & b \\ c & d \end{vmatrix} = ad - bc.$

Exercise: Find the determinant.

Answer:

(a) $\begin{bmatrix} 5 & 3 \\ 7 & -2 \end{bmatrix}$

$\begin{vmatrix} 5 & 3 \\ 7 & -2 \end{vmatrix} = 5(-2) - 3(7) = -10 - 21 = -31$

(b) $\begin{bmatrix} 5 & 2 \\ 10 & -4 \end{bmatrix}$

$\begin{vmatrix} 5 & -2 \\ 10 & -4 \end{vmatrix} = -20 + 20 = 0$

(c) $\begin{bmatrix} 3 & a \\ 4 & 5 \end{bmatrix}$

$\begin{vmatrix} 3 & a \\ 4 & 5 \end{vmatrix} = 15 - 4a$

- **Computing the determinant of an $n \times n$ matrix.**

To define the determinant for a general $n \times n$ matrix A, we need some additional terminology. The **minor** M_{ij} of the element a_{ij} is the determinant of the matrix obtained by deleting the ith row and jth column of A. The **cofactor** A_{ij} associated with the element a_{ij} is $A_{ij} = (-1)^{i+j} M_{ij}$. The determinant of A is obtained by multiplying each element of any row or column by its cofactor, and then adding the results. For example,

$$|A| = \begin{vmatrix} a_{11} & \cdots & a_{1n} \\ \vdots & \ddots & \vdots \\ a_{n1} & \cdots & a_{nn} \end{vmatrix} = a_{11} A_{11} + a_{12} A_{12} + \cdots + a_{1n} A_{1n}$$

This is called **expanding the determinant by the first row.**

Note that the cofactor A_{ij} is just the minor M_{ij} multiplied by either 1 or -1, depending on whether $i + j$ is even or odd.

Also, the determinant can be found by expanding by *any* row or column in the matrix; the result is the same. It is usually easiest to use the row or column with the most zeros.

Exercise: Find the determinant.

Answer:

(a) $\begin{bmatrix} 0 & 1 & -2 \\ 4 & 0 & 3 \\ -1 & 2 & 1 \end{bmatrix}$

$\begin{vmatrix} 0 & 1 & -2 \\ 4 & 0 & 3 \\ -1 & 2 & 1 \end{vmatrix} = (-1)^{1+1} (0) \begin{vmatrix} 0 & 3 \\ 2 & 1 \end{vmatrix} + (-1)^{1+2} (1) \begin{vmatrix} 4 & 3 \\ -1 & 1 \end{vmatrix}$

$+ (-1)^{1+3} (-2) \begin{vmatrix} 4 & 0 \\ -1 & 2 \end{vmatrix}$

$= 0 - (4 + 3) - 2 (8 - 0)$

$= -7 - 16 = -23$

(b)

$\begin{bmatrix} 0 & 1 & 0 & 2 \\ 1 & 0 & 1 & 0 \\ 4 & 0 & 3 & 2 \\ -3 & 1 & 2 & 1 \end{bmatrix}$

Expanding by the first row,

$\begin{vmatrix} 0 & 1 & 0 & 2 \\ 1 & 0 & 1 & 0 \\ 4 & 0 & 3 & 2 \\ -3 & 1 & 2 & 1 \end{vmatrix} = 0 - (1) \begin{vmatrix} 1 & 1 & 0 \\ 4 & 3 & 2 \\ -3 & 2 & 1 \end{vmatrix} + 0 - (2) \begin{vmatrix} 1 & 0 & 1 \\ 4 & 0 & 3 \\ -3 & 1 & 2 \end{vmatrix}$

$= -1 \cdot (-11) - 2 \cdot (1) = 9$

since

$\begin{vmatrix} 1 & 1 & 0 \\ 4 & 3 & 2 \\ -3 & 2 & 1 \end{vmatrix} = 1 \begin{vmatrix} 3 & 2 \\ 2 & 1 \end{vmatrix} - 1 \begin{vmatrix} 4 & 2 \\ -3 & 1 \end{vmatrix} = -11$

and

$\begin{vmatrix} 1 & 0 & 1 \\ 4 & 0 & 3 \\ -3 & 1 & 2 \end{vmatrix} = 1 \begin{vmatrix} 0 & 3 \\ 1 & 2 \end{vmatrix} - 0 + 1 \begin{vmatrix} 4 & 0 \\ -3 & 1 \end{vmatrix} = 1$

Computing determinants of $n \times n$ matrices can be tedious, but it is usually possible to simplify the calculations. If you add a nonzero multiple of one row to another row or a nonzero multiple of one column to another column, the determinant is unchanged. Sometimes these operations can add zeros to a row or column, and make it easier to calculate the determinant.

Exercise: Find the determinant of

$$A = \begin{bmatrix} 2 & 1 & 3 \\ 1 & 0 & -4 \\ -2 & 0 & 10 \end{bmatrix}$$ and the determinant of

the matrix obtained by replacing row three with $R_3 + 2R_2$.

Answer: Expanding by the second column (because it contains two 0s) gives

$$\begin{vmatrix} 2 & 1 & 3 \\ 1 & 0 & -4 \\ -2 & 0 & 10 \end{vmatrix} = (-1) \begin{vmatrix} 1 & -4 \\ -2 & 10 \end{vmatrix} = -1 \cdot (10 - 8) = -2.$$

$$\begin{bmatrix} 2 & 1 & 3 \\ 1 & 0 & -4 \\ -2 & 0 & 10 \end{bmatrix} \xrightarrow{R_3 + 2R_2 \to R_3} \begin{bmatrix} 2 & 1 & 3 \\ 1 & 0 & -4 \\ 0 & 0 & 2 \end{bmatrix},$$

and expanding by the third row gives

$$\begin{vmatrix} 2 & 1 & 3 \\ 1 & 0 & -4 \\ 0 & 0 & 2 \end{vmatrix} = (-1)^{3+3} (2) \begin{vmatrix} 2 & 1 \\ 1 & 0 \end{vmatrix} = 2 \cdot (0 - 1) = -2.$$

(B) Applications of determinants.

- **Testing for invertibility.**

  A square matrix with a nonzero determinant is called **invertible**. A square matrix is invertible if and only if A^{-1} exists. So, A^{-1} exists if and only if $|A| \neq 0$.

- **Exercise:** Determine whether the following matrices are invertible.

Answer:

(a) $$\begin{bmatrix} 6 & 3 \\ 14 & 7 \end{bmatrix}$$

$$\begin{vmatrix} 6 & 3 \\ 14 & 7 \end{vmatrix} = 42 - 42 = 0.$$ The determinant is 0 so the given matrix does not have an inverse.

(b) $$\begin{bmatrix} 1 & 0 & 5 \\ 2 & 1 & -1 \\ -1 & 2 & 1 \end{bmatrix}$$

$$\begin{vmatrix} 1 & 0 & 5 \\ 2 & 1 & -1 \\ -1 & 2 & 1 \end{vmatrix} = 1 \begin{vmatrix} 1 & -1 \\ 2 & 1 \end{vmatrix} - 0 \begin{vmatrix} 2 & -1 \\ -1 & 1 \end{vmatrix} + 5 \begin{vmatrix} 2 & 1 \\ -1 & 2 \end{vmatrix}$$

$$= 1 (3) + 0 + 5 (5) = 28$$

The determinant is nonzero, so the given matrix is invertible.

• **Solving systems of equations with Cramer's Rule.**

The solutions of a system of linear equations can sometimes be expressed using determinants. Consider the linear system corresponding to the matrix equation

$$\begin{bmatrix} a_{11} & \cdots & a_{1n} \\ \vdots & \ddots & \vdots \\ a_{n1} & \cdots & a_{nn} \end{bmatrix} \cdot \begin{bmatrix} x_1 \\ \vdots \\ x_n \end{bmatrix} = \begin{bmatrix} b_1 \\ \vdots \\ b_n \end{bmatrix}$$

Let D be the coefficient matrix and let D_{x_i} be the matrix obtained from D by replacing the ith column of D by the numbers $b_1, b_2, \ldots, b_n$. Cramer's Rule states that if $|D| \neq 0$, then

$$x_1 = \frac{|D_{x_1}|}{|D|}, x_2 = \frac{|D_{x_2}|}{|D|}, \ldots, x_n = \frac{|D_{x_n}|}{|D|}$$

Exercise: Use Cramer's Rule to solve the matrix equation

$$\begin{bmatrix} 1 & 0 & 2 \\ 2 & 7 & 6 \\ 2 & -1 & 3 \end{bmatrix} \cdot \begin{bmatrix} x \\ y \\ z \end{bmatrix} = \begin{bmatrix} 0 \\ 2 \\ 3 \end{bmatrix}.$$

Answer: $D = \begin{bmatrix} 1 & 0 & 2 \\ 2 & 7 & 6 \\ 2 & -1 & 3 \end{bmatrix}$, and we calculate

$$D_x = \begin{bmatrix} 0 & 0 & 2 \\ 2 & 7 & 6 \\ 3 & -1 & 3 \end{bmatrix}, D_y = \begin{bmatrix} 1 & 0 & 2 \\ 2 & 2 & 6 \\ 2 & 3 & 3 \end{bmatrix}, D_z = \begin{bmatrix} 1 & 0 & 0 \\ 2 & 7 & 2 \\ 2 & -1 & 3 \end{bmatrix}.$$

To find $|D|$ we expand by row 1 to get

$$\begin{vmatrix} 1 & 0 & 2 \\ 2 & 7 & 6 \\ 2 & -1 & 3 \end{vmatrix} = 1 \begin{vmatrix} 7 & 6 \\ -1 & 3 \end{vmatrix} - 0 + 2 \begin{vmatrix} 2 & 7 \\ 2 & -1 \end{vmatrix}$$

$$= (21 + 6) + 2(-2 - 14) = -5.$$

To find $|D_x|$ we expand by row 1 to get

$$\begin{vmatrix} 0 & 0 & 2 \\ 2 & 7 & 6 \\ 3 & -1 & 3 \end{vmatrix} = 0 - 0 + 2 \begin{vmatrix} 2 & 7 \\ 3 & -1 \end{vmatrix} = 2(-2 - 21) = -46.$$

To find $|D_y|$ we expand by row 1 to get

$$\begin{vmatrix} 1 & 0 & 2 \\ 2 & 2 & 6 \\ 2 & 3 & 3 \end{vmatrix} = 1 \begin{vmatrix} 2 & 6 \\ 3 & 3 \end{vmatrix} - 0 + 2 \begin{vmatrix} 2 & 2 \\ 2 & 3 \end{vmatrix} = -8.$$

To find $|D_z|$ we expand by row 1 to get

$$\begin{vmatrix} 1 & 0 & 0 \\ 2 & 7 & 2 \\ 2 & -1 & 3 \end{vmatrix} = 1 \begin{vmatrix} 7 & 2 \\ -1 & 3 \end{vmatrix} - 0 + 0 = 21 + 2 = 23.$$

So $x = \frac{|D_x|}{|D|} = \frac{-46}{-5} = \frac{46}{5}, y = \frac{|D_y|}{|D|} = \frac{-8}{-5} = \frac{8}{5}$, and

$$z = \frac{|D_z|}{|D|} = \frac{23}{-5} = -\frac{23}{5}.$$

● **Finding areas of triangles.**

If a triangle in the coordinate plane has vertices (a_1, b_1), (a_2, b_2), and (a_3, b_3), then its area is

$$A = \pm \frac{1}{2} \begin{vmatrix} a_1 & b_1 & 1 \\ a_2 & b_2 & 1 \\ a_3 & b_3 & 1 \end{vmatrix}$$

Exercise: The vertices of a triangle are $(-3, -2)$, $(-4, 7)$, and $(5, 2)$. Find the area of the triangle.

Answer: Using the formula above, we have

$$A = \pm \frac{1}{2} \begin{vmatrix} -3 & -2 & 1 \\ -4 & 7 & 1 \\ 5 & 2 & 1 \end{vmatrix}.$$ Expanding by the third column, we have

$$A = \pm \frac{1}{2} \left[1 \begin{vmatrix} -4 & 7 \\ 5 & 2 \end{vmatrix} - 1 \begin{vmatrix} -3 & -2 \\ 5 & 2 \end{vmatrix} + 1 \begin{vmatrix} -3 & -2 \\ -4 & 7 \end{vmatrix} \right]$$

$$= \pm \tfrac{1}{2} \left[(-8 - 35) - (-6 + 10) + (-21 - 8) \right]$$

$$= \pm \tfrac{1}{2} (-43 - 4 - 29) = \pm \tfrac{1}{2} (-76) = 38$$

▶ **Core Exercises**

3, 17, 25, 29, 43, 49, 57

10.8 Partial Fractions

▼ Concepts

(A) Partial fraction decomposition.

- Distinct linear factors.
- Repeated linear factors.
- Distinct irreducible quadratic factors.
- Repeated irreducible quadratic factors.
- The use of long division to prepare for partial fractions.

▼ Definitions

- **Partial fraction decomposition:** Transforming a quotient of polynomials into the sum of such quotients, with the terms of the new sum having simpler denominators.

▼ Questions to Ask Your Teacher

- Some of the later cases in the text involving repeated irreducible quadratic factors are quite complex. To what extent should we be able to do these decompositions by hand, and about how quickly should we be able to do one under test conditions?

▼ Review

(A) Partial fraction decomposition.

Let r be the proper rational function $r(x) = \dfrac{P(x)}{Q(x)}$, where the degree of $P(x)$ is less than the degree of $Q(x)$. The goal is to express the proper rational function $r(x)$ as a sum of simpler functions called **partial fractions**. Every polynomial with real coefficients can be factored completely into linear factors and irreducible quadratic factors, that is, into factors of the form $ax + b$ and of the form $ax^2 + bx + c$, where a, b, and c are real numbers. There are four possible types of factors:

---

1. Each linear factor $ax + b$ that occurs exactly once in the denominator leads to a term of the form $\dfrac{A}{ax+b}$, where $A \neq 0$.

2. Each irreducible quadratic factor $ax^2 + bx + c$ that occurs exactly once in the denominator leads to a term of the form $\dfrac{Ax + B}{ax^2 + bx + c}$ where A and B are not both zero.

3. When the linear factor $ax + b$ is repeated k times in the denominator, then the corresponding partial fraction decomposition contains the terms
$$\frac{A_1}{ax+b} + \frac{A_2}{(ax+b)^2} + \cdots + \frac{A_k}{(ax+b)^k}.$$

4. When the irreducible quadratic factor $ax^2 + bx + c$ is repeated m times in the denominator, then the corresponding partial fraction decomposition contains the terms $\dfrac{A_1 x + B_1}{ax^2 + bx + c} + \dfrac{A_2 x + B_2}{(ax^2 + bx + c)^2} + \cdots + \dfrac{A_m x + B_m}{(ax^2 + bx + c)^m}.$

The numbers k and m in cases 3 and 4 above are called the **multiplicity** of the factor.

The technique of partial fraction decomposition can be broken down into the following steps:

1. If the rational function is not proper, then perform long division before applying partial fraction decomposition to the (proper) remainder.

2. Factor the denominator into linear and irreducible quadratic factors. Express the rational function in its partial fraction decomposition with unknown constants as discussed above.

3. Multiply both sides by the common denominator. Simplify the result by grouping like terms.

4. Equate corresponding coefficients. That is, the coefficients of x^k on either side of the equation must be equal, leading to a system of linear equations based on these coefficients.

5. Solve the resulting system of linear equations.

• **Distinct linear factors.**

Exercise: Find the partial fraction decomposition of the rational function
$$\frac{-2x + 15}{x^2 - x - 12}.$$

Answer: Since $x^2 - x - 12 = (x - 4)(x + 3)$, we have
$$\frac{-2x + 15}{x^2 - x - 12} = \frac{A}{x - 4} + \frac{B}{x + 3}.$$ Multiplying both sides by $(x + 3)(x - 4)$ gives

$$-2x + 15 = A(x + 3) + B(x - 4) = Ax + 3A + Bx - 4B$$
$$= (A + B)x + (3A - 4B)$$

and equating coefficients gives $\begin{cases} -2 = A + B \\ 15 = 3A - 4B \end{cases}$ To

solve, we multiply the first equation by 4 and add it to the second, giving $7 = 7A \Leftrightarrow A = 1$. Thus, $-2 = 1 + B \Leftrightarrow B = -3$. Thus, $\frac{-2x + 15}{x^2 - x - 12} = \frac{1}{x - 4} + \frac{-3}{x + 3}$.

• **Repeated linear factors.**

Exercise: Find the partial fraction decomposition of the rational function
$$\frac{3x^3 + 2x^2 - 3x - 3}{x^4 + x^3}.$$

Answer: Since $x^4 + x^3 = x^3(x + 1)$, the factor x has multiplicity 3. Thus, we have

$$\frac{3x^3 + 2x^2 - 3x - 3}{x^4 + x^3} = \frac{A}{x} + \frac{B}{x^2} + \frac{C}{x^3} + \frac{D}{x + 1}$$

Multiplying both sides by $x^3(x + 1)$ gives

$$3x^3 + 2x^2 - 3x - 3$$
$$= Ax^2(x + 1) + Bx(x + 1) + C(x + 1) + Dx^3$$

Simplifying, this is equal to

$$Ax^3 + Ax^2 + Bx^2 + Bx + Cx + C + Dx^3$$
$$= (A + D)\, x^3 + (A + B)\, x^2 + (B + C)\, x + C$$

Equating coefficients gives

$$\begin{cases} 3 = A \qquad\qquad\quad + D \\ 2 = A + B \\ -3 = \qquad B + C \\ -3 = \qquad\qquad C \end{cases}$$

From the last equation,

$C = -3$, and back-substitution gives $B = 0$, $A = 2$, and

$D = 1$. Thus, $\dfrac{3x^3 + 2x^2 - 3x - 3}{x^4 + x^3} = \dfrac{2}{x} + \dfrac{-3}{x^3} + \dfrac{1}{x+1}$.

- **Distinct irreducible quadratic factors.**

Exercise: Find the partial fraction decomposition of the rational function $\dfrac{3x^3 - x^2 + 15x - 1}{x^4 + 8x^2 + 15}$.

Answer: The denominator factors as

$x^4 + 8x^2 + 15 = (x^2 + 3)\,(x^2 + 5)$. Since both factors are irreducible quadratics, we must have

$$\frac{3x^3 - x^2 + 15x - 1}{x^4 + 8x^2 + 15} = \frac{Ax + B}{x^2 + 3} + \frac{Cx + D}{x^2 + 5}.$$ Multiplying

both sides by $x^4 + 8x^2 + 15$ gives

$$3x^3 - x^2 + 15x - 1$$
$$= (Ax + B)\,(x^2 + 5) + (Cx + D)\,(x^2 + 3)$$
$$= Ax^3 + Bx^2 + 5Ax + 5B + Cx^3 + Dx^2 + 3Cx + 3D$$
$$= (A + C)\, x^3 + (B + D)\, x^2 + (5A + 3C)\, x + (5B + 3D)$$

Set the coefficients equal and solve:

The x^3 terms give	$3 = A$	$+ C$
The x^2 terms give	$-1 = \quad B$	$+ D$
The x terms give	$15 = 5A$	$+ 3C$
The constant terms give	$-1 = \quad 5B$	$+ 3D$

Subtracting 3 times the first equation from the third equation

gives $6 = 2A \iff A = 3$, and back-substitution gives

$C = 0$, $B = 1$, and $D = -2$. Thus,

$$\frac{3x^3 - x^2 + 15x - 1}{x^4 + 8x^2 + 15} = \frac{3x + 1}{x^2 + 3} + \frac{-2}{x^2 + 5}.$$

Exercise: Find the partial fraction decomposition of the rational function
$$\frac{3x^3 + 13x^2 + 18x + 14}{(x^2 + 3x + 4)(x - 1)(x + 2)}.$$

Answer: The denominator is already factored and $x^2 + 3x + 4$ is an irreducible quadratic, so we have
$$\frac{3x^3 + 13x^2 + 18x + 14}{(x^2 + 3x + 4)(x - 1)(x + 2)}$$
$$= \frac{Ax + B}{x^2 + 3x + 4} + \frac{C}{x - 1} + \frac{D}{x + 2}$$

Multiplying both sides by $(x^2 + 3x + 4)(x - 1)(x + 2)$ gives

$3x^3 + 13x^2 + 18x + 14$
$$= (Ax + B)(x - 1)(x + 2) + C(x^2 + 3x + 4)(x + 2)$$
$$+ D(x^2 + 3x + 4)(x - 1)$$

We use substitution to find the values of A, B, C, and D.

If $x = 1$, then the LHS of the equation gives

$3(1)^3 + 13(1)^2 + 18(1) + 14 = 48$ and the RHS gives

$(A + B)(0)(3) + C(8)(3) + D(8)(0) = 24C$, so $48 = 24C$

$\Leftrightarrow \quad C = 2$.

If $x = -2$, then the LHS of the equation gives

$3(-2)^3 + 13(-2)^2 + 18(-2) + 14 = 6$ and the RHS gives

$(-2A + B)(-3)(0) + C(2)(0) + D(2)(-3) = -6D$, so

$6 = -6D \quad \Leftrightarrow \quad D = -1$.

If $x = 0$, then the LHS of the equation gives

$3(0)^3 + 13(0)^2 + 18(0) + 14 = 14$ and the RHS gives

$(0 + B)(-1)(2) + C(4)(2) + D(4)(-1) = -2B + 8C - 4D$,

so $14 = -2B + 8C - 4D$.

Substituting the values for C and D, we have

$14 = -2B + 8(2) - 4(-1) \quad \Leftrightarrow \quad 14 = -2B + 20 \quad \Leftrightarrow$

$B = 3$.

If $x = -1$, then the LHS of the equation gives

$3(-1)^3 + 13(-1)^2 + 18(-1) + 14 = 6$ and the RHS gives

$(-A + B)(-2)(1) + C(2)(1) + D(2)(-2)$
$$= 2A - 2B + 2C - 4D$$

so $6 = 2A - 2B + 2C - 4D$.

Substituting the values for B, C, and D, we have

$6 = 2A - 2(3) + 2(2) - 4(-1) \quad \Leftrightarrow \quad 6 = 2A + 2 \quad \Leftrightarrow$

$A = 2$.

Thus,
$$\frac{3x^3 + 13x^2 + 18x + 14}{(x^2 + 3x + 4)(x - 1)(x + 2)}$$
$$= \frac{2x + 3}{x^2 + 3x + 4} + \frac{2}{x - 1} + \frac{-1}{x + 2}$$

- **Repeated irreducible quadratic factors.**

  Even simple problems of this type involve many equations in many unknowns.   (See Example 4 in your text.)
  Realistically, problems of this nature are usually solved with the aid of technology.

- **The use of long division to prepare for partial fractions.**

Exercise: Find the partial fraction decomposition of the rational function

$$\frac{x^5}{x^4 - 5x^2 + 4}.$$

Answer: Since $\dfrac{x^5}{x^4 - 5x^2 + 4}$ is not a proper fraction, we must first do long division to get a proper fraction.

$$\frac{x^5}{x^4 - 5x^2 + 4} = x + \frac{5x^3 - 4x}{x^4 - 5x^2 + 4} \text{ and }$$

$$x^4 - 5x^2 + 4 = \left(x^2 - 1\right)\left(x^2 - 4\right)$$

$$= (x - 1)(x + 1)(x - 2)(x + 2)$$

so

$$\frac{5x^3 - 4x}{x^4 - 5x^2 + 4} = \frac{A}{x - 1} + \frac{B}{x + 1} + \frac{C}{x - 2} + \frac{D}{x + 2}.$$

Multiplying both sides by $x^4 - 5x^2 + 4$ gives

$$5x^3 - 4x = A(x + 1)\left(x^2 - 4\right) + B(x - 1)\left(x^2 - 4\right)$$

$$+ C(x + 2)\left(x^2 - 1\right) + D(x - 2)\left(x^2 - 1\right)$$

Instead of writing a system of 4 equations in 4 unknowns, we use substitution to find the values of A, B, C, and D.

If $x = 1$, then $5\left(1\right)^3 - 4\left(1\right) = A\left(2\right)\left(-3\right)$ ⇔

$5 - 4 = -6A$ ⇔ $A = -\frac{1}{6}$.

If $x = -1$, then $5\left(-1\right)^3 - 4\left(-1\right) = B\left(-2\right)\left(-3\right)$ ⇔

$-5 + 4 = 6B$ ⇔ $B = -\frac{1}{6}$.

If $x = 2$, then $5\left(2\right)^3 - 4\left(2\right) = +C\left(4\right)\left(3\right)$ ⇔

$40 - 8 = 12C$ ⇔ $C = \frac{32}{12} = \frac{8}{3}$.

If $x = -2$, then $5\left(-2\right)^3 - 4\left(-2\right) = D\left(-4\right)\left(3\right)$ ⇔

$-40 + 8 = -12D$ ⇔ $D = \frac{32}{12} = \frac{8}{3}$.

Thus, $\dfrac{x^5}{x^4 - 5x^2 + 4} = x + \dfrac{-\frac{1}{6}}{x - 1} + \dfrac{-\frac{1}{6}}{x + 1} + \dfrac{\frac{8}{3}}{x - 2} + \dfrac{\frac{8}{3}}{x + 2}.$

◤ Core Exercises

1, 15, 17, 37, 43

10.9   Systems of Inequalities

Concepts

(A) **Graphing inequalities.**

- Sketching the graph of an inequality using test points.
- Solving systems of inequalities.

Definitions

- **Test point:** When we graph an inequality, we divide the plane into regions. To determine which regions are included in our solution, we test a point in each region. These points are called **test points**.

Hints and Tips

- Imagine graphing several equations at once. (You've done this before.) Notice how the plane is now divided into regions, with the curves serving as the borders of the regions. When we replace equations with inequalities, we are now looking at the regions, not the borders, and determining which regions satisfy the inequalities.

Review

(A) **Graphing inequalities.**

The graph of an inequality, in general, consists of a region in the plane whose boundary is the graph of the equation obtained by replacing the inequality sign by an equal sign. The points on the boundary are included when the inequality sign is $\geq$ or $\leq$, and this is indicated by a solid boundary. When the inequality sign is $>$ or $<$, the points on the boundary are excluded, and this is indicated by a broken boundary.

- **Sketching the graph of an inequality using test points.**

To sketch the graph of an inequality, we first sketch the boundary of the region by replacing the inequality with an equals sign. To determine which region of the plane gives the solution set of the inequality, pick a point *not on the curve*, called a **test point**, and test to see if it satisfies the inequality. If it does, then every point in the same region satisfies the inequality. If it does not, then every point in the same region does not.

Exercise: Graph the inequality $x + y^2 \geq 4$.

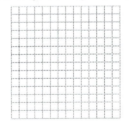

Answer: $x + y^2 = 4$ $\Leftrightarrow$ $x = -y^2 + 4$ is a parabola with vertex at $(4, 0)$ opening to the left. Using the point $(0, 0)$ as a test point, we get find that $0 + (0)^2 \ngeq 4$. Since $(0, 0)$ does not satisfy the inequality and $(0, 0)$ is inside the region, the solution lies outside of the parabola. Because the inequality is not strict, we graph the parabola with a solid line.

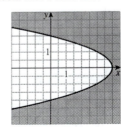

Exercise: Graph the inequality $4x^2 + 9y^2 < 36$.

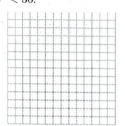

Answer: $4x^2 + 9y^2 < 36$ is an ellipse. Using the point $(0,0)$ as a test point, we have find that $4(0)^2 + 9(0)^2 < 36$. Since $(0,0)$ satisfies the inequality, the point $(0,0)$ is inside the region, and the solution lies inside the ellipse. Because the inequality is strict, we graph the ellipse with a broken line.

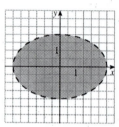

Exercise: Graph the inequality $xy > 4$.

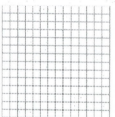

Answer: $xy > 4 \iff y = 4/x$. We use the test points $(-5, -5)$, $(0,0)$, and $(5,5)$: $(-5)(-5) > 4$, so $(-5, -5)$ is inside the region; $(0)(0) \not> 4$, so $(0,0)$ is outside; and $(5)(5) > 4$, so $(5,5)$ is inside.

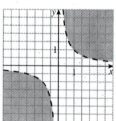

• **Solving systems of inequalities**

The solution to a system of inequalities consists of the intersection of the graphs of each inequality in the system. The **vertices** of a solution set are points where the boundaries of the graphs of two (or more) inequalities intersect. Vertices may or may not be part of the solution, but they show where the corners of the solution set lie. An inequality is **linear** if the corresponding equation is that of a line; that is, if it can be written in the form $ax + by \vartriangle c$, where $\vartriangle$ is any of the four inequality symbols.

Exercise: Graph the solution set of the

system of inequalities $\begin{cases} x + y < 3 \\ 3x + 2y \geq 5 \end{cases}$

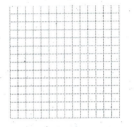

Answer:

Graph $x + y = 3$ with a broken line and graph $3x + 2y = 5$
with a solid line. The vertex is obtained by solving the system

of equations $\begin{cases} x + y = 3 \\ 3x + 2y = 5 \end{cases}$ Solving the first equation for y

gives $y = 3 - x$, and substitution into the second equation gives
$3x + 2(3 - x) = 5 \iff 3x + 6 - 2x = 5 \iff x = -1$.
Thus $y = 3 - (-1) = 4$, and the vertex is $(-1, 4)$. We use the
test point $(0, 0)$ and find that $0 + 0 < 3$, but $3(0) + 2(0) \not\geq 5$.
Thus $(0, 0)$ satisfies $x + y < 3$, but does not satisfy
$3x + 2y \geq 5$, so the solution set consists of the region lying on
the same side as $(0, 0)$ of the line $x + y = 3$, and on the
opposite side of $3x + 2y = 5$.

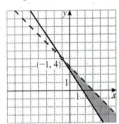

Exercise: Graph the solution set of the

system of inequalities $\begin{cases} x^2 + 4x + y \leq -6 \\ x + y > -6 \end{cases}$

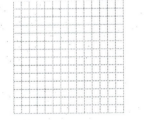

Answer: Graph $x^2 + 4x + y = -6$ with a solid line and graph
$x + y = -6$ with a broken line. We solve the system by
subtracting the second equation from the first, to obtain
$x^2 + 3x = 0 \iff x = 0$ or $x = -3$. If $x = 0$ then $y = -6$,
and if $x = -3$ then $y = -3$. So the vertices are $(0, -6)$ and
$(-3, -3)$. We use $(0, 0)$ as a test point, and find that
$0^2 + 4 \cdot 0 + 0 \not\leq -6$, but $0 + 0 > -6$. Thus, the solution set
lies on the opposite side of the parabola and the same side of the
line.

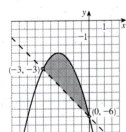

Exercise: Graph the solution set of the system of inequalities.
$$\begin{cases} 3x + y > 9 \\ 2x + y \geq 8 \\ -x + 3y \geq 3 \end{cases}$$

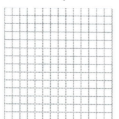

Answer: Graph $2x + y = 8$ and $-x + 3y = 3$ with a solid line and $3x + y = 9$ with a broken line. The graphs of $3x + y = 9$ and $2x + y = 8$ intersect at $(1, 6)$, and the graphs of $2x + y = 8$ and $-x + 3y = 3$ intersect at $(3, 2)$. We use $(0, 0)$ as a test point: $3 \cdot 0 + 0 \not> 9$, $2 \cdot 0 + 0 \not\geq 8$, and $-0 + 3 \cdot 0 \not\geq 3$, so the solution set lies on the opposite side of each line from the origin. Note that the vertex $(1, 6)$ does not satisfy the system because it lies on the line $3x + y = 9$, but the vertex $(3, 2)$ satisfies all three inequalities.

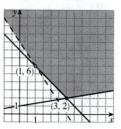

◢ Core Exercises

13, 15, 23, 35, 39, 45, 47, 49

11 Analytic Geometry

11.1 Parabolas

Concepts

(A) **Parabolas.**

- Finding an equation of a parabola.
- Finding the focus and directrix of a parabola.
- Using technology to sketch a horizontal parabola.

Definitions

- **Parabola:** A set of points in the plane equidistant from a fixed point F (called the focus) and a fixed line l (called the directrix).

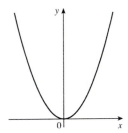

Hints and Tips

- By this point in the course you have sketched several quadratic functions $y = ax^2 + bx + c$. These are all parabolas; some have a vertex other than the origin.

Review

(A) **Parabolas.**

Graphs of equations of the form $y = ax^2 + bx + c$ are **parabolas**. A parabola is the set of points in a plane equidistant from a fixed point (the **focus**) and a fixed line l (the **directrix**). The **vertex** is the point where the parabola changes direction. The parabola is symmetric about its **axis**.

A parabola with focus $F(0, p)$ and directrix $y = -p$ is given by the equation $x^2 = 4py$. Its vertex is $(0, 0)$ and it is symmetric about the y-axis.

A parabola with focus $F(p, 0)$ and directrix $x = -p$ is given by the equation $y^2 = 4px$. Its vertex is $(0, 0)$ and it is symmetric about the x-axis.

The **latus rectum** is the line segment inside the parabola that runs through the focus perpendicular to the axis. Its length is called the **focal diameter** and is $|4p|$.

• **Finding an equation of a parabola.**

Exercise: Find an equation for each parabola whose graph is shown.

Answer:

(a)

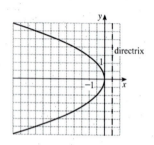

Since the directrix of this parabola is $x = \frac{1}{2}$, we have $p = -\frac{1}{2}$. The parabola opens horizontally, so we use the form

$$y^2 = 4px = 4\left(-\frac{1}{2}\right)x = -2x.$$

(b)

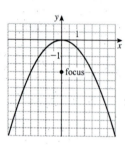

Since the focus of this parabola is $(0, -2)$, we have $p = -2$. The parabola opens vertically, so we use the form

$$x^2 = 4py = 4(-2)y = -8y.$$

Exercise: Find an equation for the parabola that has its vertex at the origin and satisfies the given condition.

Answer:

(a) Focus $F(0, -5)$

Since its vertex is the origin and the focus is below the x-axis, this parabola opens downward and we seek a parabola of the form $x^2 = 4py$. Since $p = -5$, we have

$$x^2 = 4(-5)y = -20y.$$

(b) Directrix $x = -7$

Since the directrix is the vertical line $x = -7$, this parabola is of the form $y^2 = 4px$. The directrix of this type of parabola is at $x = -p$, so $-p = -7 \quad \Leftrightarrow \quad p = 7$. Thus an equation is

$$y^2 = 4(7)x = 28x.$$

● **Finding the focus and directrix of a parabola.**

Exercise: Find the focus, directrix, and focal diameter of the parabola $-4x^2 = y$, and sketch its graph.

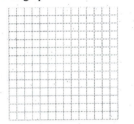

Answer: First solve for the squared term x^2: $-4x^2 = y$ $\Leftrightarrow$ $x^2 = -\frac{1}{4}y$, so $4p = -\frac{1}{4}$ $\Leftrightarrow$ $p = -\frac{1}{16}$. The focus is $\left(0, -\frac{1}{16}\right)$, the directrix is $y = -\left(-\frac{1}{16}\right) = \frac{1}{16}$, and the focal diameter is $|4p| = \left|4\left(-\frac{1}{16}\right)\right| = \frac{1}{4}$.

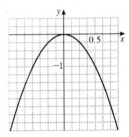

Exercise: Find the focus, directrix, and focal diameter of the parabola $20x - y^2 = 0$, and sketch its graph.

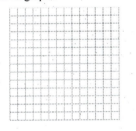

Answer: First solve for the squared term y^2: $20x - y^2 = 0$ $\Leftrightarrow$ $y^2 = 20x$, so $4p = 20$ $\Leftrightarrow$ $p = 5$. The focus is $(5, 0)$, the directrix is $x = -5$, and the focal diameter is $|4p| = |4 \cdot 5| = 20$.

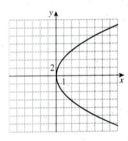

● **Using technology to sketch a horizontal parabola.**

Your graphing calculator can probably graph only functions, and a horizontal parabola is not a function. It is possible, however, to divide the horizontal parabola into two functions representing the top and bottom halves. So given the equation of a horizontal parabola (say, $x = y^2 - 1$) we solve for $y = \pm\sqrt{x+1}$ and graph each half as a separate function (the top half $y = \sqrt{x+1}$ and the bottom half $y = -\sqrt{x+1}$).

Exercise: Use a graphing device to graph $x = \dfrac{y^2}{2} + 2$.

Answer: We solve for y to obtain $y = \pm\sqrt{2x - 4}$. Then we graph $y = \sqrt{2x - 4}$ and $y = -\sqrt{2x - 4}$.

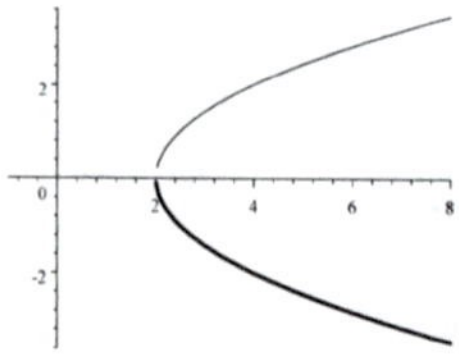

Due to limitations on the resolution of your calculator, the two halves of the parabola might not touch in your graph. If you were copying the picture to make a hand sketch, the top and bottom halves should touch.

Exercise: Use a graphing device to sketch $x = y^2 + 4y + 1$.

Answer: It is now much tougher to solve for y! We complete the square to obtain

$x = (y^2 + 4y + 4) + 1 - 4 = (y + 2)^2 - 3 \quad \Leftrightarrow$

$x + 3 = (y + 2)^2 \quad \Leftrightarrow \quad y = -2 \pm \sqrt{x + 3}$, and we graph $y = -2 + \sqrt{x + 3}$ and $y = -2 - \sqrt{x + 3}$.

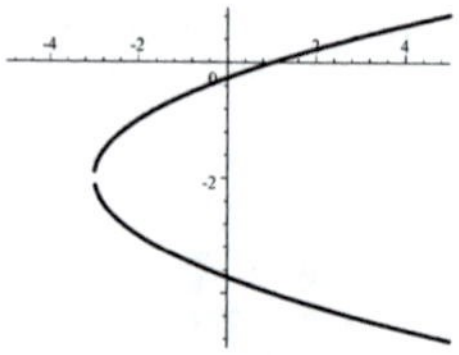

Core Exercises

11, 23, 31, 41, 43

11.2   Ellipses

Concepts

(A) Ellipses.

- Finding an equation of an ellipse.
- Finding the foci and vertices of an ellipse.
- Using technology to sketch an ellipse.

Definitions

- **Ellipse:** A set of points in the plane whose total distance from two points F_1 and F_2 (called the **foci**) is constant.

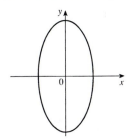

Hints and Tips

- The eccentricity of an ellipse measures how different it is from a circle. Study Figure 8 to see how varying the eccentricity affects the shape of an ellipse.

Review

(A) Ellipses.

An **ellipse** is the set of points in the plane the sum of whose distance from two fixed points (**foci**) is a constant.

Type of ellipse	Horizontal major axis	Vertical major axis
Equation ($a > b$)	$\dfrac{x^2}{a^2} + \dfrac{y^2}{b^2} = 1$	$\dfrac{x^2}{b^2} + \dfrac{y^2}{a^2} = 1$
Vertices	$(\pm a, 0)$	$(0, \pm a)$
Major axes	$(-a, 0)$ to $(a, 0)$	$(0, -a)$ to $(0, a)$
Length of major axis	$2a$	$2a$
Minor axes	$(0, -b)$ to $(0, b)$	$(-b, 0)$ to $(b, 0)$
Length of minor axis	$2b$	$2b$
Foci	$(\pm c, 0)$, where $a^2 = b^2 + c^2$	$(0, \pm c)$, where $a^2 = b^2 + c^2$
Eccentricity	$\dfrac{c}{a}$, where $a^2 = b^2 + c^2$	$\dfrac{c}{a}$, where $a^2 = b^2 + c^2$

- **Finding an equation of an ellipse.**

Exercise: Find an equation for an ellipse that satisfies the given conditions.

Answer:

(a) Foci are $(\pm 5, 0)$, vertices are $(\pm 13, 0)$

Since the foci are on the x-axis, we use the form $\dfrac{x^2}{a^2} + \dfrac{y^2}{b^2} = 1$. We are given $a = 13$ and $c = 5$, so $b^2 = a^2 - c^2 = 169 - 25 = 144$. Thus, an equation of the ellipse is $\dfrac{x^2}{169} + \dfrac{y^2}{144} = 1$.

(b) Length of major axis is 8, length of minor axis is 5, foci are on the x-axis

Since the foci are on the x-axis, the major axis is horizontal. We are given $2a = 8$ and $2b = 5$, so $a = 4$ and $b = \frac{5}{2}$. Thus, an equation of the ellipse is $\dfrac{x^2}{4^2} + \dfrac{y^2}{\left(\frac{5}{2}\right)^2} = 1 \;\Leftrightarrow\; \dfrac{x^2}{16} + \dfrac{4y^2}{25} = 1$.

(c) Length of major axis is 34, foci are $(0, \pm 8)$

Since the foci are on the y-axis, the major axis is vertical. We are given $c = 8$ and $2a = 34$, so $a = 17$. Substituting, we have $17^2 = b^2 + 8^2 \;\Leftrightarrow\; b^2 = 225$. Thus, an equation of the ellipse is $\dfrac{x^2}{225} + \dfrac{y^2}{289} = 1$.

- **Finding the foci and vertices of an ellipse.**

Exercise: Find the vertices, foci, and eccentricity of the ellipse $\dfrac{x^2}{49} + \dfrac{y^2}{64} = 1$, determine the lengths of its major and minor axes, and sketch its graph.

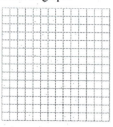

Answer: First, express the ellipse in the form $\dfrac{x^2}{7^2} + \dfrac{y^2}{8^2} = 1$. Since $8 > 7$, the major axis of this ellipse is vertical. The vertices are $(0, -8)$ and $(0, 8)$, the major axis has length $2\,(8) = 16$, and the minor axis has length $2\,(7) = 14$. To find the foci and eccentricity, we need to solve the equation $a^2 = b^2 + c^2$ for c: $8^2 = 7^2 + c^2 \;\Leftrightarrow\; 64 = 49 + c^2 \;\Leftrightarrow\; c^2 = 15 \Rightarrow\; c = \sqrt{15}$. Thus the foci are $\left(0, -\sqrt{15}\right)$ and $\left(0, \sqrt{15}\right)$ and the eccentricity is $\frac{\sqrt{15}}{8}$.

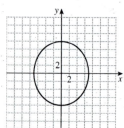

Exercise: Find the vertices, foci, and eccentricity of the ellipse $4x^2 + 25y^2 = 1$, determine the lengths of its major and minor axes, and sketch its graph.

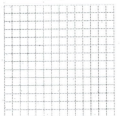

Answer: We first need to express the equation in the form $\frac{x^2}{a^2} + \frac{y^2}{b^2} = 1$. Since $w = \frac{1}{1/w}$, we can express $4x^2 + 25y^2 = x^2/\left(\frac{1}{4}\right) + y^2/\left(\frac{1}{25}\right)$. So the equation of the ellipse can be expressed as $x^2/\left(\frac{1}{2}\right)^2 + y^2/\left(\frac{1}{5}\right)^2 = 1$. Thus, $a = \frac{1}{2}$ and $b = \frac{1}{5}$. Since $\frac{1}{2} > \frac{1}{5}$, the major axis of this ellipse is horizontal. The vertices are $\left(-\frac{1}{2}, 0\right)$ and $\left(\frac{1}{2}, 0\right)$, the major axis has length $2\left(\frac{1}{2}\right) = 1$, and the minor axis has length $2\left(\frac{1}{5}\right) = \frac{2}{5}$. To find the foci and eccentricity, we need to solve the equation $a^2 = b^2 + c^2$ for c: $\left(\frac{1}{2}\right)^2 = \left(\frac{1}{5}\right)^2 + c^2 \Leftrightarrow \frac{1}{4} = \frac{1}{25} + c^2 \Leftrightarrow 25 = 4 + 100c^2 \Leftrightarrow 100c^2 = 21 \Leftrightarrow c^2 = \frac{21}{100} \Leftrightarrow c = \sqrt{\frac{21}{100}} = \frac{\sqrt{21}}{10}$. The foci are $\left(-\frac{\sqrt{21}}{10}, 0\right)$ and $\left(\frac{\sqrt{21}}{10}, 0\right)$ and the eccentricity is $\frac{\sqrt{21}/10}{1/2} = \frac{\sqrt{21}}{5}$.

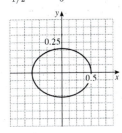

• **Using technology to sketch an ellipse.**

Your graphing calculator can probably graph only functions, and an ellipse is not a function. It is possible, however, to divide the ellipse into two functions representing the top and bottom halves. So given the equation of an ellipse (say, $\frac{x^2}{49} + \frac{y^2}{64} = 1$) we solve for $y = \pm\sqrt{64 - \frac{64}{49}x^2}$ and graph each half as a separate function (the top half $y = \sqrt{64 - \frac{64}{49}x^2}$ and the bottom half $y = -\sqrt{64 - \frac{64}{49}x^2}$).

Exercise: Use a graphing device to graph $\frac{x^2}{4} + \frac{y^2}{9} = 1$.

Answer: We solve for y to obtain $y = \pm\sqrt{9 - \frac{9}{4}x^2}$. Then we graph $y = \sqrt{9 - \frac{9}{4}x^2}$ and $y = -\sqrt{9 - \frac{9}{4}x^2}$ on the same axes.

Due to limitations on the resolution of your calculator, the two halves of the ellipse might not touch in your graph. If you were copying the picture to make a hand sketch, the left and right halves should touch.

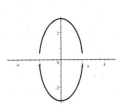

◤ Core Exercises

5, 17, 23, 33, 49, 53

11.3   Hyperbolas

▚ Concepts

(A) **Hyperbolas.**

- Finding an equation of a hyperbola.
- Finding the foci, vertices and asymptotes of an hyperbola.
- Using technology to sketch a hyperbola.

▚ Definitions

- **Hyperbola:** A set of points in the plane whose distances from two points F_1 and F_2 (the **foci**) differ by a constant.

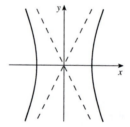

▚ Review

(A) **Hyperbolas.**

A **hyperbola** is a set of points in the plane whose distances from two fixed points differ by a constant. A hyperbola consists of two parts called **branches**. Each branch approaches the **asymptotes** $y = \pm \left(\frac{b}{a} \right) x$. There are two types of standard hyperbolas, those with a horizontal transverse axis and those with a vertical transverse axis. A convenient way to sketch a hyperbola is to draw a box $2a$ by $2b$ centered at the origin. The diagonals of the box are the asymptotes and the points where the sides intersect the x or y axes are the vertices.

Type of Hyperbola	Horizontal transverse axis	Vertical transverse axis
Equation	$\dfrac{x^2}{a^2} - \dfrac{y^2}{b^2} = 1$	$\dfrac{y^2}{b^2} - \dfrac{x^2}{a^2} = 1$
Vertices	$(\pm a, 0)$	$(0, \pm b)$
Asymptotes	$y = \pm \dfrac{b}{a} x$	$y = \pm \dfrac{b}{a} x$
Foci	$(\pm c, 0)$	$(0, \pm c)$

● **Finding an equation of a hyperbola.**

Exercise: Find an equation for the
hyperbola that satisfies the given conditions.

Answer:

(a) Foci are $(\pm15, 0)$, vertices are $(\pm9, 0)$

This hyperbola has a horizontal transverse axis, and we are
given $a = 9$ and $c = 15$. Next we find b: $15^2 = 9^2 + b^2 \iff$
$225 = 81 + b^2 \iff b^2 = 144 \iff b = 12$. Thus, an
equation of the hyperbola is $\dfrac{x^2}{9^2} - \dfrac{y^2}{12^2} = 1 \iff$
$\dfrac{x^2}{81} - \dfrac{y^2}{144} = 1.$

(b) Vertices are $(0, \pm10)$, asymptotes are
$y = \pm\frac{5}{3}x.$

This hyperbola has a vertical transverse axis, and we are given
$b = 10$ and $\dfrac{b}{a} = \dfrac{5}{3}$. So $\dfrac{10}{a} = \dfrac{5}{3} \iff a = 6$. Thus, an
equation of the hyperbola is $\dfrac{y^2}{10^2} - \dfrac{x^2}{6^2} = 1 \iff$
$\dfrac{y^2}{100} - \dfrac{x^2}{36} = 1.$

● **Finding the foci, vertices and asymptotes of an hyperbola.**

Exercise: Find the vertices, foci, and
asymptotes of the hyperbola $\dfrac{x^2}{64} - \dfrac{y^2}{36} = 1.$
Then sketch its graph.

Answer: Since x^2 has a positive coefficient in the equation
$\dfrac{x^2}{64} - \dfrac{y^2}{36} = 1$, this hyperbola has a horizontal transverse axis
(that is, it opens horizontally). The vertices are $(-8, 0)$ and
$(8, 0)$. To find the foci, we solve for c: $c^2 = 64 + 36 = 100$
$\Rightarrow c = 10$. The foci are $(-10, 0)$ and $(10, 0)$ and the
asymptotes are $y = \pm\frac{6}{8}x = \pm\frac{3}{4}x.$

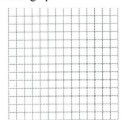

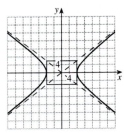

Exercise: Find the vertices, foci, and asymptotes of the hyperbola $4x^2 + 36 = 9y^2$. Then sketch its graph.

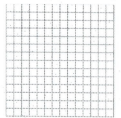

Answer: We first express this equation in a standard form. Isolate the constant term. $4x^2 + 36 = 9y^2$ $\Leftrightarrow$ $36 = 9y^2 - 4x^2$. Next divide by 36 to make the constant 1, and find a and b: $1 = \dfrac{y^2}{4} - \dfrac{x^2}{9} \Leftrightarrow 1 = \dfrac{y^2}{2^2} - \dfrac{x^2}{3^2}$. This hyperbola opens vertically. Its vertices are $(0, -2)$ and $(0, 2)$. To find the foci, we need to solve for c: $c^2 = 4 + 9 = 13 \implies c = \sqrt{13}$. The foci are $\left(0, -\sqrt{13}\right)$ and $\left(0, \sqrt{13}\right)$ and the asymptotes are $y = \pm\frac{2}{3}x$.

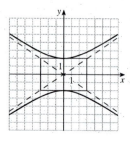

- **Using technology to sketch a hyperbola.**

Your graphing calculator can probably graph only functions, and a hyperbola is not necessarily a function. It is possible, however, to divide the hyperbola into two functions representing the top and bottom halves. So given the equation of a hyperbola (say, $\dfrac{x^2}{49} - \dfrac{y^2}{64} = 1$) we solve for $y = \pm\sqrt{\frac{64}{49}x^2 - 64}$ and graph each half as a separate function (the top half $y = \sqrt{\frac{64}{49}x^2 - 64}$ and the bottom half $y = -\sqrt{\frac{64}{49}x^2 - 64}$).

Exercise: Use a graphing device to sketch $\dfrac{x^2}{4} - \dfrac{y^2}{9} = 1$.

Answer: We solve for y to obtain $y = \pm\sqrt{\frac{9}{4}x^2 - 9}$. Then we graph $y = \sqrt{\frac{9}{4}x^2 - 9}$ and $y = -\sqrt{\frac{9}{4}x^2 - 9}$ on the same axes.

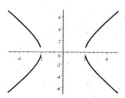

Due to limitations on the resolution of your calculator, the two halves of the hyperbola might not touch in your graph. If you were copying the picture to make a hand sketch, the top and bottom halves should touch.

▰ Core Exercises

5, 21, 45

11.4 Shifted Conics

Concepts

(A) **Equations of shifted conics.**

- Sketching shifted conics.
- Sketching degenerate conics.

Definitions

- **Degenerate conic:** If the form of an equation leads us to believe that its graph is a conic section, but that turns out not to be the case, we call the curve a **degenerate conic**.

Hints and Tips

- Once upon a time, we learned to solve quadratic equations by completing the square. (This occurred in Chapter 1). Before starting this section, it would be a good idea to review that process.

Review

(A) **Equations of shifted conics.**

When a conic is shifted h units in the x-direction and k units in the y-direction, then we can find the equation of the conic by replacing x with $x - h$ and y with $y - k$. The vertex of the parabola is now (h, k) and the center of the ellipse and hyperbola is (h, k). The standard forms of the equations of these conics are summarized in the table below.

Conic	Equation (horizontal)	Equation (vertical)
Parabola with vertex at (h, k)	$(y - k)^2 = 4p(x - h)$	$(x - h)^2 = 4p(y - k)$
Ellipse centered at (h, k)	$\dfrac{(x - h)^2}{a^2} + \dfrac{(y - k)^2}{b^2} = 1$	$\dfrac{(x - h)^2}{b^2} + \dfrac{(y - k)^2}{a^2} = 1$
Hyperbola centered at (h, k)	$\dfrac{(x - h)^2}{a^2} - \dfrac{(y - k)^2}{b^2} = 1$	$\dfrac{(y - k)^2}{b^2} - \dfrac{(x - h)^2}{a^2} = 1$

The general equation of a conic section is $Ax^2 + Cy^2 + Dx + Ey + F = 0$, where A and C are not both zero, representing a conic or a degenerate conic. The nondegenerate cases are as follows:

> The graph of $Ax^2 + Cy^2 + Dx + Ey + F = 0$ is
>
> 1. a parabola if A or C is zero.
> 2. an ellipse if A and C have the same sign (a circle if $A = C$).
> 3. a hyperbola if A and C have opposite signs.

To determine the center or vertex of the conic and aid in graphing, we **complete the square**.

● **Sketching shifted conics.**

Exercise: Find the center, foci, and vertices of the ellipse given by

$$\frac{(x-3)^2}{9} + \frac{(y+2)^2}{16} = 1,$$ and determine

the lengths of its major and minor axes. Then sketch its graph.

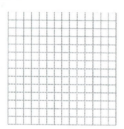

Answer: The equation is in standard form. The center is $(3, -2)$ since $y + 2 = y - (-2)$. $a = 3$ and $b = 4$, so the major axis is vertical. The vertices are $(h, k \pm b) = (3, -6)$ and $(3, 2)$, the major axis has length $2b = 8$, and the minor axis has length $2a = 6$. To find the foci, we calculate c by solving

$$4^2 = 3^2 + c^2 \quad \Leftrightarrow \quad 16 = 9 + c^2 \quad \Leftrightarrow \quad c^2 = 7 \quad \Leftrightarrow$$

$c = \sqrt{7}$. So the foci are at $(h, k \pm c) = (3, -2 - \sqrt{7})$ and $(3, -2 + \sqrt{7})$.

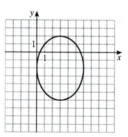

Exercise: Find the vertex, focus, and directrix of the parabola given by

$(x + 2)^2 = 2y + 3$. Then sketch its graph.

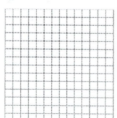

Answer: $h = -2$ and $k = -\frac{3}{2}$, so the vertex is $\left(-2, -\frac{3}{2}\right)$. Since $4p = 2$, we have $p = \frac{1}{2}$. The focus lies at $(h, k + p) = (-2, -1)$, and the directrix is

$$y = k - p = -\frac{3}{2} - \frac{1}{2} = -2.$$

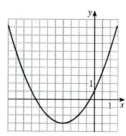

Exercise: Find the center, foci, vertices, and asymptotes of the hyperbola given by $\dfrac{(x+4)^2}{4} - \dfrac{(y-1)^2}{9} = 1$. Then sketch the graph.

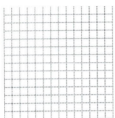

Answer: This hyperbola has a horizontal transverse axis, that is, it opens horizontally. The center is $(-4, 1)$ and the vertices are $(h \pm a, k) = (-6, 1)$ and $(-2, 1)$. To find the foci, we first need to find c. $c^2 = 2^2 + 3^2 = 4 + 9 = 13 \;\Rightarrow\; c = \sqrt{13}$, so the foci are $(h \pm c, k) = \left(-4 - \sqrt{13}, 1\right)$ and $\left(-4 + \sqrt{13}, 1\right)$.

The asymptotes are $y - k = \pm \dfrac{b}{a}(x - h)$ or

$y - 1 = \pm \frac{3}{2}(x + 4)$.

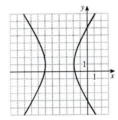

Exercise: Complete the square to determine whether the equation represents an ellipse, a parabola, a hyperbola, or a degenerate conic.

(a) $x^2 + 4y^2 - 8x + 16y + 16 = 0$

Answer:

Since the x^2 and y^2 terms have the same sign, this is the equation of an ellipse. The following steps are used in completing the square: **1.** Group the x and y terms and move the constant term to the other side:

$\left(x^2 - 8x\right) + \left(4y^2 + 16y\right) = -16$

2. Factor out the coefficients of the x^2 and y^2 terms:

$\left(x^2 - 8x\right) + 4\left(y^2 + 4y\right) = -16$

3. Complete the square in x and in y:

$\left(x^2 - 8x + \underline{\quad}\right) + 4\left(y^2 + 4y + \underline{\quad}\right)$

$\qquad\qquad = -16\left(x^2 - 8x + 16\right) + 4\left(y^2 + 4y + 4\right)$

$\qquad\qquad = -16 + 16 + 4\,(4) \quad\Leftrightarrow$

$(x - 4)^2 + 4(y + 2)^2 = 16$

4. Finish by dividing both sides by 16:

$\dfrac{(x - 4)^2}{16} + \dfrac{(y + 2)^2}{4} = 1$. This is an ellipse with horizontal major axis, centered at $(4, -2)$, with vertices at $(0, -2)$ and $(8, -2)$.

(b) $x + 3y^2 + 4y + 7 = 0$

Since this equation contains no x^2 term, it represents a parabola. The following steps are used in completing the square.

1. Group the y terms and move the x and constant terms to the other side: $3y^2 + 4y = -x - 7$

2. Factor out the coefficient of the y^2 term:
$3\left(y^2 + \frac{4}{3}y\right) = -x - 7$

3. Complete the square in y:

$3\left(y^2 + \frac{4}{3}y + \underline{}\right) = -x - 73\left(y^2 + \frac{4}{3}y + \frac{4}{9}\right)$
$$= -x - 7 + 3\left(\tfrac{4}{9}\right)3\left(y + \tfrac{2}{3}\right)^2 = -x - \tfrac{17}{3}$$

4. Finish by dividing both sides by 3:

$\left(y + \frac{2}{3}\right)^2 = -\frac{1}{3}\left(x + \frac{17}{3}\right)$. So the vertex is at $\left(-\frac{17}{3}, -\frac{2}{3}\right)$.

• **Sketching degenerate conics.**

Exercise: Graph the following.

Answer:

(a) $9x^2 - 36x + 436 + 16y^2 - 128y = 0$

Since the x^2 and y^2 terms have the same sign, this is the equation of an ellipse. We use the following steps to complete the square:

1. Group the x and y terms and move the constant term to the other side: $\left(9x^2 - 36x\right) + \left(16y^2 - 128y\right) = -436$

2. Factor out the coefficients of the x^2 and y^2 terms:
$9\left(x^2 - 4x\right) + 16\left(y^2 - 8y\right) = -436$

3. Complete the square in x and y:

$9\left(x^2 - 4x + \underline{}\right) + 16\left(y^2 - 8y + \underline{}\right) = -436 \quad \Leftrightarrow$
$9\left(x^2 - 4x + 4\right) + 16\left(y^2 - 8y + 16\right)$
$$= -436 + 9\left(4\right) + 16\left(16\right) \quad \Leftrightarrow$$
$9\left(x - 2\right)^2 + 16\left(y - 4\right)^2 = -144.$

4. Finish by dividing both sides by 144:

$$\frac{\left(x - 2\right)^2}{16} + \frac{\left(y - 4\right)^2}{9} = -1.$$

But we can't add two positive numbers together to get negative one! So there is *no* point that satisfies this equation, and the graph is empty. The conic is degenerate.

(b) $x^2 - 2x - 2y^2 - 4y - 1 = 0$

Since the x^2 and y^2 terms have opposite sign, this is the equation of a hyperbola. We use the following steps to complete the square:

1. Group the x terms and y terms and move the constant term to the other side: $(x^2 - 2x) + (-2y^2 - 4y) = 1$

2. Factor out the coefficients of the x^2 and y^2 terms:

$(x^2 - 2x) - 2(y^2 + 2y) = 1$

3. Complete the square in x and y:

$(x^2 - 2x + \_\_) - 2(y^2 + 2y + \_\_) = 1 \quad \Leftrightarrow$

$(x^2 - 2x + 1) - 2(y^2 + 2y + 1) = 1 + 1 - 2(1) \quad \Leftrightarrow$

$(x - 1)^2 - 2(y + 1)^2 = 0.$

But this equation is satisfied only if

$(x - 1)^2 = 2(y + 1)^2 = 0.$ So the solution consists of the graphs of two lines: $x - 1 = +\sqrt{2}(y + 1)$ and

$x - 1 = -\sqrt{2}(y + 1)$. The conic is degenerate.

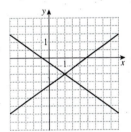

◢ Core Exercises

1, 21, 27, 35

11.5 Rotation of Axes

Concepts

Ⓐ **General equation of a conic.**

- Eliminating the xy term.
- Sketching rotated conics.

Review

Ⓐ **General equation of a conic.**

The general equation of a conic section is $Ax^2+Bxy+Cy^2+Dx+Ey+F=0$. To eliminate the xy-term in this equation, rotate the axes through the acute angle ϕ that satisfies $\cot 2\phi = \dfrac{A-C}{B}$. This produces a new coordinate system on the plane, called the **XY-plane**.

- **Eliminating the xy term.**

The coordinates (x,y) and (X,Y) of the xy- and XY-planes are related by the following **Rotation of Axes Formulas:**

$x = X\cos\phi - Y\sin\phi$	$X = x\cos\phi + y\sin\phi$
$y = X\sin\phi + Y\cos\phi$	$Y = -x\sin\phi + y\cos\phi$

Exercise: If the coordinate axes are rotated through $\phi = \cos^{-1}\frac{4}{5}$, find the XY-coordinates of the point with xy-coordinates $(-2,3)$.

Answer: We use the Rotation of Axes Formulas with $x=-2$, $y=3$, and $\phi=\cos^{-1}\frac{4}{5} \Rightarrow \cos\phi = \frac{4}{5} \Rightarrow$ $\sin\phi = \sqrt{1-\cos^2\phi} = \sqrt{1-\left(\frac{4}{5}\right)^2} = \frac{3}{5}$. Thus $X = x\cos\phi + y\sin\phi = -2\cdot\frac{4}{5}+3\cdot\frac{3}{5} = \frac{1}{5}$ and $Y = -x\sin\phi + y\cos\phi = -(-2)\cdot\frac{3}{5}+3\cdot\frac{4}{5} = \frac{18}{5}$. The XY-coordinates are $\left(\frac{1}{5},\frac{18}{5}\right)$.

Exercise: Determine an equation of the conic section $2x^2+y^2-4y=4$ in XY-coordinates when the coordinate axes are rotated through $\phi = 60^\circ$.

Answer: Using the Rotation of Axes Formulas with $\phi = 60^\circ$, we obtain $x = X\cos 60^\circ - Y\sin 60^\circ = \frac{1}{2}X - \frac{\sqrt{3}}{2}Y$ and $y = X\sin 60^\circ + Y\cos 60^\circ = \frac{\sqrt{3}}{2}X + \frac{1}{2}Y$. Substituting,

$$2\left(\tfrac{1}{2}X - \tfrac{\sqrt{3}}{2}Y\right)^2 + \left(\tfrac{\sqrt{3}}{2}X + \tfrac{1}{2}Y\right)^2 - 4\left(\tfrac{\sqrt{3}}{2}X + \tfrac{1}{2}Y\right) = 4$$

$$\Leftrightarrow\ 2\left(\tfrac{1}{4}X^2 - \tfrac{\sqrt{3}}{2}XY + \tfrac{3}{4}Y^2\right) + \tfrac{3}{4}X^2 + \tfrac{\sqrt{3}}{2}XY + \tfrac{1}{4}Y^2 - 2\sqrt{3}X - 2Y = 4$$

$$\Leftrightarrow\ \tfrac{5}{4}X^2 - \tfrac{\sqrt{3}}{2}XY + \tfrac{7}{4}Y^2 - 2\sqrt{3}X - 2Y = 4$$

● **Sketching rotated conics**

In the equation $Ax^2 + Bxy + Cy^2 + Dx + Ey + F = 0$, the quantity $B^2 - 4AC$ is called the **discriminant** of the equation.

> The graph of $Ax^2 + Bxy + Cy^2 + Dx + Ey + F = 0$ is either a conic in a degenerate conic.
>
> In the nondegenerate case, the graph is
>
> a parabola if $B^2 - 4AC = 0$ an ellipse if $B^2 - 4AC < 0$ a hyperbola if $B^2 - 4AC > 0$

Exercise: Use the discriminant of the equation $x^2 - 2xy + y^2 - 2x + 4y = 0$ to determine the shape of the graph of the equation. Use rotation of axes to eliminate the xy-term, and sketch the graph of the equation.

Answer: $B^2 - 4AC = (-2)^2 - 4 \cdot 1 \cdot 1 = 0$, so the graph is a parabola. Since $\cot 2\phi = \frac{1-1}{-2} = 0$, we have $\cot 2\phi = 0 \;\Leftrightarrow\; 2\phi = \frac{\pi}{2} \;\Leftrightarrow\; \phi = \frac{\pi}{4}$. Thus, $\cos\phi = \frac{\sqrt{2}}{2}$ and $\sin\phi = \frac{\sqrt{2}}{2}$.

Substituting

$$x = X\cos\phi - Y\sin\phi = \tfrac{\sqrt{2}}{2}X - \tfrac{\sqrt{2}}{2}Y = \tfrac{\sqrt{2}}{2}(X - Y) \text{ and}$$

$$y = X\sin\phi + Y\cos\phi = \tfrac{\sqrt{2}}{2}(X + Y), \text{ we get}$$

$$\left[\tfrac{\sqrt{2}}{2}(X - Y)\right]^2 - 2\left[\tfrac{\sqrt{2}}{2}(X - Y)\right]\left[\tfrac{\sqrt{2}}{2}(X + Y)\right]$$

$$+ \left[\tfrac{\sqrt{2}}{2}(X + Y)\right]^2 - 2\left[\tfrac{\sqrt{2}}{2}(X - Y)\right] + 4\left[\tfrac{\sqrt{2}}{2}(X + Y)\right] = 0$$

$$\Leftrightarrow$$

$$\tfrac{1}{2}\left(X^2 - 2XY + Y^2 - 2X^2 + 2Y^2 + X^2 + 2XY + Y^2\right)$$

$$- \sqrt{2}X + \sqrt{2}Y + 2\sqrt{2}X + 2\sqrt{2}Y = 0$$

$$\Leftrightarrow \;\; \tfrac{1}{2}\left(4Y^2\right) + \sqrt{2}X + 3\sqrt{2}Y = 0 \;\; \Leftrightarrow$$

$$Y^2 + \tfrac{3\sqrt{2}}{2}Y = -\tfrac{\sqrt{2}}{2}X. \text{ Completing the square, we get}$$

$$\left(Y + \tfrac{3\sqrt{2}}{4}\right)^2 = -\tfrac{\sqrt{2}}{2}\left(X - \tfrac{9\sqrt{2}}{8}\right).$$

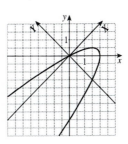

Exercise: Use the discriminant of the equation $14x^2 + 8xy - y^2 + 6 = 0$ to determine the shape of the graph of the equation. Use rotation of axes to eliminate the xy-term, and sketch the graph of the equation.

Answer: $B^2 - 4AC = 8^2 - 4(14)(-1) = 120$, so the graph is a hyperbola. We sketch a right triangle with

$$\cot 2\phi = \frac{14 - (-1)}{8} = \frac{15}{8}.$$

We see that $\cos 2\phi = \frac{15}{17}$, and using the half-angle formulas,

$$\cos \phi = \sqrt{\frac{1 + \frac{15}{17}}{2}} = \frac{4}{\sqrt{17}} \text{ and } \sin \phi = \sqrt{\frac{1 - \frac{15}{17}}{2}} = \frac{1}{\sqrt{17}}.$$

Making the substitutions

$$x = X \cos \phi - Y \sin \phi = \frac{4}{\sqrt{17}} X - \frac{1}{\sqrt{17}} Y = \frac{4X - Y}{\sqrt{17}} \text{ and}$$

$$y = X \sin \phi + Y \cos \phi = \frac{1}{\sqrt{17}} X + \frac{4}{\sqrt{17}} Y = \frac{X + 4Y}{\sqrt{17}}, \text{ we get}$$

$$14 \left(\frac{4X - Y}{\sqrt{17}} \right)^2 + 8 \left(\frac{4X - Y}{\sqrt{17}} \right) \left(\frac{X + 4Y}{\sqrt{17}} \right)$$

$$- \left(\frac{X + 4Y}{\sqrt{17}} \right)^2 + 6 = 0 \quad \Leftrightarrow$$

$$\tfrac{14}{17} \left(16X^2 - 8XY + Y^2 \right) + \tfrac{8}{17} \left(4X^2 + 15XY - 4Y^2 \right)$$

$$- \tfrac{1}{17} \left(X^2 + 8XY + 16Y^2 \right) = -6 \quad \Leftrightarrow$$

$$224X^2 - 112XY + 14Y^2 + 32X^2 + 120XY - 32Y^2$$

$$- X^2 - 8XY - 16Y^2 = -102 \quad \Leftrightarrow$$

$$255X^2 - 34Y^2 = -102 \quad \Leftrightarrow \quad 15X^2 - 2Y^2 = -6 \quad \Leftrightarrow$$

$$\frac{Y^2}{3} - \frac{5X^2}{2} = 1.$$

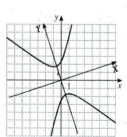

◤ **Core Exercises**

1, 7, 11, 15, 17, 21, 27, 37

11.6   Polar Equations of Conics

▼ Concepts

(A) Polar equations of conics.

- Finding a polar equation for a conic.

- Identifying and sketching a conic.

- Rotating a conic.

▼ Hints and Tips

- Notice that while the polar form of a conic section looks less familiar than the rectangular form, there are some real advantages. You can use a polar equation to graph a conic on your calculator with one equation instead of two. Furthermore, rotation of conics in polar coordinates is relatively easy, whereas in rectangular coordinates it can be tedious.

▼ Review

(A) Polar equations of conics.

Conics can be described by a positive number e called the **eccentricity**. For any point P on the conic, the eccentricity is the ratio of the distance from P to a fixed point F (the **focus**) to the distance from P to a fixed line l (called the **directrix**). If the focus is at the origin and the directrix is parallel or perpendicular to the polar axis, d units from the pole, then we have the following:

A polar equation of the form

$$r = \frac{ed}{1 \pm e\cos\theta} \qquad \text{or} \qquad r = \frac{ed}{1 \pm e\sin\theta}$$

represents a conic with eccentricity e. The conic is

1. a parabola if $e = 1$.

2. an ellipse if $e < 1$.

3. a hyperbola if $e > 1$.

The denominator $1 \pm e\cos\theta$ is used when the directrix is $x = \pm d$, the negative sign for $x = -d$ and the positive sign for $x = d$. The denominator $1 \pm e\sin\theta$ is used when the directrix is $y = \pm d$, the negative sign for $y = -d$ and the positive sign for $y = d$.

● **Finding a polar equation for a conic.**

Exercise: Write a polar equation of a conic with focus at the origin that satisfies the given conditions.

Answer:

(a) Ellipse, eccentricity 0.15, directrix
$x = -4$

$e = 0.15$, $d = 4$. Since the directrix is $x = -4$, we use $\cos\theta$ and a negative sign in the denominator. Thus, an equation is

$$r = \frac{0.15 \cdot 4}{1 - 0.15\cos\theta} = \frac{0.6}{1 - 0.15\cos\theta}.$$

(b) Hyperbola, eccentricity 3, directrix $y = 4$

$e = 3$ and $d = 4$. The directrix is $y = 4$, so we use $\sin\theta$ and a plus sign in the denominator. Thus, an equation is

$$r = \frac{3 \cdot 4}{1 + 3\sin\theta} = \frac{12}{1 + 3\sin\theta}.$$

(c) Parabola, directrix $x = 5$

$e = 1$ (because the conic is a parabola) and $d = 5$. The directrix is $x = 5$, so we use $\cos\theta$ and a plus sign in the denominator. An equation is $r = \dfrac{1 \cdot 5}{1 + 1\cos\theta} = \dfrac{5}{1 + \cos\theta}.$

● **Identifying and sketching a conic.**

Exercise: Find the eccentricity, directrix, and type of the conic with polar equation
$r = \dfrac{4}{5 - 3\sin\theta}$, and sketch its graph.

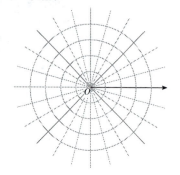

Answer: We make the leading term of the denominator 1 by multiplying the right-hand side of the equation by $\dfrac{1/5}{1/5}$:

$$r = \frac{4}{5 - 3\sin\theta} \cdot \frac{1/5}{1/5} = \frac{\frac{4}{5}}{1 - \frac{3}{5}\sin\theta}.$$ Thus, $e = \frac{3}{5}$ and the conic is an ellipse. Then $ed = \frac{4}{5} \iff \frac{3}{5}d = \frac{4}{5} \iff d = \frac{4}{3}$. Because the sign in the denominator is $-$ and the trigonometric function is sine, the directrix is $y = -\frac{4}{3}$.

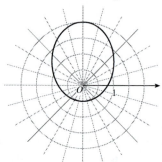

Exercise: Find the eccentricity, directrix, and type of the conic with polar equation $r = \dfrac{2}{2 + 5\cos\theta}$, and sketch its graph.

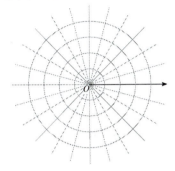

Answer: We make the leading term of the denominator 1 by multiplying the right-hand side of the equation by $\dfrac{1/2}{1/2}$:

$$r = \frac{2}{2 + 5\cos\theta} \cdot \frac{1/2}{1/2} = \frac{1}{1 + \frac{5}{2}\cos\theta}.$$ Thus, $e = \frac{5}{2}$ and the conic is a hyperbola. Then $ed = 1 \iff \frac{5}{2}d = 1 \iff d = \frac{2}{5}$. Because the sign in the denominator is $+$ and the trigonometric function is cosine, the directrix is $x = \frac{2}{5}$.

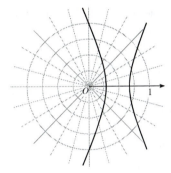

• **Rotating a conic**

Exercise: Suppose the ellipse with equation $r = \dfrac{4}{5 - 3\sin\theta}$ is rotated through an angle $45°$ about the origin. Find a polar equation for the resulting ellipse and sketch its graph.

Answer: We replace θ by $\theta - \frac{\pi}{4}$ to obtain

$$r = \frac{4}{5 - 3\sin\left(\theta - \frac{\pi}{4}\right)}.$$ We rotate the graph of

$$r = \frac{4}{5 - 3\sin\theta}$$ through $45°$ to obtain the new graph.

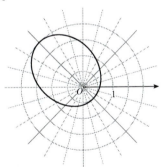

⬛ Core Exercises

7, 17, 21, 23, 31

11.7   Plane Curves and Parametric Equations

Concepts

(A) Parametric equations.

- Sketching parametric curves.
- Eliminating the parameter.
- Using a graphing device to sketch parametric curves.

(B) The relationship between parametric curves and polar coordinates.

- Converting a polar equation into a set of parametric equations.

Definitions

- **Parametric equations:** A method of specifying curves by describing how the x- and y-coordinates change with respect to an arbitrary parameter.

Hints and Tips

- It helps me to think about the toy called Etch A Sketch$^{\circledR}$. This toy allows you to draw pictures by controlling the horizontal motion of a stylus with one knob, and the vertical motion with a different knob. The stylus moves around, leaving a trail, and a picture is drawn. Parametric equations work the same way: we specify the horizontal position of a point with one equation, the vertical with a second equation. The path taken by the point traces out the curve.

Review

(A) Parametric equations.

x- and y-coordinates of points on a curve can be given separately as functions of an additional variable t, called the **parameter**:

$$x = f(t) \qquad y = g(t)$$

These are called **parametric equations** for the curve. Substituting a value of t into both equations determines the coordinates of a point $(x(t), y(t))$. As t varies, the point varies and traces out the curve. All of the graphing we have done up to this point could also have been done using parametric equations.

- **Sketching parametric curves.**

Exercise: Sketch the curve with parametric equations $x = t^3$, $y = t - 1$.

Answer: We generate points and draw the graph.

t	x	y
-2	-8	-3
-1	-1	-2
0	0	-1
1	1	0
2	8	1
3	27	2

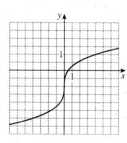

Exercise: Sketch the curve with parametric equations $x = \sin t$, $y = 1 - 2\sin^2 t$.

Answer: We generate points and draw the graph.

t	x	y
0	0	1
$\frac{\pi}{4}$	$\frac{\sqrt{2}}{2}$	0
$\frac{\pi}{2}$	1	-1
$\frac{3\pi}{4}$	$\frac{\sqrt{2}}{2}$	0
π	0	1
$\frac{5\pi}{4}$	$-\frac{\sqrt{2}}{2}$	0
$\frac{3\pi}{2}$	-1	-1
$\frac{7\pi}{4}$	$-\frac{\sqrt{2}}{2}$	0
2π	0	1

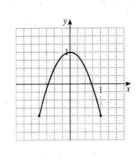

- **Eliminating the parameter.**

Often we can **eliminate the parameter** from a pair of parametric equations by solving one of the equations for t and substituting into the other.

Exercise: Eliminate the parameter from the parametric equations $x = t^3$, $y = t - 1$.

Answer: We solve the second equation for $t = y + 1$. Substituting back into the first equation, we get $x = (y + 1)^3$.

Exercise: Eliminate the parameter from the parametric equations $x = \sin t$,
$y = 1 - 2\sin^2 t$.

Answer: We substitute x for $\sin t$ and get $y = 1 - 2x^2$, a parabola. Note, however, that because $-1 \le \sin t \le 1$, we have the restriction $-1 \le x \le 1$.

- **Exercise:** Eliminate the parameter from the parametric equations $x = 3t - 9$,
$y = \sin\left(\frac{t}{3}\right)$.

Answer: Our goal is to solve one of the equations for t. In this case we can write $t = \dfrac{x + 9}{3}$ and then substitute into the other equation to obtain $y = \sin\left(\dfrac{x + 9}{9}\right)$.

- **Using graphing devices to sketch parametric curves.**

Most graphing calculators can be used to graph parametric equations.

Exercise: Use a graphing device to graph the curve with parametric equations
$x = 2t\cos t$, $y = \sin^2 t$.

Answer: Since $x = 2t\cos t$ varies without bound and $0 \le y = \sin^2 t \le 1$, we use the viewing rectangle $[-10, 10]$ by $[0, 1]$ with $t \in [-15, 15]$.

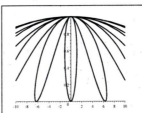

Exercise: Use a graphing device to graph the curve with parametric equations
$x = 4\cos 3t$, $y = 3\cos 4t$.

Answer: We can generate the complete graph with $t \in [0, \pi]$ because

$$
\begin{aligned}
(x(t+\pi), y(t+\pi)) &= (4\cos 3(t+\pi), 3\cos 4(t+\pi)) \\
&= (-4\cos 3t, 3\cos 4t) \\
&= (x(\pi - t), y(\pi - t))
\end{aligned}
$$

We use the viewing rectangle $[-4, 4]$ by $[-3, 3]$.

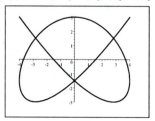

(B) The relationship between parametric curves and polar coordinates.

- **Converting a polar equation into a set of parametric equations.**

The graph of the polar equation $r = f(\theta)$ is equivalent to the graph of the parametric equations $x = f(t)\cos t$, $y = f(t)\sin t$.

Exercise: Rewrite the polar equation $r = 1 - \sin 3\theta$ as a set of parametric equations.

Answer: $x = f(t)\cos t$, $y = f(t)\sin t$. So we write $x = (1 - \sin 3t)\cos t$, $y = (1 - \sin 3t)\sin t$.

▰ Core Exercises

1, 5, 25, 33, 35, 61

12 Sequences and Series

12.1 Sequences and Summation Notation

▲ Concepts

(A) Sequences.

- Listing the terms of a sequence described explicitly.
- Listing the terms of a sequence described recursively.
- Finding the nth term of a sequence.

(B) Partial sums.

- Finding the partial sums of a sequence.

(C) Sigma notation.

- Using sigma notation.
- Understanding the properties of sums.

▲ Definitions

- **Sequence:** An infinite ordered list of numbers, such as $2, 4, 6, 8, 10, \ldots$ or $3, 1, 4, 1, 5, 9, \ldots$. Technically, a function whose domain is the set of natural numbers.
- **Fibonacci sequence**: The sequence beginning $0, 1, 1, 2, 3, 5, 8, 13, \ldots$ in which each term is the sum of the previous two terms.
- **Partial sums of a sequence:** The ith partial sum S_i of a sequence is the sum of the first i terms of that sequence.
- **Sigma Notation**: $\sum_{i=1}^{k} a_i = a_1 + a_2 + \cdots + a_k$

▲ Hints and Tips

- One of the nice things about sequences is that the mathematical and English definitions of the word "sequence" are similar: an ordered list.
- If you have a computer science background, you can think of sigma notation as the mathematical way of writing this computer program:

```
for i = 1 to k do
    sum = sum + a[i]
end
```

◤ Review

Ⓐ Sequences.

A **sequence** is a set of numbers written in a specific order: $a_1, a_2, a_3, a_4, \ldots, a_n, \ldots$. The number a_1 is called the *first term*, a_2 is the *second term*, and in general, a_n is the *nth term*. A sequence can also be defined as a function f whose domain is the set of positive integers. In this form the values $f(1), f(2), f(3), \ldots$ are called the **terms**, where $f(n) = a_n$ is the *n*th term. Some sequences have formulas that give the *n*th term directly; these sequences are called **explicit**. Other sequences have formulas that give the *n*th term based on the values of previous terms; these sequences are called **recursive**.

- **Listing the terms of a sequence described explicitly.**

Exercise: Find the first five terms, the 10th term, and the 100th term of the sequence.

Answer:

(a) $a_n = \dfrac{n-1}{2n+1}$

For $a_n = \dfrac{n-1}{2n+1}$:

$$a_1 = \frac{1-1}{2(1)+1} = 0$$

$$a_2 = \frac{2-1}{2(2)+1} = \frac{1}{5}$$

$$a_3 = \frac{3-1}{2(3)+1} = \frac{2}{7}$$

$$a_4 = \frac{4-1}{2(4)+1} = \frac{3}{9} = \frac{1}{3}$$

$$a_5 = \frac{5-1}{2(5)+1} = \frac{4}{11}$$

$$a_{10} = \frac{10-1}{2(10)+1} = \frac{9}{21} = \frac{3}{7}$$

$$a_{100} = \frac{100-1}{2(100)+1} = \frac{99}{201} = \frac{33}{67}$$

(b) $a_n = 2^n + (-1)^n$

For $a_n = 2^n + (-1)^n$:

$$a_1 = 2^{(1)} + (-1)^{(1)} = 2 - 1 = 1$$

$$a_2 = 2^{(2)} + (-1)^{(2)} = 4 + 1 = 5$$

$$a_3 = 2^{(3)} + (-1)^{(3)} = 8 - 1 = 7$$

$$a_4 = 2^{(4)} + (-1)^{(4)} = 16 + 1 = 17$$

$$a_5 = 2^{(5)} + (-1)^{(5)} = 32 - 1 = 31$$

$$a_{10} = 2^{(10)} + (-1)^{(10)} = 1024 + 1 = 1025$$

$$a_{100} = 2^{(100)} + (-1)^{(100)} = 2^{100} + 1$$

(c) $a_n = \dfrac{2n - 3^n}{4^n}$

For $a_n = \dfrac{2n - 3^n}{4^n}$:

$a_1 = \dfrac{2(1)-3^1}{4^1} = \dfrac{2-3}{4} = -\dfrac{1}{4}$

$a_2 = \dfrac{2(2)-3^2}{4^2} = \dfrac{4-9}{16} = -\dfrac{5}{16}$

$a_3 = \dfrac{2(3)-3^3}{4^3} = \dfrac{6-27}{64} = -\dfrac{21}{64}$

$a_4 = \dfrac{2(4)-3^4}{4^4} = \dfrac{8-81}{256} = -\dfrac{73}{256}$

$a_5 = \dfrac{2(5)-3^5}{4^5} = \dfrac{10-243}{1024} = -\dfrac{233}{1024}$

$a_{10} = \dfrac{2(10)-3^{10}}{4^{10}} = \dfrac{20-59{,}049}{1{,}048{,}576} = -\dfrac{59{,}029}{1{,}048{,}576}$

$a_{100} = \dfrac{2(100)-3^{100}}{4^{100}}$

• **Listing the terms of a sequence described recursively.**

Exercise: Find the first twelve terms of the Fibonacci sequence.

Answer:
The Fibonacci sequence is defined recursively by $F_1 = 1$, $F_2 = 1$, and $F_n = F_{n-1} + F_{n-2}$:
$F_3 = F_2 + F_1 = 1 + 1 = 2$, $F_4 = F_3 + F_2 = 2 + 1 = 3$,
$F_5 = F_4 + F_3 = 3 + 2 = 5$, $F_6 = F_5 + F_4 = 5 + 3 = 8$,
$F_7 = F_6 + F_5 = 8 + 5 = 13$, $F_8 = F_7 + F_6 = 13 + 8 = 21$,
$F_9 = F_8 + F_7 = 21 + 13 = 34$,
$F_{10} = F_9 + F_8 = 34 + 21 = 55$,
$F_{11} = F_{10} + F_9 = 55 + 34 = 89$,
$F_{12} = F_{11} + F_{10} = 89 + 55 = 144$

• **Finding the nth term of a sequence.**

Exercise: Find the 100th term of the sequence $a_n = 6n - 5(-1)^n$.

Answer: $a_{100} = 6(100) - 5(-1)^{100} = 595$

Exercise: Find the formula for the nth term of the sequence.

(a) $\dfrac{2}{3}, \dfrac{8}{3}, \dfrac{32}{3}, \dfrac{128}{3}, \dfrac{512}{3}, \ldots$

Some facts we can observe about this sequence is that the denominator of each term is 3, and the numerators are all powers of 2. We determine the exponent of 2 in each numerator and correlate it with the number n of the term:

n	1	2	3	4	5
Numerator of a_n	2	8	32	128	512
Exponent of 2	1	3	5	7	9

Thus, a formula for the exponents of the numerator is $2n - 1$, and the numerator of a_n is 2^{2n-1}. Putting it all together, we get

$a_n = \dfrac{2^{2n-1}}{3}$.

(b) $\frac{2}{3}, \frac{4}{9}, \frac{6}{27}, \frac{8}{81}, \ldots$

The numerators are all multiples of 2. By trial and error, we find that each numerator is $2n$. The denominators are all powers of 3, and by trial and error we find that each denominator is 3^n.

Thus, $a_n = \dfrac{2n}{3^n}$.

B **Partial sums.**

For the sequence $a_1, a_2, a_3, \ldots, a_n, \ldots$, the **partial sums** are

$$S_1 = a_1$$
$$S_2 = a_1 + a_2$$
$$S_3 = a_1 + a_2 + a_3$$
$$S_4 = a_1 + a_2 + a_3 + a_4$$
$$\vdots$$
$$S_n = a_1 + a_2 + a_3 + a_4 + \cdots + a_n$$
$$\vdots$$

S_1 is called the **first partial sum**, S_2 the **second partial sum**, and so on. S_n is called the nth **partial sum**. The sequence $S_1, S_2, S_3, S_4, \ldots, S_n, \ldots$ is called the **sequence of partial sums**. The sequence of partial sums is useful in detecting a pattern to find the sum.

• **Finding the partial sums of a sequence.**

Exercise: Find the first five partial sums of the sequence given by $a_n = 2^n + (-1)^n$.

Answer: We found the first five terms of this sequence above; they are $a_1 = 1$, $a_2 = 5$, $a_3 = 7$, $a_4 = 17$, and $a_5 = 31$. Thus,
$S_1 = a_1 = 1$, $S_2 = a_1 + a_2 = 1 + 5 = 6$,
$S_3 = a_1 + a_2 + a_3 = 1 + 5 + 7 = 13$,
$S_4 = a_1 + a_2 + a_3 + a_4 = 1 + 5 + 7 + 17 = 30$, and
$S_5 = a_1 + a_2 + a_3 + a_4 + a_5 = 1 + 5 + 7 + 17 + 31 = 61$.

Exercise: Find the first five partial sums of the sequence given by $a_n = \dfrac{n-1}{2n+1}$.

Answer: We found the first five terms of this sequence above; they are $a_1 = 0$, $a_2 = \frac{1}{5}$, $a_3 = \frac{2}{7}$, $a_4 = \frac{1}{3}$, and $a_5 = \frac{4}{11}$. Thus,
$S_1 = a_1 = 0$, $S_2 = a_1 + a_2 = 0 + \frac{1}{5} = \frac{1}{5}$,
$S_3 = a_1 + a_2 + a_3 = 0 + \frac{1}{5} + \frac{2}{7} = \frac{17}{35}$,
$S_4 = a_1 + a_2 + a_3 + a_4 = 0 + \frac{1}{5} + \frac{2}{7} + \frac{1}{3} = \frac{86}{105}$, and
$S_5 = a_1 + a_2 + a_3 + a_4 + a_5 = 0 + \frac{1}{5} + \frac{2}{7} + \frac{1}{3} + \frac{4}{11} = \frac{1366}{1155}$

Exercise: Find the first six partial sums of the sequence given by $a_n = \dfrac{2n}{n+3}$.

Answer: $S_1 = a_1 = \frac{2}{4} = \frac{1}{2}$, $S_2 = a_1 + a_2 = \frac{1}{2} + \frac{4}{5} = \frac{13}{10}$,
$S_3 = a_1 + a_2 + a_3 = \frac{1}{2} + \frac{4}{5} + \frac{6}{6} = \frac{23}{10}$,
$S_4 = a_1 + a_2 + a_3 + a_4 = \frac{1}{2} + \frac{4}{5} + \frac{6}{6} + \frac{8}{7} = \frac{241}{70}$,
$S_5 = a_1 + a_2 + a_3 + a_4 + a_5 = \frac{1}{2} + \frac{4}{5} + \frac{6}{6} + \frac{8}{7} + \frac{10}{8} = \frac{657}{140}$,
and
$S_6 = a_1 + a_2 + a_3 + a_4 + a_5 + a_6$
$\quad = \frac{1}{2} + \frac{4}{5} + \frac{6}{6} + \frac{8}{7} + \frac{10}{8} + \frac{12}{9} = \frac{2531}{420}$

Exercise: Find the first four partial sums and then find a formula for the nth partial sum of the sequence given by

$$a_n = \frac{1}{n+1} - \frac{1}{n+3}.$$

Answer: Start by finding the first several partial sums until we find a pattern:

$$S_1 = \left(\tfrac{1}{2} - \tfrac{1}{4}\right)$$

$$S_2 = \left(\tfrac{1}{2} - \tfrac{1}{4}\right) + \left(\tfrac{1}{3} - \tfrac{1}{5}\right)$$
$$= \left(\tfrac{1}{2} + \tfrac{1}{3}\right) - \left(\tfrac{1}{4} + \tfrac{1}{5}\right)$$

$$S_3 = \left(\tfrac{1}{2} - \tfrac{1}{4}\right) + \left(\tfrac{1}{3} - \tfrac{1}{5}\right) + \left(\tfrac{1}{4} - \tfrac{1}{6}\right)$$
$$= \left(\tfrac{1}{2} + \tfrac{1}{3}\right) - \left(\tfrac{1}{5} + \tfrac{1}{6}\right)$$

At this point a pattern is starting to emerge.

$$S_4 = \left(\tfrac{1}{2} - \tfrac{1}{4}\right) + \left(\tfrac{1}{3} - \tfrac{1}{5}\right) + \left(\tfrac{1}{4} - \tfrac{1}{6}\right) + \left(\tfrac{1}{5} - \tfrac{1}{7}\right)$$
$$= \left(\tfrac{1}{2} + \tfrac{1}{3}\right) - \left(\tfrac{1}{6} + \tfrac{1}{7}\right)$$

This type of sum is called a telescoping sum because the middle terms collapse. A formula for the partial sums is

$$S_n = \left(\frac{1}{2} + \frac{1}{3}\right) - \left(\frac{1}{n+2} + \frac{1}{n+3}\right).$$

(C) Sigma notation.

Sigma notation is a way of expressing the sum of the first n terms of the sequence. $\sum_{k=1}^{n} a_k$ is read as "the sum of a_k from $k = 1$ to $k = n$" and $\sum_{k=1}^{n} a_k = a_1 + a_2 + a_3 + a_4 + \cdots + a_n$. The letter k is called the **index of summation** or the **summation variable** and the idea is to find the sum as k takes on the values $1, 2, 3, 4, \ldots, n$. Any letter can be used for the index of summation, and the index need not start at 1.

- **Using sigma notation.**

Exercise: Find the sum.

(a) $\displaystyle\sum_{k=1}^{4} \frac{1}{k^2}$

(b) $\displaystyle\sum_{j=3}^{7} (-1)^j j^3$

Answer:

$$\sum_{k=1}^{4} \frac{1}{k^2} = \tfrac{1}{1^2} + \tfrac{1}{2^2} + \tfrac{1}{3^2} + \tfrac{1}{4^2} = 1 + \tfrac{1}{4} + \tfrac{1}{9} + \tfrac{1}{16} = \tfrac{205}{144}$$

$$\sum_{j=3}^{7} (-1)^j j^3$$
$$= (-1)^3 \, 3^3 + (-1)^4 \, 4^3 + (-1)^5 \, 5^3 + (-1)^6 \, 6^3 + (-1)^7 \, 7^3$$
$$= -27 + 64 - 125 + 216 - 343 = -215$$

Exercise: Write $\tfrac{2}{3} + \tfrac{3}{4} + \tfrac{4}{5} + \tfrac{5}{6} + \tfrac{6}{7}$ in sigma notation.

Answer: We find the sequence associated with the terms $\tfrac{2}{3}, \tfrac{3}{4}$, $\tfrac{4}{5}, \tfrac{5}{6}$, and $\tfrac{6}{7}$. Both the numerator and denominator increase by 1, and the denominator is 1 greater than the numerator. So the numerator is k, the denominator is $k + 1$, and the index of summation runs from $k = 2$ to $k = 6$. Thus, one formula for the sum is $\displaystyle\sum_{k=2}^{6} \frac{k}{k+1}$. Note there are many possible representations; another is $\displaystyle\sum_{j=1}^{5} \frac{j+1}{j+2}$.

- **Understanding the properties of sums.**

  Sums have the following properties:

  > Let $a_1, a_2, a_3, a_4, \ldots$ and $b_1, b_2, b_3, b_4, \ldots$ be sequences. Then for any positive integer n and any constant c, the following properties hold.
  >
  > 1. $\sum_{k=1}^{n} (a_k + b_k) = \left(\sum_{k=1}^{n} a_k\right) + \left(\sum_{k=1}^{n} b_k\right)$
  >
  > 2. $\sum_{k=1}^{n} (a_k - b_k) = \left(\sum_{k=1}^{n} a_k\right) - \left(\sum_{k=1}^{n} b_k\right)$
  >
  > 3. $\sum_{k=1}^{n} (ca_k) = c \left(\sum_{k=1}^{n} a_k\right)$

Core Exercises

13, 25, 33, 63, 71, 79

12.2   Arithmetic Sequences

Concepts

(A) **Arithmetic sequences.**

- Finding terms of an arithmetic sequence.
- Finding partial sums of arithmetic sequences.

Definitions

- **Arithmetic sequence:** A sequence of the form $a_n = a + (n-1)d$.

Hints and Tips

- You can tell a sequence is arithmetic by taking the difference of neighboring terms. In an arithmetic sequence, this difference is always the constant d.

Review

(A) **Arithmetic sequences.**

- **Finding terms of an arithmetic sequence.**

  An arithmetic sequence is a sequence of the form $a, a + d, a + 2d, a + 3d, \ldots$. The number a is the **first term** and the number d is called the **common difference**. Any two consecutive terms of an arithmetic sequence differ by d. The nth term of the arithmetic sequence is given by the formula $a_n = a + (n-1)d$.

Exercise: Write the first five terms of an arithmetic sequence for which 9 is the first term and 5 is the common difference.	**Answer:** $a_n = a + (n-1)d \Rightarrow$ $a_1 = 9$, $a_2 = 9 + (2-1)5 = 9 + 5 = 14$, $a_3 = 9 + (3-1)5 = 9 + 10 = 19$, $a_4 = 9 + (4-1)5 = 9 + 15 = 24$, and $a_5 = 9 + (5-1)5 = 9 + 20 = 29$.

Exercise: The first four terms of an arithmetic sequence are 6, 13, 20, and 27. Find a and d, and write a formula for the nth term.

Answer: a is the first term, so $a = 6$. d is the common difference between any two terms, so using the first and second term, we have $d = 13 - 6 = 7$. So $a_n = 6 + (n - 1)\,7$.

Exercise: The fifth term of an arithmetic sequence is 34 and the eighth term is 43. Find a, d, and the formula for the nth term.

Answer: Use the formula for the nth term, $a_n = a + (n - 1)\,d$. When $n = 5$, we have $a_5 = a + (5 - 1)\,d$, so $34 = a + 4d$. When $n = 8$, we have $a_8 = a + (8 - 1)\,d$, so $43 = a + 7d$. Thus, we have the system

$$\begin{cases} 34 = a + 4d \\ 43 = a + 7d \end{cases}$$ Subtracting the first equation from the

second equation, we get $9 = 3d$, so $d = 3$. Back-substituting then gives $34 = a + 4\,(3)$ $\Leftrightarrow$ $34 = a + 12$ $\Leftrightarrow$ $a = 22$. So $a = 22$, $d = 3$, and $a_n = 22 + (n - 1)\,3$.

• **Finding partial sums of arithmetic sequences.**
For the arithmetic sequence $a_k = a + (k - 1)\,d$, the **nth partial sum** (the sum of the first n terms) is $S_n = \sum_{k=1}^{n} [a + (k - 1)\,d] = a + [a + d] + [a + 2d] + \cdots + [a + (n - 1)\,d]$. This sum is given by $S_n = \dfrac{n}{2}\,[2a + (n - 1)\,d] = n\left(\dfrac{a + a_n}{2}\right)$.

Exercise: Find the sum of the first 40 positive multiples of 3.

Answer: This is the sequence $3, 6, 9, 12, \ldots$, which has $a = 3$ and $d = 3$. So $a_{40} = 3 + (40 - 1)\,3 = 120$. Then $S_{40} = \sum_{k=1}^{k} a_k = 40\left(\frac{3 + 120}{2}\right) = 2460$.

Exercise: Find the sum of the first five terms of the arithmetic sequence with $a = -2$ and $d = 8$.

Answer: Using $S_n = \frac{n}{2}\,[2a + (n - 1)\,d]$ with $n = 5$ and $a = -2$, we get $S_5 = \frac{5}{2}\,[2\,(-2) + (5 - 1)\,8] = \frac{5}{2}\,(-4 + 32) = \frac{5}{2}\,(28) = 70$.

Exercise: Find the partial sum S_{10} of the arithmetic sequence with $a_1 = 0.31$ and $a_{10} = 8.35$.

Answer: This time use the formula $S_n = n\left(\dfrac{a + a_n}{2}\right)$ with $n = 10$ to find the partial sum. So $S_{10} = 10\left(\dfrac{0.31 + 8.35}{2}\right) = 10\left(\dfrac{8.66}{2}\right) = 43.3$.

◥ Core Exercises

11, 21, 31, 33, 45, 57, 63

12.3 Geometric Sequences

Concepts

(A) **Geometric sequences.**

- Finding terms of a geometric sequence.
- Finding partial sums of a geometric sequence.
- Summing an infinite geometric series.

Definitions

- **Geometric sequence:** A sequence of the form $a_n = a\left(r^{n-1}\right)$.
- **Infinite series:** An expression of the form $a_1 + a_2 + a_3 + \cdots$ or $\sum_{i=1}^{\infty} a_i$

Hints and Tips

- You can tell a sequence is geometric by taking the *ratio* of neighboring terms. In a geometric sequence, this ratio is always the constant d.

Review

(A) **Geometric sequences.**

A **geometric sequence** is a sequence of the form $a, ar, ar^2, ar^3, ar^4, \ldots$. The number a is the **first term** and r is the **common ratio**. The ratio of any two consecutive terms is r, so $r = \dfrac{a_{k+1}}{a_k}$ for any nonnegative integer k. The nth term of the sequence is given by the formula $a_n = ar^{n-1}$.

- **Finding the terms of a geometric sequence.**

Exercise: Write the first five terms of a sequence where 9 is the first number and 4 is the common ratio.	**Answer:** $a_n = ar^{n-1}$, so $a_1 = 9 \cdot 4^{1-1} = 9$, $a_2 = 9 \cdot 4^{2-1} = 9 \cdot 4 = 36$, $a_3 = 9 \cdot 4^{3-1} = 9 \cdot 16 = 144$, $a_4 = 9 \cdot 4^{4-1} = 9 \cdot 64 = 576$, and $a_5 = 9 \cdot 4^{5-1} = 9 \cdot 256 = 2304$.

- **Finding partial sums of a geometric sequence.**

For the geometric sequence $a_k = ar^{k-1}$, the nth **partial sum** is $S_n = \sum_{k=1}^{n} ar^{k-1} = a + ar + ar^2 + ar^3 + \cdots + ar^n$. For $r \neq 1$, this sum is given by $S_n = a\left(\dfrac{1 - r^n}{1 - r}\right)$.

Exercise: Find the partial sum of the first 20 terms of the geometric sequence with $a = 5.6$ and $r = 2$.	**Answer:** Using the formula $S_n = a\left(\dfrac{1 - r^n}{1 - r}\right)$ and substituting the values $n = 20$, $a = 5.6$, and $r = 2$, we get $S_{20} = 5.6\left(\dfrac{1 - 2^{20}}{1 - 2}\right) = 5.6 \cdot 1{,}048{,}575 = 5{,}872{,}020$.

Exercise: Find $\sum_{k=6}^{15} (0.3)^{k-1}$.

Answer: To use the formula we first express the sum with the index starting at 1. Since

$$\sum_{k=1}^{5} (0.3)^{k-1} + \sum_{k=6}^{15} (0.3)^{k-1} = \sum_{k=1}^{15} (0.3)^{k-1}, \text{ we get}$$

$$\sum_{k=6}^{15} (0.3)^{k-1} = \sum_{k=1}^{15} (0.3)^{k-1} - \sum_{k=1}^{5} (0.3)^{k-1}. \text{ Now}$$

$$\sum_{k=1}^{15} (0.3)^{k-1} = 1 \left(\frac{1-(0.3)^{15}}{1-(0.3)} \right) = \frac{0.999999986}{0.7} = 1.4286 \text{ and}$$

$$\sum_{k=1}^{5} (0.3)^{k-1} = 1 \left(\frac{1-(0.3)^{5}}{1-(0.3)} \right) = \frac{0.99757}{0.7} = 1.4251, \text{ so}$$

$$\sum_{k=6}^{15} (0.3)^{k-1} = 1.4286 - 1.4251 = 0.0035.$$

- **Summing an infinite geometric series.**

A sum of the form $a + ar + ar^2 + ar^3 + \cdots + ar^{n-1} + \cdots$ is an **infinite series**. From above, the nth partial sum of such a series is $S_n = a \left(\dfrac{1-r^n}{1-r} \right)$ for $r \neq 1$. If $|r| < 1$, then r^n gets close to 0 as n gets large, so S_n gets close to $\dfrac{a}{1-r}$. That is, if $|r| < 1$, then the sum of the infinite series is $a + ar + ar^2 + ar^3 + \cdots + ar^{n-1} + \cdots = \dfrac{a}{1-r}$.

Exercise: Find the sum of the infinite geometric series $\frac{4}{3} + \frac{4}{9} + \frac{4}{27} + \frac{4}{81} + \cdots$.

Answer: We need to find the values of a and r, then substitute these values into the formula. a is the first term, so $a = \frac{4}{3}$. r is the common ratio, found by dividing two consecutive terms, so

$$r = \frac{4/9}{4/3} = \frac{1}{3}. \text{ Since } |r| < 1, \ S = \frac{a}{1-r} = \frac{4/3}{1-(1/3)} = 2.$$

Exercise: Find the sum of the infinite geometric series $27 + 18 + 12 + 8 + \cdots$.

Answer: We need to find the values of a and r, then substitute these values into the formula. a is the first term, so $a = 27$. r is the common ratio, found by dividing two consecutive terms,

$$\text{so } r = \frac{18}{27} = \frac{2}{3}. \text{ Since } |r| < 1, \ S = \frac{a}{1-r} = \frac{27}{1-(2/3)} = 81.$$

Exercise: Find the sum of the infinite geometric series $\frac{3}{\sqrt{5}} - \frac{6}{5} + \frac{12}{5\sqrt{5}} - \frac{24}{25} + \cdots$.

Answer: We need to find the values a and r, then substitute these values into the formula. a is the first term, so $a = \frac{3}{\sqrt{5}}$. r is the common ratio, found by dividing two consecutive terms,

$$\text{so } r = \frac{a_2}{a_1} = \frac{-\frac{6}{5}}{\frac{3}{\sqrt{5}}} = -\frac{2}{\sqrt{5}}. \text{ Since } |r| < 1,$$

$$S = \frac{a}{1-r} = \frac{\frac{3}{\sqrt{5}}}{1 + \frac{2}{\sqrt{5}}} = \frac{\frac{3}{\sqrt{5}}}{\frac{\sqrt{5}+2}{\sqrt{5}}} = \frac{3}{\sqrt{5}+2}$$

$$= \frac{3}{\sqrt{5}+2} \cdot \frac{\sqrt{5}-2}{\sqrt{5}-2} = \frac{3\left(\sqrt{5}-2\right)}{5-4} = 3\sqrt{5} - 6$$

Exercise: Express the repeating decimal $0.090909\ldots$ as a fraction.

Answer: First express as an infinite geometric sum. $0.09090909\ldots = 0.09 + 0.0009 + 0.000009 + \cdots$. The first term is $a = 0.09$ and the common ratio is $\frac{0.0009}{0.09} = 0.01$. Since $|r| < 1, \ S = \frac{a}{1-r} = \frac{0.09}{1-0.01} = \frac{0.09}{0.99} = \frac{1}{11}$.

Exercise: Express the repeating decimal $0.3\overline{45}$ as a fraction.

Answer: First express as an infinite geometric sum.

$0.3\overline{45} = 0.3 + 0.045 + 0.00045 + 0.0000045 + \cdots$. The geometric series involved is

$0.045 + 0.00045 + 0.0000045 + \cdots$. Its first term is $a = 0.045$ and its common ratio is $r = \frac{0.00045}{0.045} = 0.01$. Since $|r| < 1$, the sum of the geometric series is

$S = \frac{a}{1-r} = \frac{0.045}{1-0.01} = \frac{0.045}{0.99} = \frac{1}{22}$. Thus,

$0.3\overline{45} = 0.3 + 0.045 + 0.00045 + 0.0000045 + \cdots$
$$= \frac{3}{10} + \frac{1}{22} = \frac{38}{110} = \frac{19}{55}$$

Core Exercises

3, 19, 39, 45, 63, 67

12.4   Mathematics of Finance

Concepts

(A) **Annuities and installment buying**

- Working with future values of annuities.
- Working with present values of annuities and installment payments.

Definitions

- **Annuity:** A sum of money that is paid in regular, equal payments.
- **Amount of an annuity:** The sum of all the individual payments made from an annuity.
- **Periodic rent:** The amount of the regular annuity payment.
- **Present value of an annuity:** The amount of money you would have to invest now to withdraw regular payments that are the equivalent of a given annuity.

Hints and Tips

- An investment advisor described annuities to me once in a way that made a lot of sense. He pointed out that an insurance policy is where you give a company a little bit of money every year, and then they give your beneficiaries a large amount of money at some point in the future. An annuity is the opposite: you give the company a large amount of money up front and then they give you a little bit of money every year.

Review

(A) **Annuities and installment buying**

- **Working with future values of annuities.**
  An **annuity** is a sum of money that is paid into an account (or fund) in regular equal intervals. The amount A_f of an annuity is the sum of all the individual payments from the beginning until the last payment is made, including interest. This is a partial sum of a geometric sequence and results in the formula $A_f = R\dfrac{(1+i)^n - 1}{i}$, where R is the amount of each regular payment, i is the interest per compounding period, and n is the number of payments. Since the value of the money is paid in the future, this is called the **future value**.

Exercise: New parents put away $50 a month into a saving account paying 3%, compounded monthly. How much money will accumulate in this account in 16 years?

Answer: Apply the formula $A_f = R\dfrac{(1+i)^n - 1}{i}$, with $R = 50$, $n = 16 \cdot 12 = 192$ payments, and $i = \frac{0.03}{12} = 0.0025$:

$$A_f = 50 \cdot \frac{(1 + 0.0025)^{192} - 1}{0.0025} = 12{,}302.13.$$ There will be $12,302.13 in the account in 16 years.

Exercise: A small company decides to save money to purchase a photocopier that costs $9,500. How much do they need to deposit each month in an account that pays 6% compounded monthly if they need to save this money over a 9 month period?

Answer: Substitute the values $A_f = 9500$, $n = 9$, and $i = \frac{0.06}{12} = 0.005$ into $A_f = R\dfrac{(1+i)^n - 1}{i}$ and solve for R:

$$9500 = R\frac{(1.005)^9 - 1}{0.005} = 9.182115828R$$

$\Leftrightarrow \quad R = 1034.62$. They must deposit $1034.62 monthly.

- **Working with present values of annuities and installment payments.**
  The **present value**, PV, is today's value of an amount to be paid in n compounded periods in the future. $PV = A(1+i)^{-n}$, where A is the amount to be paid in the future, i is the interest per compounding period, and n is the number of compounding periods. The **present value of an annuity**, A_p, consisting of n regular equal payments of size R with interest rate i per compounding period, is given by $A_p = R\dfrac{1 - (1+i)^{-n}}{i}$. This formula can be solved for R to obtain the amount of a regular payment on an installment loan: $R = \dfrac{iA_p}{1 - (1+i)^{-n}}$.

  Although it is not possible to solve the formula algebraically for i, it is possible to use a graphing device to find i.

Exercise: Suppose you plan to purchase a DVD player from an electronics superstore for $300. If you put the $300 on a charge card at 19.8% interest compounded monthly and pay it off in 10 equal payments, how much will each payment be?

Answer: Substitute the values $A_p = 300$, $n = 10$, and $i = \frac{0.198}{12} = 0.0165$ into $R = \dfrac{iA_p}{1 - (1+i)^{-n}}$ and calculate R:

$$R = \frac{0.0165 \cdot 300}{1 - (1.0165)^{-10}} = \frac{4.95}{0.150963872} = 32.79.$$ Each payment will be $32.79.

Exercise: Suppose you wish to purchase a car and you have $125 a month that you can spend on payments for the next 4 years. If the current interest rate on used cars is 6.6%, what is the highest price you can pay?

Answer: Substitute the values $R = 125$, $n = 4 \cdot 12 = 48$, and $i = \frac{0.066}{12} = 0.0055$ into $A_p = R\dfrac{1 - (1 + i)^{-n}}{i}$:

$$A_p = 125\frac{1 - (1.0055)^{-48}}{0.0055} = 5260.69.$$ You can pay no more than $5260.69.

Exercise: A used car dealership advertise a car for $9595 (plus tax and license) or $229.59 a month for 48 months (plus tax and license). What interest rate are they charging?

Answer: Let x be the annual interest rate. Then $i = \frac{x}{12}$. So we can graph the function $R(x) = \dfrac{\frac{x}{12}(9595)}{1 - \left(1 + \frac{x}{12}\right)^{-48}}$ and find where it intersects the line $y = 229.59$.

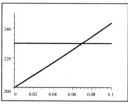

From this graph, we see that $x \approx 0.070$, so the annual interest rate is 7%.

◥ Core Exercises

1, 9, 13, 15, 23

◤ 12.5   Mathematical Induction

◥ Concepts

(A) Principle of mathematical induction

- Using mathematical induction to complete proofs.

◥ Definitions

- **Mathematical induction:** A method used to prove that a given statement is true for all natural numbers. (Technically, all natural numbers after a certain base number.)

◥ Questions to Ask Your Teacher

- Teachers vary widely in the format they require for proofs in general, and induction proofs in particular. My precalculus teacher, for example, wanted the phrase, "Therefore, by the principle of mathematical induction," to be in every induction proof, verbatim. After your teacher has done an example or two, it would probably be good to clarify your teacher's expectations for your own proofs.

- Think of induction in terms of an infinitely long line of dominos. How can you convince yourself logically that the 1,343,729th domino in the line will fall if you push the first one? Well, the first domino falls; and if some arbitrary kth domino falls, it certainly causes the $(k + 1)$th domino to fall. So by mathematical induction, all of the dominos will fall.

◤ Review

(A) Principle of mathematical induction

Mathematical induction is a type of proof used to show that a statement that depends only on n is true for all natural numbers n. Let $P(n)$ be a statement that depends on n. The principle of mathematical induction is based on showing that the following two conditions are true:

> **1.** $P(1)$ is true.
>
> **2.** For every natural number k, *if* $P(k)$ is true, *then* $P(k+1)$ is true.

If these two conditions are met, then it follows that $P(n)$ is true for all natural numbers n.

Proof by mathematical induction involves the following steps.

> **1.** Basis of induction: *Prove* that the statement $P(n)$ is true for some initial value (usually 1).
>
> **2.** Induction hypothesis: Assume that the statement $P(k)$ is true.
>
> **3.** The induction step: Prove that $P(k+1)$ is true using the hypothesis above.
>
> **4.** Conclusion: Since the two conditions have been shown true, by the principle of mathematical induction, the statement is true for all natural numbers.

● **Using mathematical induction to complete proofs.**

Exercise: Prove that $n^3 - n + 3$ is divisible by 3 for all natural numbers n.

Answer: Here $P(n)$ is the statement "$n^3 - n + 3$ is divisible by 3. "

Basis of induction: For $n = 1$, we calculate
$(1)^3 - (1) + 3 = 3$ which is divisible by 3. So $P(1)$ is true.

Induction hypothesis: Assume that the statement $P(k)$ is true for some $k \geq 1$, that is, assume that the statement "$k^3 - k + 3$ is divisible by 3" is true.

Induction step: Show that $P(k+1)$ is true, that is, "$(k+1)^3 - (k+1) + 3$ is divisible by 3": Now
$(k+1)^3 - (k+1) + 3$
$$= \left(k^3 + 3k^2 + 3k + 1\right) - k - 1 + 3$$
$$= \left(k^3 - k + 3\right) + \left(3k^2 + 3k\right)$$
$$= \left(k^3 - k + 3\right) + 3\left(k^2 + k\right)$$

But $k^3 - k + 3$ is divisible by 3 (by the induction hypothesis), and $3\left(k^2 + k\right)$ is divisible by 3, since it is the product of 3 and an integer. Hence the sum $\left(k^3 - k + 3\right) + 3\left(k^2 + k\right)$ is also divisible by 3.

Conclusion: $P(1)$ is true and $P(k)$ implies $P(k+1)$, so by mathematical induction, the statement is true for all natural numbers n.

Exercise: Let $\{a_n\}$ be the sequence recursively defined by

$a_n = a_{n-1} + 2a_{n-2} - 6$, where $a_1 = 4$ and $a_2 = 5$. Find the first 5 terms of the sequence and prove that $a_n = 2^{n-1} + 3$.

Answer: $a_1 = 4$; $a_2 = 5$;
$a_3 = a_2 + 2a_1 - 6 = 5 + 2(4) - 6 = 7$;
$a_4 = a_3 + 2a_2 - 6 = 7 + 2(5) - 6 = 11$;
$a_5 = a_4 + 2a_3 - 6 = 11 + 2(7) - 6 = 19$.

Basis of induction: $n = 1$, $n = 2$, and $n = 3$. Note that all cases are needed since a_n is based on the terms a_{n-1} and a_{n-2}, and a_3 is the first term defined by the recursive definition.
$a_1 = 2^{1-1} + 3 = 4$, $a_2 = 2^{2-1} + 3 = 5$, and
$a_3 = 2^{3-1} + 3 = 7$. So $P(1)$, $P(2)$, and $P(3)$ are true.

Induction hypothesis: Note that we must start at $k \geq 3$ because a_2 is not given by the formula $a_j = a_{j-1} + 2a_{j-2} - 6$. Also note that since the terms a_{j-1} and a_{j-2} are both used in the definition of a_j, we need an induction hypotheses that includes both terms. So assume that
$a_k = a_{k-1} + 2a_{k-2} - 6 = 2^{k-1} + 3$ for $k \geq 3$.

Induction step: We must show that
$a_{k+1} = 2^{(k+1)-1} + 3 = 2^k + 3$. By the recursive definition, $a_{k+1} = a_k + 2a_{k-1} - 6$, and by the induction hypothesis,
$a_k = 2^{k-1} + 3$ and $a_{k-1} = 2^{k-2} + 3$. Substituting,

$$a_{k+1} = a_k + 2a_{k-1} - 6 = \left(2^{k-1} + 3\right) + 2\left(2^{k-2} + 3\right) - 6$$
$$= 2^{k-1} + 3 + 2 \cdot 2^{k-2} + 6 - 6 = 2^{k-1} + 2^{k-1} + 3$$
$$= 2 \cdot 2^{k-1} + 3 = 2^k + 3$$

Conclusion: By the principle of mathematical induction, $a_n = a_{n-1} + 2a_{n-2} - 6 = 2^{n-1} + 3$, where $a_1 = 4$ and $a_2 = 5$, is true for all natural numbers n.

Core Exercises

1, 5, 13, 21, 25, 35

12.6   The Binomial Theorem

Concepts

(A) **Expanding binomials.**

- Using Pascal's triangle to expand powers of binomials.

(B) **Binomial coefficients and their properties.**

- Calculating binomial coefficients.
- Using the Binomial Theorem to expand powers of binomials.
- Finding a particular term in a binomial expansion.

Definitions

- **Pascal's triangle**: A triangular array of numbers in which each term is the sum of the two entries above it. The terms on the sides of the triangle are always 1.
- **Factorial**: n factorial, written $n!$, is given by $1 \cdot 2 \cdot 3 \cdot 4 \cdots \cdots n$.
- **Binomial coefficient**: The binomial coefficient is denoted $\binom{n}{r}$ and is equal to $\dfrac{n!}{r!\,(n-r)!}$.

Review

(A) **Expanding binomials.**

An expression of the form $a + b$ is called a **binomial**. When $(a + b)^n$ is expanded, the following pattern emerges:

> **1.** There are $n + 1$ terms, the first being $a^n b^0$ and the last $a^0 b^n$.
>
> **2.** The exponents of a *decrease* by 1 from term to term while the exponents of b *increase* by 1.
>
> **3.** The sum of the exponents of a and b in each term is n.

Pascal's triangle is the following triangular array. It gives the coefficients of the terms of a binomial expansion.

$$
\begin{array}{cc}
(a+b)^0 & 1 \\
(a+b)^1 & 1 \quad 1 \\
(a+b)^2 & 1 \quad 2 \quad 1 \\
(a+b)^3 & 1 \quad 3 \quad 3 \quad 1 \\
(a+b)^4 & 1 \quad 4 \quad 6 \quad 4 \quad 1 \\
(a+b)^5 & 1 \quad 5 \quad 10 \quad 10 \quad 5 \quad 1 \\
(a+b)^6 & 1 \quad 6 \quad 15 \quad 20 \quad 15 \quad 6 \quad 1
\end{array}
$$

Every entry (other than a 1) is the sum of the two entries diagonally above it.

• **Using Pascal's triangle to expand powers of binomials.**

Exercise: Use Pascal's triangle above to expand $(3 + 2x)^5$.

Answer:
$(a+b)^5 = a^5 + 5a^4 b^1 + 10a^3 b^2 + 10a^2 b^3 + 5a^1 b^4 + b^5$, so

$$
\begin{aligned}
(3 + 2x)^5 &= (3)^5 + 5\,(3)^4\,(2x) + 10\,(3)^3\,(2x)^2 \\
&\quad + 10\,(3)^2\,(2x)^3 + 5\,(3)\,(2x)^4 + (2x)^5 \\
&= 243 + 810x + 1080x^2 + 720x^3 + 240x^4 + 32x^5
\end{aligned}
$$

Exercise: Use Pascal's triangle above to expand $(3x - 5y)^3$.

Answer: $(a+b)^3 = a^3 + 3a^2 b^1 + 3a^1 b^2 + b^3$, so

$$
\begin{aligned}
(3x - 5y)^3 &= (3x)^3 + 3\,(3x)^2\,(-5y) + 3\,(3x)\,(-5y)^2 + (-5y)^3 \\
&= 27x^3 - 135x^2 y + 225xy^2 - 125y^3
\end{aligned}
$$

(B) **Binomial coefficients and their properties.**

The product of the first n natural numbers is denoted by $n!$ and is called n **factorial**. So $n! = 1 \cdot 2 \cdot 3 \cdot \cdots \cdot (n-1) \cdot n$ and we define $0! = 1$. For $0 \le r \le n$, the **binomial coefficient** is denoted by $\binom{n}{r} = \dfrac{n!}{r!\,(n-r)!}$.

Two important relations between binomial coefficients are the following:

$$
\binom{n}{r} = \binom{n}{n-r}
$$

$$
\binom{k}{r-1} + \binom{k}{r} = \binom{k+1}{r} \quad \text{for any nonnegative integers } r \text{ and } k \text{ with } r \le k
$$

Now **Pascal's triangle** can also be written as

$$\binom{0}{0}$$
$$\binom{1}{0} \quad \binom{1}{1}$$
$$\binom{2}{0} \quad \binom{2}{1} \quad \binom{2}{2}$$
$$\binom{3}{0} \quad \binom{3}{1} \quad \binom{3}{2} \quad \binom{3}{3}$$
$$\binom{4}{0} \quad \binom{4}{1} \quad \binom{4}{2} \quad \binom{4}{3} \quad \binom{4}{4}$$
$$\binom{5}{0} \quad \binom{5}{1} \quad \binom{5}{2} \quad \binom{5}{3} \quad \binom{5}{4} \quad \binom{5}{5}$$
$$\binom{6}{0} \quad \binom{6}{1} \quad \binom{6}{2} \quad \binom{6}{3} \quad \binom{6}{4} \quad \binom{6}{5} \quad \binom{6}{6}$$

The **Binomial Theorem** states that

$$(a+b)^n = \binom{n}{0} a^n b^0 + \binom{n}{1} a^{n-1} b^1 + \binom{n}{2} a^{n-2} b^2 + \cdots + \binom{n}{n-1} a^1 b^{n-1} + \binom{n}{n} a^0 b^n$$

The term that contains a^r in the expansion of $(a+b)^n$ is $\binom{n}{r} a^r b^{n-r}$.

- **Calculating binomial coefficients.**

Exercise: Compute $\binom{7}{2}$.

Answer: $\binom{7}{2} = \dfrac{7!}{2!\,5!} = \dfrac{7 \cdot 6}{2} = 21$

- **Using the Binomial Theorem to expand powers of binomials.**

Exercise: Use the binomial theorem to expand $(3x - 5y)^3$.

Answer:

$$(3x - 5y)^3 = \binom{3}{0} (3x)^3 + \binom{3}{1} (3x)^2 (-5y)$$
$$+ \binom{3}{2} (3x) (-5y)^2 + \binom{3}{3} (-5y)^3$$
$$= 27x^3 - 135x^2 y + 225xy^2 - 125y^3$$

Exercise: Find the first three terms in the expansion of $(4 - x)^9$.

Answer: The first three terms are $\binom{9}{0} 4^9 (-x)^0$, $\binom{9}{1} 4^8 (-x)^1$, and $\binom{9}{2} 4^7 (-x)^2$, or $262{,}144$, $-589{,}824x$, and $589{,}824x^2$.

Exercise: Find the last three terms in the expansion of $(7 + y)^{11}$.

Answer: The last three terms are $\binom{11}{9} 7^2 (y)^9$, $\binom{11}{10} 7^1 (y)^{10}$, and $\binom{11}{11} 7^0 (y)^{11}$, or $2695 y^9$, $77 y^{10}$, and y^{11}.

- **Finding a particular term in a binomial expansion.**

Exercise: Find the term containing x^8 in the expansion of $(2x - 3)^{15}$.

Answer: The term containing a^r is $\binom{n}{r} a^r b^{n-r}$, so the term containing x^8 is

$$\binom{15}{8} (2x)^8 (-3)^{15-8} = 6435 \cdot 2^8 \cdot x^8 \cdot (-3)^7$$
$$= -3{,}602{,}776{,}320 x^8.$$

▷ Core Exercises

1, 11, 21, 43, 49

13 Counting and Probability

13.1 Counting Principles

◢ Concepts
(A) Tree diagrams.
(B) The Fundamental Counting Principle.
  • Counting with cards and dice.

◢ Definitions
• **The Fundamental Counting Principle:** If we can break a process up into a series of independent choices, the number of ways to complete the process is the product of the number of ways to make each choice.

◢ Hints and Tips
• In theory, we could solve every counting problem by drawing a tree diagram. If a problem involves 8 possibilities, then drawing a tree diagram may be the easiest way to solve it. However, if a problem involves 262,144 possibilities, then we need to use another technique — such as the Fundamental Counting Principle.

◢ Review
(A) **Tree diagrams.**
A **tree diagram** is a systematic method of determining and listing all the possibilities of a counting or probability problem. At each vertex, all possible choices are drawn as edges and labeled. To create a list of possible outcomes to a counting problem, start at an end vertex and trace your way back to the initial vertex.

Exercise: A family is planning on having three children. List all possible birth orders of birth (boy-girl-boy, girl-girl-boy, etc.).

Answer: Here we use a tree diagram to help us count. Let B represent a boy and G represent a girl. This leads to the tree diagram below.

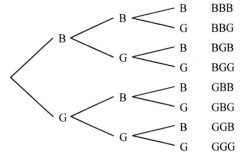

So the 8 possible birth orders are GGG, GGB, GBG, GBB, BGG, BGB, BBG, BBB.

Exercise: A pet food manufacturer tests two types of dog food to see which food is preferred by a dog. A dog is observed at most five times or until it chooses the same food three times. How many different outcomes are possible? Draw a tree diagram to count this question.

Answer: Although this count can be made with other methods still to come, using a tree diagram is the easiest. Let A and B represent the two types of dog food. This leads to the tree diagram below.

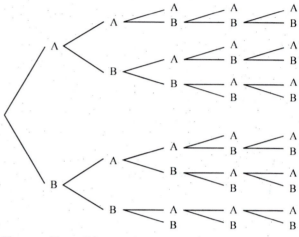

There are 20 possible outcomes.

(B) The Fundamental Counting Principle.

This principle states that if two events occur in order and if the first event can occur in m ways and the second in n ways (after the first event has occurred), then the two events can occur in order in $m \times n$ ways. When the events $E_1, E_2, \ldots, E_k$ occur in order and E_1 can occur in n_1 ways, E_2 in n_2 ways, and so on, then the events can occur in order in $n_1 \times n_2 \times \cdots \times n_k$ ways.

Exercise: For security purposes, a company classifies each employee according to five hair colors, four eye colors, six weight categories, five height categories, and two sex categories. How many classifications are possible?

Answer: Using the Fundamental Counting Principle there are $5 \cdot 4 \cdot 6 \cdot 5 \cdot 2 = 1200$ classifications.

Exercise: The Uniform Pricing Code (the bar code on a package used to price the item) consists of a 10 digit number. How many different bar codes are possible?

Answer: Using the Fundamental Counting Principle there are ten choices for each of the ten digits. As a result there are $10 \cdot 10 \cdot 10 \cdot 10 \cdot 10 \cdot 10 \cdot 10 \cdot 10 \cdot 10 \cdot 10 = 10^{10}$ Uniform Pricing Codes.

Exercise: A car manufacturer makes its lowest-priced car in five exterior colors, four interior colors, a choice of manual or automatic transmission, and three levels of options. How many models are available?

Answer: Using the Fundamental Counting Principle there are $5 \cdot 4 \cdot 2 \cdot 3 = 120$ models available.

- **Counting with cards and dice.**

Many counting questions involve a **standard deck of cards**. A standard deck of cards consists of 52 cards broken up into 4 **suits** of 13 cards each. The suits **diamonds** and **hearts** are red, while the suits **clubs** and **spades** are black. There are 13 **kinds** in a deck: Ace, 2, 3, ..., 10, Jack, Queen, King. For example, the cards 3 *of diamonds*, 3 *of hearts*, 3 *of clubs*, and 3 *of spades* are of the same **kind**. Other common counting questions involve a **die** (or **dice**, the plural of die) which is a six-sided cube with the numbers 1 through 6 painted on the faces. When a pair of dice is rolled there are 36 possible outcomes that can occur. Also, a common "experiment" is to look at the sum of the faces. The table below shows the 36 possible outcomes along with the corresponding sum of the faces.

		Die 2				
	1	2	3	4	5	6
1	$(1,1)$ 2	$(1,2)$ 3	$(1,3)$ 4	$(1,4)$ 5	$(1,5)$ 6	$(1,6)$ 7
2	$(2,1)$ 3	$(2,2)$ 4	$(2,3)$ 5	$(2,4)$ 6	$(2,5)$ 7	$(2,6)$ 8
3	$(3,1)$ 4	$(3,2)$ 5	$(3,3)$ 6	$(3,4)$ 7	$(3,5)$ 8	$(3,6)$ 9
4	$(4,1)$ 5	$(4,2)$ 6	$(4,3)$ 7	$(4,4)$ 8	$(4,5)$ 9	$(4,6)$ 10
5	$(5,1)$ 6	$(5,2)$ 7	$(5,3)$ 8	$(5,4)$ 9	$(5,5)$ 10	$(5,6)$ 11
6	$(6,1)$ 7	$(6,2)$ 8	$(6,3)$ 9	$(6,4)$ 10	$(6,5)$ 11	$(6,6)$ 12

(Die 1 labels the rows; Die 2 labels the columns.)

Core Exercises

3, 9, 13, 21, 23, 37, 41

13.2   Permutations and Combinations

Concepts

(A) **Factorial notation.**

(B) **Permutations.**

- Distinguishable permutations.

(C) **Combinations.**

- Difference between permutations and combinations.

Definitions

- **n factorial:** $n! = 1 \cdot 2 \cdot 3 \cdot 4 \cdots \cdot n$.
- **Permutation:** A set of objects taken from a larger set, in which order counts.
- **Combination:** A set of objects taken from a larger set, in which order does not count.
- **Partition:** A partition of n is a set of numbers $n_1, n_2, ..., n_k$ such that $n_1 + n_2 + \cdots + n_k = n$.

◤ Review

(A) **Factorial notation.**

The product of the first n natural numbers is denoted by $n!$ and is called n **factorial**. $n! = 1 \cdot 2 \cdot 3 \cdot 4 \cdots (n-1) \cdot n$ for $n \geq 1$, and $0! = 1$. $n!$ can also be defined in a recursive form as $n! = n \cdot (n-1)!$ for $n \geq 1$. For example $6! = 6 \cdot (5!)$.

Exercise: A woman sits down to make four sales calls. In how many ways can she make her calls?	**Answer:** She has 4 choices for the first call, 3 choices for the call, 2 choices for the third call, and 1 choice for the last call. So there are $4 \cdot 3 \cdot 2 \cdot 1 = 4!$ ways for her to make her sales calls.

(B) **Permutations.**

A **permutation** of a set of distinct objects is an ordering or arrangement of these objects.   The number of ways n objects can be arranged is $n!$.   When only r of the n objects of the set are arranged, then **the number of permutations of n objects taken r at a time** is denoted by $P(n, r)$ and is given by the formula $P(n, r) = n \cdot (n-1) \cdot (n-2) \cdots (n-r+1) = \dfrac{n!}{(n-r)!}$.

Permutations are used when (1) order is important, and (2) there is no repetition.

Exercise: A club of 20 people gather for a group photo standing in a row. In how many ways can the photo be taken?	**Answer:** Here we are after an arrangement of all 20 members of the group. So there are $$P(20, 20) = 20! \approx 2.432902008 \times 10^{18}$$ ways to arrange the group.

Exercise: A volleyball team has nine members and six members are on the court at any one time. In how many ways can the six be selected and arranged on the court?	**Answer:** $$P(9, 6) = \frac{9!}{(9-6)!} = \frac{9!}{3!} = 9 \cdot 8 \cdot 7 \cdot 6 \cdot 5 \cdot 4 \cdot \frac{3!}{3!} = 60{,}480$$

Exercise: A new car dealership has ten different new car models and five pedestals in front of the dealership to display the cars. In how many ways can the dealership display the cars?	**Answer:** $$P(10, 5) = \frac{10!}{(10-5)!} = \frac{10!}{5!} = 10 \cdot 9 \cdot 8 \cdot 7 \cdot 6 \cdot \frac{5!}{5!} = 30{,}240$$

• **Distinguishable permutations.**

When considering a set of objects, some of which are of the same kind, then two permutations are **distinguishable** if one cannot be obtained from the other by interchanging the positions of objects of the same kind. If a set of n objects consists of k different kinds of objects with n_1 objects of the first kind, n_2 objects of the second kind, n_3 objects of the third kind, and so on, where $n_1 + n_2 + n_3 + \cdots + n_k = n$, then the number of distinguishable permutations of these objects is $\dfrac{n!}{n_1! \, n_2! \, n_3! \cdots n_k!}$. The numbers $n_1, n_2, n_3, \ldots, n_k$ are called a **partition** of n when $n_1 + n_2 + n_3 + \cdots + n_k = n$.

Exercise: An instructor makes three versions of an exam for her chemistry class, which contains 30 students. If she makes 10 copies of each version, in how many ways can she distribute the exams in her class?

Answer: Since all 30 exams are distributed to the class, this is a distinguishable permutation. The number of ways is then

$$\frac{30!}{10! \, 10! \, 10!} = 5.550996791 \times 10^{12}.$$

Exercise: How many ways can the letters of the word "MOTORSPORTS" be arranged?

Answer: Since MOTORSPORTS has 11 letters (1 M, 3 Os, 2 Ts, 2 R's, 2 S's, and 1 P), there

are $\dfrac{11!}{1! \, 3! \, 2! \, 2! \, 2! \, 1!} = 831{,}600$ ways to arrange these letters.

Exercise: An ice chest contains six colas, four orange sodas, and eight fruit-flavored waters. In how many ways can these be distributed to 18 people at a picnic?

Answer: Since we distribute all 18 items in the ice chest, this is a distinguishable permutation. So there are

$$\frac{18!}{6! \, 4! \, 8!} = 9{,}189{,}180 \text{ ways to distribute the soft drinks.}$$

Ⓒ **Combinations.**

A **combination** of r elements of a set is any subset of r elements from the set without regard to order. If the set has n elements ($n \geq r$), then the **number of combinations of n elements taken r at a time** is denoted by $C(n, r)$ and is given by the formula $C(n, r) = \dfrac{n!}{r! \, (n - r)!}$. Combinations are used when (1) order is not important and (2) there is no repetition. A set with n elements has 2^n subsets.

Exercise: A drawer contains 28 pairs of socks. In how many ways can a person select ten pairs of socks to pack for a vacation?

Answer: Order is not important and there are no repeats. So there are $C(28, 10) = \frac{28!}{10! \, (28-10)!} = \frac{28!}{10! \, 18!} = 13123110$ ways to selected the pairs of socks.

Exercise: Twelve cars are stopped at a fruit inspection checkpoint operated by the state of California. In how many ways can the inspectors choose three of the twelve cars to inspect?

Answer: Here order is not important, because the inspectors simply want a group of cars to inspect. There is no repetition, since they will not inspect the same car twice. So there are

$$C(12, 3) = \frac{12!}{3! \, (12 - 3)!} = \frac{12!}{3! \, 9!} = 220 \text{ ways to select the}$$

cars to inspect.

Exercise: A group of 12 friends are at a party.

(a) In how many ways can 3 people be selected to go and get more pizza?

Answer:

Since a group is selected, order is not important, and obviously there is no repetition. As a result, there are

$$C(12, 3) = \frac{12!}{3! \, (12 - 3)!} = \frac{12!}{3! \, 9!} = 220 \text{ ways to select the}$$

three people.

(b) If there are 6 females and 6 males at a party, in how many ways can two males and one female be selected to go and get more pizza?

This is still a combination. Since the males can be selected only from the six males and the female can be selected only from the six females, there are

$$C(6, 2) \cdot C(6, 1) = \frac{6!}{2! \, 4!} \cdot \frac{6!}{1! \, 5!} = 15 \cdot 6 = 90 \text{ ways to select}$$

the three people.

Exercise: Pauline's Pizza Palace has 14 different kinds of toppings. Pauline's Special Combination is a large pizza with any three different toppings. How many Pauline's Special Combinations are possible?

Answer: Since the toppings must be different and order is not important, this is a combination problem. The number of

Pauline's Special Combinations is $C(14, 3) = \dfrac{14!}{3! \, 11!} = 364$.

Exercise: A hamburger chain advertises that they can make a hamburger in 256 ways. How many different toppings do they need to meet this claim?

Answer: Since changing a topping will change the type of hamburger, a subset of toppings will represent a different hamburger. Let n be the number different toppings they have. So we need to solve the equation $2^n = 256$. Thus $n = \log_2 256 = 8$.

Exercise: A *full house* is a poker hand consisting of 3 cards of one kind and 2 cards of another kind. How many full houses are possible?

Answer: To count this, we first pick 1 of the 13 kinds and 3 of the 4 cards of that kind, then pick another kind from the 12 remaining kinds, and 2 of the 4 cards from that kind. So the number of possible full houses is

$$\begin{pmatrix} \text{pick the} \\ \text{first kind} \end{pmatrix} \begin{pmatrix} \text{pick 3 of} \\ \text{4 cards} \end{pmatrix} \begin{pmatrix} \text{pick the} \\ \text{second kind} \end{pmatrix} \begin{pmatrix} \text{pick 2 of} \\ \text{4 cards} \end{pmatrix}$$

$$= C(13, 1) \cdot C(4, 3) \cdot C(12, 1) \cdot C(4, 2)$$
$$= \frac{13!}{1! \, 12!} \cdot \frac{4!}{3! \, 1!} \cdot \frac{12!}{1! \, 11!} \cdot \frac{4!}{2! \, 2!}$$
$$= 13 \cdot 4 \cdot 12 \cdot 6$$
$$= 3744$$

• **Difference between permutations and combinations.**

Both permutations and combinations use a set of n distinct objects where r objects are selected at random *without repetitions* (or replacements). The key difference between permutations and combinations is *order*. When order is important, permutations are used and when order is not important, combinations are used. Other key words that help to distinguish combinations are: *group, sets, subsets*, etc. Key words that distinguish permutations are: *arrange, order, arrangements*, etc.

Exercise: Fifteen people are selected for a study of diet pills. Eight will be given a new diet pill and seven will be given a placebo. In how many ways can the eight be selected?

Answer: This is a combination problem because (1) there is no repetition (a person can be selected for only one group) and (2) we are interested in a group only. So there are

$$C(15, 8) = \frac{15!}{8! \, 7!} = 6435 \text{ ways to select the test group.}$$

Exercise: Five couples line up outside a movie theater. In how many ways can they line up?

Answer: First treat each couple as one item and then arrange the people in the couple. There are $P(5,5) = 5!$ ways to arrange the couples, and there are $P(2,2) = 2!$ ways to arrange each couple. So there are $5!\,(2!)^5$ ways to arrange the couples in line at the movie theater. Since $2! = 2$ we get $5!\,(2!)^5 = 5!\,2^5 = 3840$.

Exercise: A newly formed Ad Hoc Committee on Student Retention plans to study ten new students during the Fall semester in order to identify ways to improve student retention. Suppose next fall's incoming class contains 112 new female and 108 new male students.

Answer:

(a) In how many different ways can 10 new students be chosen at random (with no regard to sex) for this study?

Since there are 220 new students expected, we pick the 10 students from this group. So there are $C(220,10) = \dfrac{220!}{10!\,210!}$ ways to pick the study group.

(b) How many possible studies in part (a) have 5 females and 5 males?

The five female students must be picked from among the 112 female students, and the five male students must be picked from among the 108 male students. So there are $C(112,5) \cdot C(108,5) = \dfrac{112!}{5!\,107!} \cdot \dfrac{108!}{5!\,103!}$ ways to form the study group.

Exercise: A college theater group plans on doing 5 different shows during the year: *Cats*, *The Tempest*, *Annie*, *A Streetcar Named Desire*, and *Guys and Dolls*. If *Cats* must be scheduled last, and *Annie* must be either first or second, how many different schedules are possible?

Answer: There are five slots to fill, and *Cats* must be placed in the last slot, while *Annie* can be placed in either first or second slot. So there are 2 ways to place *Annie*, 1 way to place *Cats* (in the last slot), and there are three remaining slots to fill. Since there are 3! ways to fill these remaining slots, there are a total of $2 \cdot 3! \cdot 1 = 12$ ways to schedule the shows.

Exercise: In poker, a *flush* consists of five cards of the same suit. How many different flushes are possible?

Answer: Here we first pick the suit and then 5 cards from that suit. There are $C(4,1)$ ways to pick the suit, and there are $C(13,5)$ ways to pick the 5 cards from the suit. So there are $C(4,1) \cdot C(13,5) = 5148$ flushes.

Core Exercises

3, 9, 13, 31, 43, 49, 53, 59

13.3 Probability

Concepts

(A) **Sample spaces and probability.**

- Complements.
- Mutually exclusive events and unions of events.
- Probability of the union of two sets.
- Independent events.

Definitions

- **Sample space:** The sample space of an experiment is the set of all possible outcomes that we are considering. For example, if we toss a coin and let it land on the floor, the sample space is {Heads, Tails}.
- **Probability:** The probability of an event occuring as the outcome of an experiment is the number of ways in which the event could occur, divided by the total number of possible outcomes.
- **Complement:** The complement of an event is the set of all possible outcomes in which the event does *not* occur.
- **Mutually Exclusive Events:** Two events are mutually exclusive if there is no possible outcome in which both events occur.
- **Independent events:** Two events are independent if the occurance of one event does not affect the probability of the other event.

Review

(A) **Sample spaces and probability.**

An **experiment** is a process that has a finite set of definite outcomes. The set of all possible outcomes of an experiment is called the **sample space**, and we use the letter S to denote it. An **event** is any subset of the sample space. When each individual outcome has an equal chance of occurring, then the **probability** that event E occurs, written as $P(E)$, is $P(E) = \dfrac{n(E)}{n(S)} = \dfrac{\text{number of elements in event } E}{\text{number of elements in sample space } S}$. As a result, $0 \le P(E) \le 1$. The closer $P(E)$ is to 1, the more likely the event E is to happen, whereas the closer to zero the less likely. If $P(E) = 1$, then E is called the **certain event**, which means *that it is definitely going to happen*. When $P(E) = 0$, the event is called an **impossible event**.

Exercise: A pair of dice are rolled and the sum of the faces are recorded.

(a) Find the probability that the sum of the numbers showing is 3.

(b) Find the probability that the sum of the faces is 7.

Answer:

Using the table at the end of Section 13.1 in this book, we see that the sum is 3 in only 2 of the 36 squares. Thus,
$$P(\text{sum is 3}) = \tfrac{2}{36} = \tfrac{1}{18}.$$

Again using the table, there are 7 squares where the sum is 7. Thus, $P(\text{sum is 7}) = \tfrac{6}{36} = \tfrac{1}{6}$.

(c) Find the probability that the sum of the faces is greater than 9.

The sum is greater than 9 is the same as *the sum is* 10, *the sum is* 11, *or the sum is* 12. Looking at the table, there are three squares where the sum is 10, two squares where the sum is 11, and one square where the sum is 12, for a total of 6 squares. Thus, P (sum is greater than 9) $= \frac{6}{36} \approx 0.167$.

Exercise: A box of 100 ping pong balls contains 50 yellow balls and 50 white balls.

Answer:

(a) If 5 balls are selected at random, what is the probability that all 5 balls are yellow?

Since there are 100 balls, the sample space S consists of all ways to select 5 balls without regard to color. So $n(S) = C(100, 5) = 75{,}287{,}520$. Since there are 50 yellow balls, there are $C(50, 5) = 2{,}118{,}760$ ways to select this group. Thus,

$$P(5 \text{ yellow balls}) = \frac{C(50, 5)}{C(100, 5)} = \frac{2{,}118{,}760}{75{,}287{,}520} \approx 0.02814.$$

(b) If 5 balls are selected at random, what is the probability that 3 balls are yellow and 2 balls are white?

Pick the 3 yellow balls from the 50 yellow balls and pick the 2 white balls from the 50 white balls. There are $C(50, 3) \cdot C(50, 2) = 19600 \cdot 1225 = 24{,}010{,}000$ ways to make these selections. So the probability is

$$P(3 \text{ yellow and 2 white}) = \frac{C(50, 3) \cdot C(50, 2)}{C(100, 5)}$$
$$= \frac{24{,}010{,}000}{75{,}287{,}520} \approx 0.3189$$

• **Complements.**

The **complement** of an event E is the set of outcomes of the sample space S that are not in E, and is denoted as E'. The probability that E does not occur is $P(E') = 1 - P(E)$. Sometimes it is easier to find $P(E')$ than to find $P(E)$.

Exercise: A box of 100 calculators contains 3 defective calculators. Suppose 5 calculators are selected at random. What is the probability that at least one calculator is defective?

Answer: The complement of *at least one calculator is defective* is *no calculator is defective*. There are 97 non-defectives; choose 5. The number of ways of choosing 5 non-defective is $C(97, 5)$. So,

$$P(\text{all non-defective}) = \frac{C(97, 5)}{C(100, 5)} \approx 0.8560. \text{ Then}$$

$$P(\text{at least 1 defective}) = 1 - P(\text{all non-defective})$$
$$\approx 1 - 0.8560 \approx 0.1440.$$

Exercise: Five friends compare birth months. What is the probability that at least two were born in the same month?

Answer: Here we assume that each month is equally likely to occur. Since we seek *at least two* this means that two could have the same birth month, three could have the same birth month, two pairs could have the same birth month, etc. The complement, *no two have the same birth month*, is much easier to count. Thus,

$$P\,(\text{different birth months}) = \frac{\left(\begin{array}{c}\text{number of ways to assign}\\ \text{5 different birth months}\end{array}\right)}{\left(\begin{array}{c}\text{number of ways to assign}\\ \text{birth months}\end{array}\right)}$$

$$= \frac{12 \cdot 11 \cdot 10 \cdot 9 \cdot 8}{12 \cdot 12 \cdot 12 \cdot 12 \cdot 12} = \frac{55}{144}$$

So the probability that at least two will have the same birth month is $1 - \frac{55}{144} = \frac{89}{144} \approx 0.618$.

• **Mutually exclusive events and unions of events.**

When two events have no outcomes in common they are said to be **mutually exclusive**. The probability of the union of two mutually exclusive events E and F in a sample space S is given by $P\,(E \cup F) = P\,(E) + P\,(F)$.

Exercise: A box contains 6 red building blocks and 4 blue building blocks. An infant selects 6 blocks from the box. Consider the events
E: the infant selects 3 red & 3 blue blocks
F: the infant selects all red blocks, and
G: the infant select at least four red blocks.
Which pairs of these events are mutually exclusive?

Answer:
E and F?   Since blue building blocks are not red, an outcome cannot satisfy both events, so $E \cap F = \emptyset$, and the events are mutually exclusive.
E and G?   Since the infant cannot select four red blocks and three blue blocks, these events are mutually exclusive.
F and G?   Since it is possible for the infant to select *at least 4 red blocks* and 6 *red blocks*, these events are not mutually exclusive.

Exercise: A five-card hand is drawn at random. Consider the following events:
E: all 5 cards are red
F: the hand is a full house
G: the hand contains two pairs
H: all 5 cards are face cards (Jack, Queen, or King)
Which pairs of these events are mutually exclusive?

Answer: In this problem, events are mutually exclusive if there is no hand that belong to both events. To show that two events are not mutually exclusive, we must show that there is at least one hand that belongs to both events.
E and F?   These events are mutually exclusive because a full house contains 3 cards of one kind, and these 3 cards cannot all be the same color. So a full house has to contain at least one black card.
E and G?   These events are not mutually exclusive. There are many hands with 2 pairs and a fifth card that are all red.

E **and** *H*?   These events are not mutually exclusive. There
are three face cards per suit, so there are six red face cards.
Thus there are hands that have 5 red face cards.

F **and** *G*?   These events are not mutually exclusive; in fact,
any full house contains two pairs.

F **and** *H*?   These events are not mutually exclusive. There
are many full houses which involve only face cards.

G **and** *H*?   These events are not mutually exclusive. Since
there are three kinds of face card, each pair and the fifth card
can come from a different kind of face card.

- **Probability of the union of two sets.**

When two events are *not* mutually exclusive, the probability of *E* or *F* is given by the formula $P(E \cup F) = P(E) + P(F) - P(E \cap F)$. Remember, when the word "or" is used, you are looking for the union of the events.

Exercise: A five-card hand is drawn at
random. Find the probability that all red
cards are drawn or a full house is drawn.

Answer: The sample space is all five-card hands. The number
of elements in the sample space is $C(52, 5)$. Let *E* be the event
all 5 cards are red. Let *F* be the event *full house*. These events
are mutually exclusive (see the previous exercise). Since there
are 26 red cards from which 5 red cards are drawn,

$$P(E) = \frac{C(26, 5)}{C(52, 5)} = \frac{65{,}780}{C(52, 5)}.$$ Also, from a prior exercise,

$$P(F) = \frac{3744}{C(52, 5)}.$$ Thus,

$$P(E \cup F) = P(E) + P(F) = \frac{65{,}780}{C(52, 5)} + \frac{3744}{C(52, 5)}$$
$$= \frac{69{,}524}{C(52, 5)}$$

Exercise: A five-card hand is drawn at
random. Find the probability that a full
house is drawn or all five cards are face cards.

Answer: The sample space is all five-card hands. Since there
are 52 cards in a deck, the number of 5 card hands is $C(52, 5)$.
Let *F* be the event *full house*. Let *H* be the event *all five cards
are face cards*. In the previous exercise, we saw that these
events are not mutually exclusive. Here we have

$$P(F) = \frac{3744}{C(52, 5)}.$$ Each suit has 3 face cards and there are

4 suits, so there are 12 face cards. We choose 5; so

$$n(H) = C(12, 5) \text{ and } P(H) = \frac{C(12, 5)}{C(52, 5)} = \frac{792}{C(52, 5)}.$$

$F \cap H$ is the event that the full house is made from face cards.
Since there are 3 kinds of face cards, there are $C(3, 1)$ ways to
pick the first kind, $C(4, 3)$ ways to pick 3 of the 4 cards from
that kind; there are $C(2, 1)$ ways to pick the second kind, and
$C(4, 2)$ ways to pick 2 of the 4 cards from that kind.

$$n\left(F \cap H\right) = C\left(3, 1\right) \cdot C\left(4, 3\right) \cdot C\left(2, 1\right) \cdot C\left(4, 2\right)$$
$$= 3 \cdot 4 \cdot 2 \cdot 6 = 144$$

so $P\left(F \cap H\right) = \dfrac{144}{C\left(52, 5\right)}$. Thus,

$$P\left(F \cup H\right) = P\left(F\right) + P\left(H\right) - P\left(F \cap H\right)$$
$$= \frac{3744}{C\left(52, 5\right)} + \frac{792}{C\left(52, 5\right)} - \frac{144}{C\left(52, 5\right)} = \frac{4392}{C\left(52, 5\right)}$$

- **Independent events.**

Two events are **independent** when the occurrence of one event does not affect the probability of the other event occurring. When events E and F are independent, then $P\left(E \cap F\right) = P\left(E\right) \cdot P\left(F\right)$. Many students get *independent events* and *mutually exclusive events* confused. The difference is that when two events are mutually exclusive they have nothing in common, but when the events are independent then the events do not influence each other. If two events are independent, then they are *not* mutually exclusive. And if they are mutually exclusive, then the two events are not independent.

Exercise: A pair of dice is rolled five times and the sum of the faces is recorded. What is the probability of rolling five 7's in a row?

Answer: Since rolling a 7 on any one roll does not influence the next roll, each roll is independent. In the first exercise in this section, we found the probability of rolling a 7 is $\frac{1}{6}$. Thus, the probability of rolling five 7s is $\dfrac{1}{6} \cdot \dfrac{1}{6} \cdot \dfrac{1}{6} \cdot \dfrac{1}{6} \cdot \dfrac{1}{6} = \dfrac{1}{6^5}$.

Exercise: A driver on a certain stretch of interstate has a 0.0001 probability of getting a radar ticket. Suppose a driver drives this stretch of road 6 times. What is the probability that the driver gets at least one radar ticket?

Answer: These events are independent. Since the complement is easier to find, we find it first. The complement of *getting at least one ticket* is *getting no ticket*.

$P\left(\text{no ticket}\right) = 1 - 0.0001 = 0.9999$, So $\left(0.9999\right)^6 \approx 0.9994$ is the probability of no tickets in 6 trips. Thus,

$P\left(\text{getting a ticket}\right) = 1 - 0.9994 = 0.0006$

Core Exercises

7, 13, 21, 25, 33, 43, 57

13.4   Binomial Probability

Concepts

(A) Binomial probability.

Definitions

- **Binomial experiment:** An experiment with only two possible outcomes.

Hints and Tips

- If you are working on a problem and find yourself thinking in terms of wins and losses, or coin tosses, then you are probably dealing with binomial probability.

Review

(A) **Binomial probability.**

When an experiment has two possible outcomes, E and F, with $P(E) = p$ and $P(F) = q = 1 - p$, then the probability of obtaining exactly r successes in n *independent* trials is $P(r \text{ successes in } n \text{ trials}) = C(n, r) \cdot p^r q^{n-r}$. It is important that the trials are independent.

Exercise: A multiple-choice quiz has five choices per question. Suppose a person guesses the answers to each of ten questions.

Answer:

(a) What is the probability that the person guesses correctly on a question? That the person guesses incorrectly?

Since there are five choices, only one of which is correct, the probability that the person guesses correctly is $\frac{1}{5} = 0.2$ So the probability that the person guesses incorrectly is $\frac{4}{5} = 0.8$.

(b) What is the probability that the person guesses none of the ten questions correctly?

The trials are independent, since what the person guesses on one question does not influence whether the person guesses correctly on any other question. Here we want zero successes in ten trials:

$$P(\text{no success in 10 trials}) = C(10, 0) \cdot 0.2^0 \cdot 0.8^{10}$$
$$= 1 \cdot 1 \cdot 0.107 = 0.107$$

(c) What is the probability that the person guesses correctly on exactly 7 questions?

$$P(7 \text{ successes in 10 trials}) = C(10, 7) \cdot 0.2^7 \cdot 0.8^3$$
$$= 120 \cdot 0.0000128 \cdot 0.512 = 0.000786$$

(d) What is the probability that the person guesses correctly on *at least* 9 questions?

Since getting at least 9 correct is the same as getting 9 correct or 10 correct,

$P(\text{at least 9 successes in 10 trials})$
$$= P(9 \text{ of } 10) + P(10 \text{ of } 10)$$
$$= C(10, 9)\,(0.2^9)\,(0.8^1) + C(10, 10)\,(0.2^{10})\,(0.8^0)$$
$$= 10 \cdot 0.000000512 \cdot 0.8 + 1 \cdot 0.000000102 \cdot 1$$
$$= 0.00000410 + 0.000000102 = 0.00000420$$

Exercise: A fast food chain has a contest in which one of every contestants wins free food. Suppose you go to this fast food restaurant four times in one week.

Answer: In this problem a success constitutes winning free food. So $P(\text{success}) = \frac{1}{3}$ and $P(\text{failure}) = \frac{2}{3}$.

(a) What is the probability that you win on all four trips?

$$P(4 \text{ successes in 4 trials}) = C(4, 4) \cdot \left(\tfrac{1}{3}\right)^4 \left(\tfrac{2}{3}\right)^0$$
$$= 1 \cdot \tfrac{1}{81} \cdot 1 \approx 0.0123$$

(b) What is the probability that you do not win free food that week?

$$P\,(\text{no success in 4 trials}) = C\,(4,0) \cdot \left(\tfrac{1}{3}\right)^0 \left(\tfrac{2}{3}\right)^4$$
$$= 1 \cdot 1 \cdot \tfrac{16}{81} \approx 0.1975$$

(c) What is the probability that you win at least once during the week?

$P\,(\text{win at least once}) = 1 - P\,(\text{lose every time})$. Since

$P\,(\text{lose every time}) = C\,(4,0) \cdot \left(\tfrac{1}{3}\right)^0 \left(\tfrac{2}{3}\right)^4$ we have

$$= 1 \cdot 1 \cdot \tfrac{16}{81} = \tfrac{16}{81}$$

$P\,(\text{win at least once}) = 1 - P\,(\text{lose every time})$

$$= 1 - \tfrac{16}{81} = \tfrac{65}{81} \approx 0.8025$$

◤ Core Exercises

1, 21, 27

13.5 Expected Value

◤ Concepts

(A) **Expected Value.**

◤ Definitions

- **Expected value:** Informally, the expected value represents the average outcome one expects an experiment to produce. For example, if we spend one dollar on a state lottery ticket, on average we do not expect to get a lot of money back, and therefore the expected value will be a small number. More formally, the expected value is the sum we get when we take the value of each possible outcome, multiply by the probability of that outcome, and add all the results together.

◤ Review

(A) **Expected Value.**

The **expected value** of an experiment is a weighted average of *payoffs* based on the outcomes of the experiment. The weight applied to each payoff is the probability of the event (that corresponds to the payoff) occurring. Expected value E is $E = a_1 p_1 + a_2 p_2 + \cdots + a_n p_n$. Here a_i is the payoff and p_i is the corresponding probability.

Exercise: A game consists of rolling a die. You receive the face value in dollars if an odd number is rolled. If the game costs $1.50 to play, what is the expected value to you?

Answer: The payoffs and their respective probabilities are given in the following table.

Payoff	Event Description	Probability
$-\$1.50$	An even number is rolled	$\frac{1}{2}$
$-1.50 + 1 = -\$0.50$	A 1 is rolled	$\frac{1}{6}$
$-1.50 + 3 = \$1.50$	A 3 is rolled	$\frac{1}{6}$
$-1.50 + 5 = \$3.50$	A 5 is rolled	$\frac{1}{6}$

Thus, $E = -1.50 \cdot \frac{1}{2} - 0.50 \cdot \frac{1}{6} + 1.50 \cdot \frac{1}{6} + 3.50 \cdot \frac{1}{6} = 0.$

Exercise: A sales clerk at a car stereo store works for commission only. If 20% of the people purchase an item for which the clerk gets a $3 commission, 10% purchase an item for which the clerk gets a $5 commission, and 5% purchase an item for which the clerk gets an $8 commission, what is the expected value per customer who enters the store?

Answer: The payoffs (commissions) and their respective probabilities are given in the following table.

Payoff	Probability
$3	0.20
$5	0.10
$8	0.05
$0	the rest

$E = 3\,(0.20) + 5\,(0.10) + 8\,(0.05) + 0 = 0.6 + 0.5 + 0.4 = 1.50$

Exercise: A bowl contains eight chips which cannot be distinguished by touch alone. Five chips are marked $1 each and the remaining three chips are marked $4 each. A player is blindfolded and draws two chips at random without replacement from the bowl. The player is then paid the sum of the two chips. Find the expected payoff of this game.

Answer: The possible payoff are $2 (two $1 chips are selected), $5 (one $1 chip and one $4 chip), and $8 (two $4 chips). The probability of getting $2 is $\frac{C(5,2)}{C(8,2)} = \frac{10}{28}$, the probability of getting $5 is $\frac{C(5,1) \cdot C(3,1)}{C(8,2)} = \frac{15}{28}$, and the probability of getting $8 is $\frac{C(3,2)}{C(8,2)} = \frac{3}{28}$. Thus,

$E = 2 \cdot \frac{10}{28} + 5 \cdot \frac{15}{28} + 8 \cdot \frac{3}{28} = \frac{119}{28} = \$4.25.$

Exercise: A box of 10 light bulbs contains 3 defective light bulbs. If a random sample of 2 bulbs is drawn from the box, what is the expected number of defective bulbs?

Answer: We use the number of defective bulbs as the payoff, so a sample of two bulbs can have 0, 1, or 2 defective bulbs.

Payoff	Probability
0	$\frac{C(7,2)}{C(10,2)} = \frac{7}{15}$
1	$\frac{C(7,1) \cdot C(3,1)}{C(10,2)} = \frac{7}{15}$
2	$\frac{C(3,2)}{C(10,2)} = \frac{1}{15}$

Thus $E = 0 \cdot \frac{7}{15} + 1 \cdot \frac{7}{15} + 2 \cdot \frac{1}{15} = \frac{9}{15} = \frac{3}{5}.$

Core Exercises

13, 15, 17, 19